DRILL HALL LIBRARY
MEDWAY

Microorganisms in Foods 7

Microorganisms in Foods 7

Microbiological Testing in
Food Safety Management

ICMSF

International Commission on Microbiological Specifications for Foods.
Microbiological testing in food safety management / ICMSF ; editorial committee, R.B. Tompkin . . . [et. al.].
p. cm.—(Microorganisms in foods ; 7)
Includes bibliographical references and index.
ISBN 0-306-47262-7
1. Food—Microbiology. 2. Food—Standards. I. Tompkin, R.B. (R. Bruce) II. Title. III. Series.
QR115.I46   2001
664′.001′579—dc21
                    2001053529

Library of Congress Control Number: 2001053529

ISBN-10: 0-306-47262-7

Printed on acid-free paper.

© 2002 Springer Science+Business Media, LLC
All rights reserved. This work may not be translated or copied in whole or in part without the written permission of the publisher (Springer Science+Business Media, LLC, 233 Spring Street, New York, NY 10013, USA), except for brief excerpts in connection with reviews or scholarly analysis. Use in connection with any form of information storage and retrieval, electronic adaptation, computer software, or by similar or dissimilar methodology now known or hereafter developed is forbidden.
The use in this publication of trade names, trademarks, service marks, and similar terms, even if they are not identified as such, is not to be taken as an expression of opinion as to whether or not they are subject to proprietary rights.

Printed in the United States of America.   (EB)

9 8 7 6 5 4 3

springer.com

Table of Contents

INTRODUCTION xi
EDITORIAL COMMITTEE xiii
ICMSF MEMBERS DURING PREPARATION OF BOOK 7 xiii
CONSULTANTS xiii
CONTRIBUTORS xiii

CHAPTER 1—MICROBIOLOGICAL HAZARDS AND THEIR CONTROL 1

    1.1 Introduction 1
    1.2 History 2
    1.3 The Concept of a Food Safety Management System 4
    1.4 Historical Development 6
    1.5 Status of Foodborne Illness: Etiologic Agents or Contaminants 9
    1.6 Practices Contributing to Foodborne Illness 13
    1.7 Importance of Effective Control Measures 13
    1.8 Effectiveness of GHP and HACCP 14
    1.9 Would an FSO Improve Food Safety and Reduce Foodborne Illness? .. 14
    1.10 Use of FSOs in Food Safety Management 15
    1.11 Performance, Process, Product, and Default Criteria 16
    1.12 Establishment of Control Measures 17
    1.13 Assessment of Control of a Process 17
    1.14 Acceptance Criteria 17
    1.15 Microbiological Criteria 17
    1.16 Microbiological Tests 18
    1.17 Summary 18
    1.18 References 19

CHAPTER 2—EVALUATING RISKS AND ESTABLISHING FOOD SAFETY OBJECTIVES ... 23

    2.1 Introduction 23
    2.2 Tolerable Level of Consumer Protection 26
    2.3 Importance of Epidemiologic Data 28
    2.4 Evaluation of Risk 31
    2.5 Food Safety Objectives (FSOs) 33
    2.6 Establishment of an FSO Based on a Risk Evaluation by an Expert Panel .. 35
    2.7 Evaluation of Risk by Quantitative Risk Assessment 36
    2.8 Establishment of an FSO Based on Quantitative Risk Assessment ... 40
    2.9 Stringency of FSOs in Relation to Risk and Other Factors 41

2.10 Summary 42
2.11 References 42

CHAPTER 3—MEETING THE FSO THROUGH CONTROL MEASURES 45

3.1 Introduction 45
3.2 Control Measures 45
3.3 Confirm That the FSO Is Technically Achievable 48
3.4 Importance of Control Measures 49
3.5 Performance Criteria 54
3.6 Process and Product Criteria 59
3.7 The Use of Microbiological Sampling and Performance Criteria 59
3.8 Default Criteria 61
3.9 Process Validation 61
3.10 Monitoring and Verifying Control Measures 65
3.11 Examples of Control Options 66
3.12 Assessing Equivalency of Food Safety Management Systems 68
3.13 References 68
Appendix 3–A: Control Measures Commonly Applied to Foodborne Diseases .. 71

CHAPTER 4—SELECTION AND USE OF ACCEPTANCE CRITERIA 79

4.1 Introduction 79
4.2 Equivalence 80
4.3 Establishment of Acceptance Criteria 81
4.4 Application of Acceptance Criteria 84
4.5 Determining Acceptance by Approval of Supplier 85
4.6 Examples To Demonstrate the Process of Lot Acceptance 87
4.7 Auditing Food Operations for Supplier Acceptance 90
4.8 References 97

**CHAPTER 5—ESTABLISHMENT OF MICROBIOLOGICAL
CRITERIA FOR LOT ACCEPTANCE 99**

5.1 Introduction 99
5.2 Purposes and Application of Microbiological Criteria for Foods 101
5.3 Definition of Microbiological Criterion 101
5.4 Types of Microbiological Criteria 102
5.5 Application of Microbiological Criteria 103
5.6 Principles for the Establishment of Microbiological Criteria 104
5.7 Components of Microbiological Criteria for Foods 106
5.8 Examples of Microbiological Criteria 111
5.9 References 112

CHAPTER 6—CONCEPTS OF PROBABILITY AND PRINCIPLES OF SAMPLING 113

6.1 Introduction 113
6.2 Probability 113
6.3 Population and Sample of the Population 114

6.4 Choosing the Sample Units 115
6.5 The Sampling Plan 115
6.6 The Operating Characteristic Function 115
6.7 Consumer Risk and Producer Risk 117
6.8 Stringency and Discrimination 117
6.9 Acceptance and Rejection 118
6.10 What Is a Lot? 118
6.11 What Is a Representative Sample? 119
6.12 Confidence in Interpretation of Results 120
6.13 Practical Considerations 121
6.14 References 121

CHAPTER 7—SAMPLING PLANS 123

7.1 Introduction 123
7.2 Attributes Plans 123
7.3 Variables Plans 131
7.4 Comparison of Sampling Plans 134
7.5 References 142

CHAPTER 8—SELECTION OF CASES AND ATTRIBUTES PLANS 145

8.1 Introduction 145
8.2 Microbial Criteria: Utility, Indicator, and Pathogens 146
8.3 Factors Affecting the Risk Associated with Pathogens 147
8.4 Categorizing Microbial Hazards According to Risk 151
8.5 Definition of Cases 152
8.6 Deciding between Two-Class and Three-Class Attributes Sampling Plans .. 157
8.7 Determining Values for m and M 158
8.8 Specific Knowledge about the Lot 159
8.9 What Is a Satisfactory Probability of Acceptance? 160
8.10 Selecting n and c 161
8.11 Sampling Plan Performance of the Cases 162
8.12 References 165
Appendix 8–A: Ranking of Foodborne Pathogens or Toxins into Hazard Groups 166

CHAPTER 9—TIGHTENED, REDUCED, AND INVESTIGATIONAL SAMPLING 173

9.1 Introduction 173
9.2 Application of Tightened Sampling and Investigational Sampling 176
9.3 Tightened Sampling Plans 177
9.4 Example of the Influence of Sampling Plan Stringency in Detecting Defective Lots 180
9.5 Selecting the Sampling Plan According to Purpose 181
9.6 Reduced Sampling 182
9.7 References 182

CHAPTER 10—EXPERIENCE IN THE USE OF TWO-CLASS ATTRIBUTES PLANS FOR LOT ACCEPTANCE ... 183

    10.1 Introduction ... 183
    10.2 The Concept of Zero Tolerance ... 184
    10.3 The Need for Compromise ... 185
    10.4 Application of Two-Class Sampling Plans for Pathogens Such as *Salmonella* ... 186
    10.5 Problems in the Implementation of Stringent Sampling Plans ... 188
    10.6 Relation to Commercial Practice ... 189
    10.7 Discrepancies between Original and Retest Results ... 191
    10.8 Summary ... 196
    10.9 References ... 197

CHAPTER 11—SAMPLING TO ASSESS CONTROL OF THE ENVIRONMENT ... 199

    11.1 Introduction ... 199
    11.2 Principles of GHP ... 200
    11.3 Post-Process Contamination ... 200
    11.4 Establishment and Growth of Pathogens in the Food Processing Environment ... 202
    11.5 Measures To Control Pathogens in the Food Processing Environment ... 207
    11.6 Sampling the Processing Environment ... 210
    11.7 References ... 220

CHAPTER 12—SAMPLING, SAMPLE HANDLING, AND SAMPLE ANALYSIS ... 225

    12.1 Introduction ... 225
    12.2 Collection of Sample Units ... 226
    12.3 Intermediate Storage and Transportation ... 229
    12.4 Reception of Samples ... 231
    12.5 Sample Analysis ... 231
    12.6 Recovery of Injured Cells ... 232
    12.7 Errors Associated with Methods and Performance of Laboratories ... 233
    12.8 References ... 235

CHAPTER 13—PROCESS CONTROL ... 237

    13.1 Introduction ... 237
    13.2 Knowledge of the Degree of Variability and the Factors That Influence Variability ... 240
    13.3 Process Capability Study ... 243
    13.4 Control during Production: Monitoring and Verifying a Single Lot of Food ... 246
    13.5 Control during Production: Organizing Data from Across Multiple Lots of Food To Maintain or Improve Control ... 247
    13.6 Use of Process Control Testing as a Regulatory Tool ... 259
    13.7 Investigating and Learning from Previously Unrecognized Factors or Unforeseen Events ... 260
    13.8 References ... 261

Table of Contents ix

CHAPTER 14—AFLATOXINS IN PEANUTS **263**

    14.1 Introduction 263
    14.2 Risk Assessment 263
    14.3 Risk Management 265
    14.4 Acceptance Criteria for Final Product 269
    14.5 References 270

CHAPTER 15—*SALMONELLA* IN DRIED MILK **273**

    15.1 Introduction 273
    15.2 Risk Evaluation 273
    15.3 Risk Management 275
    15.4 Product and Process Criteria 279
    15.5 GHP and HACCP 280
    15.6 Acceptance Criteria for Final Product 281
    15.7 References 282

CHAPTER 16—*LISTERIA MONOCYTOGENES* IN
                       COOKED SAUSAGE (FRANKFURTERS) **285**

    16.1 Introduction 285
    16.2 Risk Evaluation 287
    16.3 Risk Management 293
    16.4 Process and Product Criteria 304
    16.5 GHP and HACCP 308
    16.6 Acceptance Criteria for Final Product 308
    16.7 References 309

CHAPTER 17—*E. COLI* O157:H7 IN FROZEN RAW GROUND BEEF PATTIES **313**

    17.1 Introduction 313
    17.2 Risk Evaluation 314
    17.3 Risk Management 322
    17.4 Control Measures 322
    17.5 Acceptance Criteria 325
    17.6 Statistical Implications of the Proposed Sampling Plan 327
    17.7 References 330

Appendix A—Glossary 333
Appendix B—Objectives and Accomplishments of the International Commission
                 on Microbiological Specifications for Foods 337
Appendix C—ICMSF Participants 341
Appendix D—Publications of the ICMSF 347
Appendix E—Table of Sources 351

INDEX **357**

Addenda

Since the publication of this book in 2002, a number of the concepts introduced by the International Commission on Microbiological Specifications for Foods (ICMSF), such as Food Safety Objective (FSO) and the food safety management scheme, have been further developed through the step-wise Codex Alimentarius Commission (CAC) procedure. Discussions within the Codex Committee on Food Hygiene (CCFH) led to the recognition that, in addition to the FSO, which relates to the level of a hazard at the point of consumption, there was the need to introduce a new term, Performance Objective, for the level of a hazard at earlier points within the food chain. Here are the definitions for the concepts as proposed by CCFH:

- *Food safety objective* = the maximum frequency and/or concentration of a hazard in a food at the time of consumption that provides or contributes to the appropriate level of protection (ALOP).
- *Performance objective* = the maximum frequency and/or concentration of a hazard in a food at a specified step in the food chain before the time of consumption that provides or contributes to an FSO or ALOP, as applicable.
- *Performance criterion* = the effect in frequency and/or concentration of a hazard in a food that must be achieved by the application of one or more control measures to provide or contribute to a PO or an FSO.

These concepts have now been adopted by the CAC and included in the Procedural Manual, which is updated annually and published by the Joint FAO/WHO Food Standards Programme in Rome, Italy. (The 15th edition is of 2005; ISBN: 92-5-105420-7)[1]. These and other terms may continue to evolve as the international community strives to reach a consensus on the principles and guidelines for microbiological risk management within the overall framework of risk analysis. Because the latest information should be considered when reading this book, the reader is encouraged to refer to the most recent Codex documents for current terms, their definitions and application. Codex documents can generally be accessed through the CAC website (http://www.codexalimentarius.net/web/index_en.jsp). It is anticipated that this book will be revised to reflect the final deliberations of the CCFH and adoption of the documents by the CAC.

The Editorial Committee
28 March, 2006

---

[1] ftp://ftp.fao.org/codex/Publications/ProcManuals/manual_15e.pdf).

Introduction

Microorganisms in Foods 7: Microbiological Testing in Food Safety Management was written by the International Commission on Microbiological Specifications for Foods (ICMSF/the Commission) with assistance from a limited number of consultants.

Microorganisms in Foods 7 is based upon Part I of *Microorganisms in Foods 2: Sampling for Microbiological Analysis: Principles and Specific Applications (2nd ed. 1986)*. In the 1980s, control of food safety was largely by inspection and compliance with hygiene regulations, together with end product testing. *Microorganisms in Foods 2* put such testing on a sounder statistical basis through sampling plans, which remain useful at port-of-entry when there is no information on the conditions under which a food has been produced or processed. At an early stage, the Commission recognized that no sampling plan could ensure the absence of a pathogen in food. Testing foods at ports-of-entry, or elsewhere in the food chain, could not guarantee food safety.

This led the Commission to explore the potential value of Hazard Analysis Critical Control Point (HACCP) for enhancing food safety, particularly in developing countries. *Microorganisms in Foods 4: Application of the Hazard Analysis Critical Control Point (HACCP) System to Ensure Microbiological Safety and Quality (1988)*, illustrated the procedures used to identify the microbiological hazards in a practice or a process, to identify the critical control points at which those hazards could be controlled, and to establish systems by which the effectiveness of control could be monitored. In addition, recommendations were included for the application of HACCP from production/harvest to consumption, together with examples of how HACCP could be applied at each step in the food chain.

Effective implementation of HACCP requires knowledge of the hazardous microorganisms and their response to conditions in foods (e.g., pH, a_w, temperature, preservatives). The Commission concluded that such information was not collected together in a form that could be assessed easily by food industry personnel in quality assurance, technical support, research and development, and by those in food inspection at local, state, regional, or national levels. *Microorganisms in Foods 5: Characteristics of Microbial Pathogens (1996)* is a thorough, but concise, review of the literature on growth, survival, and death responses of foodborne pathogens. It is intended as a quick reference manual to assist in making judgements on the growth, survival, or death of pathogens in support of HACCP plans and to improve food safety.

Microorganisms in Foods 6: Microbial Ecology of Food Commodities (1998) is intended for those primarily involved in applied aspects of food microbiology such as food processors, food microbiologists, food technologists, veterinarians, public health workers, and regulatory officials. Addressing 16 commodity areas, it describes the initial microbial flora and the prevalence of pathogens, the microbiological consequences of processing, typical spoilage patterns, episodes implicating those commodities with foodborne illness, and measures to control pathogens and limit spoilage.

This book, *Microorganisms in Foods 7: Microbiological Testing in Food Safety Management (2002),* illustrates how systems such as HACCP and good hygiene practices (GHP) provide greater assurance of safety than microbiological testing, but also identifies circumstances where microbiological testing still plays a useful role in systems to manage food safety. It continues to address the Commission's objectives to: (a) assemble, correlate, and evaluate evidence about the microbiological safety and quality of foods; (b) consider whether microbiological criteria would improve and ensure the microbiological safety of particular foods; (c) propose, where appropriate, such criteria; (d) recommend methods of sampling and examination; (e) provide guidance on appraising and controlling the microbiological safety of foods.

This book introduces the reader to a structured approach for managing food safety, including sampling and microbiological testing. The text outlines how to meet specific food safety goals for a food or process using GHP and the HACCP system.

The concept of a food safety objective (FSO) is recommended to industry and control authorities to translate "risk" into a definable goal for establishing food safety management systems that incorporate the principles of GHP and HACCP. FSOs provide the scientific basis for the industry to select and implement measures that control the hazard(s) of concern in specific foods or food operations, for control authorities to develop and implement inspection procedures to assess the adequacy of control measures adopted by industry, and for quantifying the equivalence of inspection procedures in different countries.

Microbiological testing can be a useful tool in the management of food safety. However, microbiological tests should be selected and applied with knowledge of their limitations, as well as their benefits and the purposes for which they are used. In many instances other means of assessment are quicker and more effective.

The need for microbiological testing varies along the food chain. Points in the food chain should be selected where information about the microbiological status of a food will prove most useful for control purposes. Similarly, in a food operation, samples may be collected from different points in a process for control purposes.

Finally, a framework is provided by which importing countries can assess whether foods from other countries have been produced in a manner that provides a level of protection equivalent to that required for domestically produced foods.

This book illustrates the insensitivity of even statistically based sampling plans and encourages a rational approach to the use of microbiological testing in systems that manage food safety through GHP and HACCP. Several new chapters are based on the experience of the food industry in controlling salmonellae, *Listeria monocytogenes* and *Escherichia coli* O157:H7, on tightened or investigational sampling, on microbiological testing of the processing environment, and the use of statistical process control to detect trends and work toward continuous improvement.

The book is intended to be useful for anyone who is engaged in setting microbiological criteria for the purpose of governmental or industry food inspection and control. For students in food science and technology it offers a wealth of information on food safety management and many references for further study.

EDITORIAL COMMITTEE

R. B. Tompkin (Chairman)
T. A. Roberts
M. van Schothorst
M. B. Cole

L. Gram
R. L. Buchanan
S. Dahms

ICMSF Members during Preparation of Book 7

Chairman   T. A. Roberts (1991–2000)   M. B. Cole (from 2000)

Secretary M. van Schothorst

Treasurer A. N. Sharpe (1989–1998)   J. M. Farber
             F. F. Busta (1998–2000)   (from 2000)

Members   A. C. Baird-Parker (retired 1999)   F. H. Grau (retired 1999)
             R. L. Buchanan   J.-L. Jouve
             J.-L. Cordier   A. M. Lammerding (from 1998)
             S. Dahms (from 1998)   S. Mendoza (retired 1998)
             M. P. Doyle (resigned 1999)   Z. Merican
             M. Eyles (resigned 1999)   J. I. Pitt
             J. Farkas (retired 1998)   F. Quevedo (retired 1998)
             R. S. Flowers   P. Teufel
             B. D. G. M. Franco (from 2000)   R. B. Tompkin
             L. Gram (from 1998)

Consultants

J. Braeunig (2000)
S. Dahms (1997, 1998)
P. Desmarchelier (1999)
J. M. Farber (1998)
B. D. G. M. Franco (1998, 1999)
W. Garthwright (1999)
L. G. M. Gorris (2000)

L. Gram (1997, 1998)
H. Kruse (1999, 2000)
A. M. Lammerding (1997, 1998)
B. Shay (1999)
K. Swanson (2000)
A. von Holy (1997)

Contributors

D. Kilsby, R. B. Smittle, J. H. Silliker

Chapter 1

Microbiological Hazards and Their Control

1.1	Introduction	**1.10**	Use of FSOs in Food Safety Management
1.2	History	**1.11**	Performance, Process, Product, and Default Criteria
1.3	The Concept of a Food Safety Management System	**1.12**	Establishment of Control Measures
1.4	Historical Development	**1.13**	Assessment of Control of a Process
1.5	Status of Foodborne Illness: Etiologic Agents or Contaminants	**1.14**	Acceptance Criteria
1.6	Practices Contributing to Foodborne Illness	**1.15**	Microbiological Criteria
		1.16	Microbiological Tests
1.7	Importance of Effective Control Measures	**1.17**	Summary
1.8	Effectiveness of GHP and HACCP	**1.18**	References
1.9	Would an FSO Improve Food Safety and Reduce Foodborne Illness?		

1.1 INTRODUCTION

The purpose of this book is to introduce the reader to a structured approach for managing food safety, including sampling and microbiological testing. In addition, the text outlines how to meet specific food safety goals for a food or process using Good Hygienic Practices (GHP) and the Hazard Analysis Critical Control Point (HACCP) systems.

The International Commission on Microbiological Specifications for Foods (ICMSF/the Commission) is recommending that industry and control authorities adopt the concept of a food safety objective (FSO). This concept translates *risk* into a definable goal for establishing food safety management systems that incorporate the principles of GHP and HACCP. FSOs provide the scientific basis for industry to select and implement measures that control the hazard(s) of concern in specific foods or food operations, for control authorities to develop and implement inspection procedures to assess the adequacy of control measures adopted by industry, and for quantifying the equivalence of inspection procedures in different countries.

Microbiological testing can be a useful tool in the management of food safety. However, microbiological tests should be selected and applied with knowledge of their limitations, as well as their benefits and the purposes for which they are used. In many instances, other means of assessment are quicker and equally effective.

The text provides guidance to the food industry on how to establish effective management systems to control specific hazards in foods. This approach will similarly be of interest to control authorities with responsibility for assessing whether industry has developed and implemented adequate systems of food safety management.

The need for microbiological testing varies along the food chain. Points should be selected in the food chain where information about the microbiological status of a food will prove most useful for control purposes. Similarly, in a food operation, samples may be collected from different points in a process for control purposes.

A framework is provided by which importing countries can assess whether foods from other countries have been produced in a manner that provides a level of protection equivalent to that required for domestically produced foods.

Guidance is provided for establishing sampling plans based on risk to consumers. In Chapter 8, 15 cases are described that take into account whether the hazard will increase, decrease, or not change between when a food is sampled and when the food is consumed. These same principles are useful for foods at port-of-entry and in domestic situations when the safety and acceptability of a food are uncertain.

1.2 HISTORY

The ICMSF was asked in 1995 by the Codex Alimentarius Committee on Food Hygiene to write a discussion paper on The Management of Pathogens in Foods in International Trade. During its meeting in 1996, the ICMSF concluded that such management should use existing Codex documents and should be in line with the requirements of the World Trade Organization Agreement on the Application of Sanitary and PhytoSanitary Measures (WTO/SPS agreement) (WTO, 1995), which stated that foods could freely be imported if they would not endanger the country's appropriate level of (consumer) protection (ALOP). In the same agreement, risk assessment was identified as a tool to determine whether a food would or would not endanger the ALOP. How an ALOP needed to be expressed or how it should be established was not elaborated on in the agreement.

A procedure for microbiological risk assessment was already described in a Codex document. Other documents in the Codex system included the General Principles of Food Hygiene, with its annex on HACCP, and the Principles for the Establishment and Application of Microbiological Criteria for Foods.

The ICMSF recognized that it would be difficult for the food industry to prove that a product would meet the ALOP of an importing country if that ALOP was expressed in terms such as "the number of illnesses per 100,000 of a population, caused by a hazard/food combination." However, the Codex tools to ensure the safety of the food were the GHP and HACCP systems.

In the HACCP system, hazards are controlled by their elimination from the food or reduction to acceptable levels. It was the Commission's understanding that these acceptable levels would not endanger the ALOP. However, as long as an ALOP was not expressed as "the level of a hazard in a food which would be acceptable," the industry would not know what the acceptable level in the HACCP system would be. The ICMSF thus felt the need to develop the concept of an FSO, following the concept of *quality objectives* in quality assurance and quality management standards (Jouve, 1992) and the introduction of FSOs in broad terms (Jouve, 1994, 1996). At that time, the same wording had been proposed at congresses to express some form of justification for sanitary measures with respect to determining equivalence, and later in refereed publications (Hathaway, 1997; Hathaway & Cook, 1997).

The intention was that the FSO would convert the ALOP or acceptable (tolerable) level of risk (number of illnesses) into the maximum frequency and/or concentration of a hazard considered to be tolerable for consumer protection. That FSO could then be translated into the performance of a food process that would ensure that, at the moment of consumption, the level of the hazard in a food would not be greater than the FSO. Risk assessment was considered to be helpful in establishing the FSO, because the risk characterization would be expressed as "the estimated number of illnesses per year," in terms similar to the ALOP. Moreover, risk assessment could be used to select control measures that would ensure that the FSO would be met.

Consequently, in 1996, the ICMSF recommended to the Codex secretariat that a stepwise procedure should be used to manage pathogens in foods. The first step would be to perform a microbiological risk assessment; the second step would be to develop an FSO. Step three should confirm that the FSO would be technically achievable by the application of GHP and HACCP. Step four was the establishment of microbiological criteria, when there was a need, and step five was establishing other acceptance criteria for foods in international trade.

In 1997, an additional step was introduced between steps one and two, i.e., the use of the newly developed concept of risk management. Through the introduction of this step, it was recognized that the establishment of an FSO was not only a scientific risk assessment exercise, but also a societal decision in which the various stakeholders (consumers, industries, etc.) should participate.

At the same time that the FSO concept was brought into discussion in the Codex Alimentarius Committee on Food Hygiene, the term *food safety objectives* was also introduced in the Codex Committee on Import and Export to deal with the outcome of all kinds of sanitary measures. This situation has led to great confusion about the nature of FSOs, why they are needed, what they should accomplish, etc. The Codex Alimentarius Commission finally decided in 2000 that the term FSO would no longer be used in the broader sense, but would only be used by the Codex Committee on Food Hygiene for its own purposes. In 2001, that Committee agreed upon a definition of an FSO as an initial step toward incorporating the concept into its future recommendations.

The ICMSF decided that in the present revision of ICMSF Book 2 (ICMSF, 1986), the FSO concept should be introduced as the basis for establishing a food safety management system. Discussions within the ICMSF of how an ALOP should be developed, the use of risk assessment within food safety management, and the use of FSOs and how they would be established concluded that establishing FSOs should be broader than just converting an ALOP into a level of a hazard in a food. It was also recognized that a full risk assessment, according to the Codex procedures, would in many cases not be necessary to determine control measures required to meet an FSO, but that an expert panel and a less detailed risk evaluation would often suffice.

Although the FSO concept was originally linked to the ALOP concept as described in the WTO/SPS agreement, it was later recognized that the Codex not only had the task of producing procedures and guidelines that could be used by WTO, but that it also should help nations to improve their food safety. In this light, the food safety objectives were not only seen as levels of hazards that were already achieved in a country, but also as levels that should be achieved as part of a food safety enhancement program. This situation has also led to confusion. On the one hand, a country cannot ask of an exporting country that the exported foods should meet higher requirements (more stringent FSOs) than are achieved in the importing country. Such FSOs would merely reflect the status quo. On the other hand, FSOs used in food safety enhancement program within a country should be regarded quite separately from FSOs used to manage foods in international trade.

The ICMSF recognizes this situation and recommends that FSOs can be used for both purposes, and that governments establish them to communicate to industries the maximum level of a hazard in a food that must not be surpassed. The ICMSF recommends that FSOs should be established on the basis of epidemiologic evidence that a certain level of a hazard in a food does not cause an intolerable public health problem. If evidence exists that a certain level of a hazard in a food is indeed unacceptable, lower levels should be set if they can be obtained by control measures that are technically achievable with acceptable costs. How FSOs for the various purposes should be established might vary from situation to situation. In this book, the ICMSF gives some general guidelines.

1.3 THE CONCEPT OF A FOOD SAFETY MANAGEMENT SYSTEM

Chapters 2–5 describe in detail a sequence of activities for establishing a comprehensive food safety system, as summarized in Figure 1–1. The respective roles of industry and government are described because it is through their collective activities that effective food safety systems are developed and verified. A series of steps is described.

Risk managers in government use epidemiologic evidence to link human illness with a microbiological agent and, if possible, a food. A decision is then made that a new policy is needed to prevent or reduce future illnesses of the same nature or to prevent deterioration of public health, e.g., through imported foods. Additional information may be required to understand better the options available for control. Depending on urgency, severity, risk, likelihood of further occurrences, the hazard-food combination, and other factors, risk managers may establish an expert panel to develop recommendations or ask risk assessors to perform a risk profile.

If possible from available information, risk managers in government, with input from other stakeholders, perform an evaluation to decide whether an FSO should be established that will reflect existing or improved control of the hazard-food combination. Using information obtained from epidemiologic studies, knowledge of the hazard, and conditions leading to foodborne illness, risk managers in government develop an FSO for the hazard-food combination, i.e., the maximum frequency and/or concentration of a microbial hazard in food considered tolerable for consumer protection.

Risk managers in industry and government assess whether the FSO, communicated in verifiable terms, is achievable with current or improved technologies, processes, and practices. FSOs define the level of a hazard at the moment of consumption. This implies that preparation and use should be an integral part of the food safety management considerations.

If the FSO is achievable, the next step is to establish criteria that can be used to assess whether the control measures will control the hazard and meet the FSO. The criteria may consist of performance criteria (e.g., 5D kill, no greater than a 100-fold increase in numbers), process criteria (e.g., 71 °C for 2.5 minutes), or product criteria (pH, a_w). The criteria are met through application of GHP and HACCP, including procedures for monitoring to ensure or check control, while taking account of process variability (see Chap-

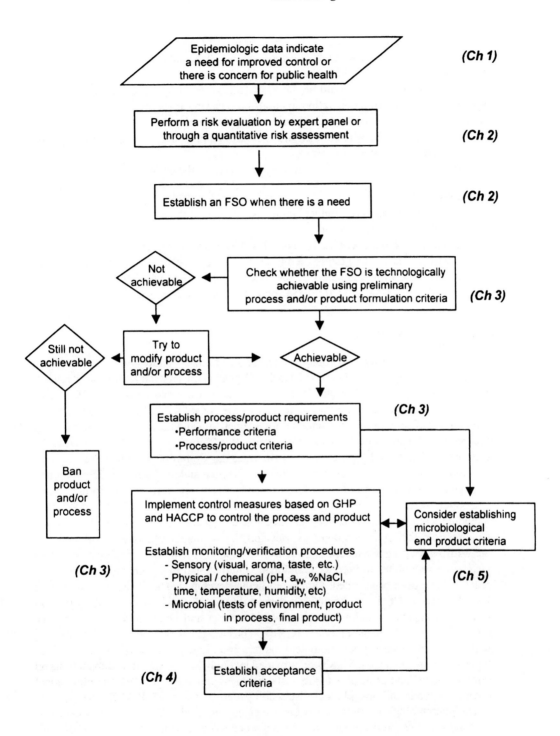

Figure 1-1 Proposed scheme for managing the microbiological safety of food.

ter 13). If knowledge is lacking and process or product criteria cannot be soundly based, default values may be used to ensure food safety. Those values are often conservative, e.g., very low pH or a_w, a severe heat treatment.

The foregoing criteria should be incorporated into auditing/inspection procedures to assess the adequacy of an operation to control the hazard(s) and meet established FSOs. In addition, lot acceptance criteria should be established that are based on parameters of the food that determine whether a hazard is controlled (e.g., pH, a_w) and, if judged appropriate, microbiological criteria.

If the FSO is judged to be *not* technically achievable, the options are to:

- reassess the FSO,
- modify the process or product and its use to achieve the FSO,
- ban the product/process, or
- establish acceptance criteria for assessing the adequacy of an operation to control the hazard(s) and meeting established FSOs and, if appropriate, establish lot acceptance criteria based on parameters of the food that determine whether a hazard is controlled (e.g., pH, a_w) and, if judged appropriate, microbiological end product criteria.

1.4 HISTORICAL DEVELOPMENT

Microbiological criteria for foods in international trade are addressed in the joint Food and Agriculture Organization/World Health Organization (FAO/WHO) food standards program, as implemented by the Codex Alimentarius Commission (CAC, 1997). That program (established in 1962, the same year as the constitution of the ICMSF) was the direct result of conflict between national food legislation and the general requirements of the main food markets of the world. Serious non-tariff obstacles to trade were caused by differing national food legislation. At that time, the Commission's objectives were to develop international food standards, Codes of Practice and guidelines, anticipating that their general adoption would help to remove and prevent non-tariff barriers to food trade.

Several aspects of food control are covered in Codex standards and Codes of Practice, including composition, labeling, additives, and hygiene. Subsidiary bodies of the Commission, the Codex committees, develop the standards and codes. Although the period from drafting a standard or code to its adoption is lengthy (normally four to five years), the system has worked well, and many international food standards and codes of practice have been established. Others are being developed. The Codex Committee on Food Hygiene (CCFH) has the major responsibility for all provisions on food hygiene in Codex documents, including microbiological criteria and codes of hygienic practice (effectively, good hygienic practices (GHP)). The CCFH needs expert advice in dealing with highly specialized microbiological matters and especially in developing microbiological criteria. Such advice has been provided by the ICMSF through its publications on sampling plans and principles for the establishment and application of microbiological criteria for foods and several other discussion papers (CAC, 1997; ICMSF, 1998b).

The microbiological safety of food is principally ensured by selection of raw materials, control at the source, product design and process control, and the application of GHP and HACCP during production, processing, distribution, storage, sale, preparation, and use. This comprehensive preventive system offers much more control than end-product testing. Examples of measures that have successfully controlled foodborne hazards are given in Table 1–1.

Table 1-1 Measures That Have Successfully Controlled Foodborne Hazards

	Hazard	Control measures	Other measures
Bacteria	B. cereus	Time-temperature control during cooking, cooling, and storage	
	brucellae	Eradication of brucellosis	Animal health
	Campylobacter (thermophilic)	Raw material selection, avoid cross-contamination	GHP
	C. botulinum (proteolytic strains)	Retorting, acidification, low water activity	HACCP
	C. botulinum (non-proteolytic strains)	Cooking, time-temperature control, pH and a_w of products	HACCP
	C. perfringens	Time-temperature control during cooking, cooling, and storage	HACCP
	E. coli (pathogenic)	Cooking, controlled fermentation and aging, chilled storage, avoid recontamination	GHP
	L. monocytogenes	Cooking, controlled fermentation and aging, chilled storage, avoid recontamination	GHP
	M. bovis	Eradication of TB in cattle, pasteurization of milk	
	Salmonella (non-typhoid)	Cooking, controlled fermentation and aging, chilled storage, avoid recontamination	GHP
	S. typhi	Personal hygiene	
	Shigella	Cooking, water quality, waste water management, personal hygiene	
	S. aureus	Cooking, controlled fermentation and aging, chilled storage, avoid recontamination	GHP
	V. cholerae	Water quality, waste water management, personal hygiene	GHP
	V. parahaemolyticus	Raw material selection, avoid eating raw fish	GHP, HACCP
	V. vulnificus	Avoid certain foods	
	Y. enterocolitica	Separation of raw from ready-to-eat, extra hygiene at slaughter	GHP

continues

Table 1-1 continued

	Hazard	Control measures	Other measures
Viruses	Hepatitis A	Water quality, personal hygiene, consumer education	Controls at harvesting
	Norwalk and other SRSVs	Water quality, personal hygiene, consumer education, cooking	Controls at harvesting
Parasites	*Trichinella spiralis*	(at farms) limit access to farms, rodent control, freezing, cooking, control at slaughter for animals at risk (outdoor farmed pigs, game, esp. wild boars), freezing, cooking	Animal health
	Toxoplasma gondii	Meat: freezing, cooking; vegetables: wash thoroughly	
Toxigenic fungi	*Aspergillus, Fusarium, Penicillium* spp.	Raw material selection, sorting, dry storage, dehydration	GHP
Seafood toxins	Ciguatera	Controls on harvesting, consumer education	GHP, HACCP
	Scombroid poisoning	Hygiene, temperature control	GHP
	Shellfish intoxication	Controls on harvesting, consumer education	

Microbiological testing is time-consuming, often lacks sensitivity and specificity, and the levels of sampling routinely applied have a low probability of detecting defective lots when the proportion of defectives within the lot is low (see Chapters 6–7). Recognizing the limitations of end-product testing to ensure microbiological safety at port-of-entry, the ICMSF proposed a system of verification based on the use of GHP in combination with the HACCP system as a more reliable means of assuring product safety in the modern food industry (ICMSF, 1988).

Much of the information needed in HACCP to judge whether microorganisms grow, survive, or die during food processing, distribution, and use is contained in the scientific literature, but this information is not organized in a way that is convenient for those in the food industry needing to use that information. Hence, the ICMSF compiled published information that was judged by experts to be reliable in a series of easily used tables (ICMSF, 1996). This was a conscious step towards promoting the newly emerging concept of food safety management systems, based on Codex documents. The ICMSF then recognized that much of the information in the ICMSF Book 3 (1980a,b) was out of date and did not consider the newly emerged pathogenic bacteria, such as *Listeria monocytogenes* and *Campylobacter jejuni/coli*, or many of the newer processes used in food manufacture. Consequently, the ICMSF updated its reviews of commodities in ICMSF Book 6 (1998a), but deliberately omitted consideration of sampling plans and microbiological criteria to emphasize that food management systems were replacing the earlier approach of end-product testing and offered better control of microbiological hazards. That evolution of management systems continues in this book, with end-product testing merely one of several components of ensuring food safety. Different types of sampling plans are considered, some more intensive than attributes plans used at port-of-entry, and intended for use when attempting to identify a problem and its source.

1.5 STATUS OF FOODBORNE ILLNESS: ETIOLOGIC AGENTS OR CONTAMINANTS

1.5.1 Bacteria

Periodically, overviews summarize trends in foodborne disease (CAST, 1994, 1998; POST, 1997; Bean et al., 1990, 1997). Diseases caused by foodborne pathogens constitute a worldwide public health problem (Tauxe, 1997; WHO, 1997) influenced by demographics, industrialization and centralization of food production and supply, travel and trade, and microbial evolution and adaptation. The symptoms of foodborne illness range from mild to severe gastroenteritis, to life-threatening disease. Foodborne disease is commonly acute, but also can become chronic with long-term sequelae. In industrialized countries, the most common causes of foodborne illness are salmonellae, *Staphylococcus aureus*, *Clostridium perfringens*, and *Vibrio parahaemolyticus*, with thermophilic *Campylobacter* spp. assuming greater importance (Altekruse et al., 1999). Until the mid-1990s in many countries, the combined annual number of outbreaks of illness attributed to *Salmonella* spp., *S. aureus*, and *C. perfringens* often represented 70–80% of the reported outbreaks due to bacteria. *C. jejuni/coli*, *E. coli* O157:H7, *L. monocytogenes* and *Cyclospora cayetanensis* were not recognized as pathogens 20 years ago (Mead et al., 1999).

Salmonellae have for many years been considered the most important foodborne pathogens worldwide. Foods implicated in outbreaks of salmonellosis include meat and poultry, eggs and egg products, milk and milk products, fresh produce (de Roever, 1998), and spices. In many countries, *Salmonella enteritidis* [*Salmonella enterica* serovar Enteritidis] in poultry and in eggs increased greatly from the mid-1980s to become a serious problem, but outbreaks and sporadic cases attributed to this pathogen have apparently declined in the UK from 1996–1997 (ACMSF, 1993, 1996, 2001).

In recent years, the incidence of disease caused by thermophilic *Campylobacter* spp. has exceeded that due to salmonellae in some industrialized countries. Outbreaks of campylobacteriosis are rare, most cases being sporadic, and commonly attributed to undercooked poultry or cross-contamination from raw poultry. Other foods, untreated water, and raw milk also have been implicated (Altekruse *et al.*, 1999).

In countries where raw fish is an important part of the diet, disease caused by *V. parahaemolyticus* is frequent. Occasional outbreaks occur in western countries, but there the vehicle of transmission is usually processed rather than raw seafoods. *Vibrio cholerae* is endemic in many tropical countries, and water plays a major role in the epidemiology of cholera.

Shigella spp. also represent an important public health problem in many developing countries. Cases of shigellosis reported in developed countries are often associated with traveling, food handlers, and day care centers. Because the reservoir of *Shigella* spp. is restricted to humans, the source of infection is food or water contaminated by human carriers.

Disease caused by *Yersinia enterocolitica* occurs worldwide and has been primarily associated with the consumption of raw or undercooked pork.

Escherichia coli strains are a common part of normal microbial flora of animals, including humans. Most strains are harmless, but some cause diarrhea. Strains carrying particularly virulent properties have emerged as a serious hazard, with consumption of even low numbers of these organisms bearing a risk for life-threatening illness. During the last two decades, enterohaemorrhagic *E. coli* (EHEC), producing verocytotoxins (VTs), also called Shiga-like toxins (SLTs), have emerged as a serious foodborne hazard. Many different serotypes of *E. coli* produce VTs or SLTs. Initially, most human outbreaks were due to *E. coli* O157:H7, although in Australia serogroup O111 is the most common cause of illness. Human illness has now been associated with many verocytotoxin-producing serotypes of *E. coli*. Human cases of EHEC were initially associated with consumption of undercooked ground beef (Bell *et al.*, 1994) and, occasionally, unpasteurized milk. Outbreaks have also been traced to unpasteurized apple juice ("apple cider" in the US), vegetable sprouts (Mahon *et al.*, 1997), yogurt, fermented sausage, water, and contact with farm animals (Doyle *et al.*, 1997).

Disease caused by *L. monocytogenes* is not frequent, but can be severe, with a high mortality rate in populations at risk, such as infants, pregnant women, and the immunocompromised. The bacterium is ubiquitous. Foods implicated in outbreaks have included products made from raw milk, butter, ready-to-eat meat products, surimi, smoked mussels and trout, and vegetables.

Botulism occurs relatively infrequently, but remains a serious concern because of the life-threatening nature of the disease and the impact on trade of the incriminated product type. For many years, commercially processed foods were less frequently involved than home-canned or home-prepared foods, but more recently several commercially prepared products have been implicated. Faulty processing and/or inappropriate storage temperatures have been the most common reasons for botulism, with home-processed foods and foods mishandled in food service establishments also responsible. For example, hot sauce containing jalapeño peppers prepared without adequate heating, potato salad prepared with foil-wrapped baked potatoes previously stored at room temperature, sautéed onions in butter stored unrefrigerated and then served on a sandwich, and tiramisu-mascarpone cheese.

With improved refrigerated storage during food distribution and use, *S. aureus* and *C. perfringens* now cause illness only when there has been temperature abuse. Also with improved refrigeration, the shelf-life of many foods has lengthened, leading to a new concern that psychrotrophic pathogens may increase to dangerous levels without spoilage being evident to the consumer. Microorganisms of most concern are nonproteolytic strains of *C. botulinum* types B, E, and F, *L. monocytogenes* and *Y. enterocolitica*, all of which cause little or no deterioration of the food supporting their growth.

Other foodborne bacterial pathogens include *Streptococcus pyogenes*, *Mycobacterium tuberculosis*, *Brucella abortus*, and *Bacillus cereus*.

1.5.2 Viruses

Hepatitis A virus and viruses known as Small Round Structured Viruses (SRSV), including Norwalk virus, are known causes of foodborne illness. Viruses are occasionally involved in large outbreaks (Weltman *et al.*, 1996), but the true extent and importance of viruses in foodborne illness has not been adequately assessed (ACMSF, 1995). Mead *et al.* (1999) estimated that viruses are more important than bacteria and protozoa as a cause of foodborne illness. As methods of detection improve and national initiatives on surveillance and ascertainment are intensified, SRSVs have been recognized as the most important cause of nonbacterial gastroenteritis throughout the world (Caul, 2000). Live bivalve molluscs are often implicated in viral foodborne disease outbreaks (Halliday *et al.*, 1991).

1.5.3 Protozoa

Foodborne protozoa have also been incriminated in large outbreaks, e.g., *Cryptosporidium parvum* from apple juice and water (Osewe, 1996), and *Cyclospora cayetanensis* from raspberries, lettuce, and water (Speer, 1997). In immunocompromised persons, diarrhea may be severe, making the illness serious and difficult to treat. Large outbreaks in North America have had a huge impact on international trade in soft fruit and salad vegetables because if present on the initial crop, these protozoa are almost impossible to eliminate. Illness due to *Toxoplasma gondii* is also much more serious in immunocompromised persons and pregnant women. Sporadic cases of protozoal infection have been linked to consumption of undercooked meat or fish products.

1.5.4 Seafood Toxins

Disease caused by histamine and other biogenic amines can arise from several foods, notably scombroid fish species. In North America, illness attributed to histamine is the second most common disease from fish, excluding shellfish (MMWR, 2000).

The principal intoxications having a microbiological origin in seafoods include paralytic shellfish poisoning (PSP), diarrhetic shellfish poisoning (DSP), neurotoxic shellfish poisoning (NSP), amnesic shellfish poisoning (ASP) (also known as domoic acid poisoning), ciguatera, and scombroid (histamine) fish poisoning. PSP, DSP, and NSP are caused by toxins produced by dinoflagellates, and ASP by a diatom. All these diseases typically result from consumption of bivalve mollusks that have been feeding on the toxigenic algae. The toxin(s) causing ciguatera are derived from toxigenic microalgae growing in and around tropical coral reefs and passed up the marine food chain through herbivorous reef fish to more far-ranging carnivorous species. Humans typically become intoxicated from eating the toxic fish. Histamine or scombroid poisoning is caused by consumption of fish containing high levels of histamine (and other biogenic amines) resulting from histidine dehydrogenase activity of bacteria multiplying on the fish after death. With the exception of histamine and other biogenic amines, toxin accumulation is passive. All the seafood toxins are resistant to heating and, therefore, cannot be destroyed by cooking. They are undetectable organoleptically (ICMSF, 1996; Liston, 2000; Whittle & Gallacher, 2000).

1.5.5 Toxigenic Fungi

Mycotoxins that have a role in human disease are toxic metabolites produced by common fungi when they grow in susceptible crops. The most important compounds are aflatoxins, produced by *Aspergillus flavus* and *A. parasiticus* in peanuts and maize, which cause primary liver cancer and are responsible for many deaths annually, especially in developing countries. A role in human disease is also probable for fumonisins, produced by *Fusarium verticilloides* and *F. proliferatum* during growth in maize, which have been implicated in human esophageal cancer. Other mycotoxins with a role in human disease include ochratoxin A, produced by *Aspergillus ochraceus* in stored foods, *A. carbonarius* in grapes and grape products, and *Penicillium verrucosum* in cereals, which probably has a role in kidney disease in wide areas of Europe. Trichothecene toxins, produced by *Fusarium graminearum* and related species in cereals, cause immunosuppresssion and consequently have an ill-defined but potentially important role in reducing disease resistance. Some evidence indicates that aflatoxins are also immunosuppressive; these compounds are widespread in foods in some countries and may have an important but largely unsuspected role in susceptibility to a wide variety of diseases.

1.5.6 Surveillance of Foodborne Disease

Although foodborne infections have been recognized as a major cause of human illness for many years, the true incidence remains unknown (Motarjemi & Käferstein, 1997; Clark *et al.*, 2000). In Canada and the US, for example, the number of cases and outbreaks that actually occur is estimated to be many times greater than the reported figures (Todd, 1996, 1997; Buzby & Roberts, 1997; Mead *et al.*, 1999). Recent studies of infectious diseases estimating the extent of underreporting include Notermans & Hoogenboom-Verdegaal (1992) for the Netherlands, and Wheeler *et al.* (1999) and FSA (2000) for England.

Such underreporting is unfortunate, for in addition to providing a continuing assessment of trends in etiologic agents and food vehicles, as indicated above, prompt reporting and investigation of outbreaks provides for (a) identification and removal of contaminated products from the market, (b) correction of faulty food-preparation practices in food-service establishments, processing plants, and homes, (c) identification and appropriate treatment of human carriers of foodborne pathogens, (d) possible detection of new agents of foodborne disease, (e) a better understanding of the effectiveness of regulatory policy and/or its implementation, and (f) improved understanding of the epidemiology of various pathogens.

Common systems for detecting and monitoring foodborne illness are essential for identifying trends in different countries. In the US, there was a decline in foodborne illness in 1999 compared with 1996–1998 (MMWR, 2000), primarily due to decreases in campylobacteriosis and shigellosis, with salmonellosis not changing. In the UK, campylobacteriosis has remained at approximately the same level since 1995, while cases due to *S. enteritidis* and *S. typhimurium* have declined dramatically since 1997 (www.phls.co.uk). Campylobacteriosis has become recognized as the most common cause of bacterial foodborne illness in many developed countries. Also, the predominant salmonellae strains vary with each country, but for the past 5–10 years the two mentioned have been of greatest concern worldwide. The data for the UK may not apply to the US or to other countries. Each country tracks all the strains of salmonellae and may show a figure for trends for the top one or two or all of them. A greater awareness worldwide of foodborne diseases and consumer concerns points to the need for more effective control measures using internationally accepted methods of risk management.

1.6 PRACTICES CONTRIBUTING TO FOODBORNE ILLNESS

A relatively low proportion of illness is traced back to a particular food. Where that has been possible, the common contributory factors are very similar in different countries. For example, inadequate thawing prior to cooking occurred on numerous occasions with large Christmas turkeys. The subsequent cooking failed to kill salmonellae in the center of the partially thawed bird, resulting in salmonellosis.

Inadequate temperature control after cooking includes preparation too far in advance of consumption; storing at warm ambient (not refrigeration) temperatures; and attempting to cool large amounts of heated food that cannot be cooled quickly, resulting in growth of surviving bacteria during the slow cooling, and consumption without reheating above ca 63 °C to kill the vegetative cells of, e.g., *Bacillus cereus* in rice or *C. perfringens* in meat or gravy.

Contamination of a cooked, safe food with microorganisms from raw food has resulted in illness, e.g., at barbecues.

1.7 IMPORTANCE OF EFFECTIVE CONTROL MEASURES

An effective food safety management system almost always comprises a number of control measures, such as raw material selection, hygienic handling prior to a process, correct application of the process, and GHP during and subsequent to processing (see Chapter 3).

At the industrial level, heat processes have been developed for low-acid canned foods to control *C. botulinum* spores. Less severe heat processes are adequate for acid or acidified foods (pH 4.5 or below) because *C. botulinum* spores are unable to multiply at those low pH values.

Spores of *C. perfringens* in beef survive most cooking processes and are able to multiply if the cooked meat is held at suitable temperatures. Growth is minimized by cooling quickly through the range of temperatures supporting growth (50–15 °C).

At the catering level, other controls can be applied. *Escherichia coli* O157:H7 occurs sporadically in ground (minced) beef used in hamburgers. Heating sufficiently will eliminate the hazard, and appropriate heat processes have been developed to ensure that the center of the meat patty reaches a temperature that is lethal to pathogenic *E. coli*, taking account of the weight, thickness, and initial temperature of the patty, the temperature of the grill, and the duration of cooking.

To prevent salmonellosis from *S. enteritidis* in eggs at the domestic level, vulnerable persons, such as the elderly and the immunocompromised, should consume only eggs that have been cooked until the yolk is solid, or use a commercially pasteurized product. Numerous cases of salmonellosis have been traced to using raw eggs in homemade mayonnaise and tiramisu. Handling raw poultry in the kitchen can spread thermophilic *Campylobacter* spp. to other working surfaces. Hands should be washed frequently and thoroughly after handling raw poultry.

1.8 EFFECTIVENESS OF GHP AND HACCP

HACCP has proven very effective as a control system for many food processes, especially those with a step that eliminates the hazard (e.g., canning of low-acid foods, cooking to eliminate salmonellae or *L. monocytogenes*), or foods that are formulated to prevent microbial multiplication (e.g., low a_w, low pH). Of greater concern are processes that contain no step that can prevent or eliminate a known hazard.

Reporting foodborne illnesses in the media and prompt distribution of information via the Internet fuel public perception that food safety is declining. In reality, today's food safety systems are stronger than ever before due to better implementation of GHP and HACCP. Where records are available, the majority of recalls are due to failures to adhere to GHP, rather than in failures in the GHP/HACCP plans. In addition improved laboratory and detection methods allow better recognition of potential problems, possible agents, and hazard-food combinations. Hence the safety of foods is improving, despite recognizing new agents and new hazard-food combinations. Some new pathogens force reconsideration of traditional control measures, e.g., VT-producing *E. coli*, present sporadically and usually in low numbers in cattle, are difficult to detect and control. When new agents are recognized, government and the food industry respond to control them, but the consumer must also recognize that it may take time to understand the conditions leading to illness and the changes necessary to control them.

The current trends in food processing—to reduce the extent of heating, to minimize the use of chemical preservatives, and to provide foods that require little or no preparation or are ready-to-eat and consequently not subjected to heating prior to consumption—all increase the likelihood of pathogens reaching the consumer. Even with the greatest care in agricultural practices, it is not possible to eliminate all pathogens from raw agricultural and seafood commodities. In ICMSF Book 4 (1988), critical control points (CCPs) were divided into those that eliminate the hazard, CCP1, and those that can reduce but cannot prevent or eliminate the hazard, CCP2. Many HACCP plans would be better understood if that differentiation were reintroduced.

1.9 WOULD AN FSO IMPROVE FOOD SAFETY AND REDUCE FOODBORNE ILLNESS?

Although HACCP plans are now more widely implemented, the main weaknesses are that the level of control needed is not stated clearly, and there is little or no guidance on what is expected of an adequately designed and implemented HACCP plan. This omission is widespread in documents prepared by Codex, many advisory groups, and in governmental regulations. An FSO would indicate the level of control needed for adequate GHP and HACCP systems.

Another issue of concern in relation to food safety management is the continued indiscriminate use of microbiological testing of the end products. That testing is usually inadequate in terms of sampling and the number of samples tested and often poorly targeted in terms of the hazard(s) most likely in the particular product. With the integration of HACCP and GHP in the 1990s, there has been no decrease in end-product testing that might have been expected with increased control. If anything, such testing has continued to increase.

No risk management options will provide absolute safety (i.e., there will always be some risk).

1.10 USE OF FSOs IN FOOD SAFETY MANAGEMENT

When considering the level of tolerable risk for a given hazard-food combination, risk managers should seek input from risk assessors and stakeholders, such as the affected industry and consumers. Risk is an estimate of the probability and severity of the adverse effects in exposed populations that may result from a hazard in a food. The ALOP is the achieved or achievable level proposed following consideration of public health impact, technological feasibility, economic implications, and comparison with other risks in everyday life.

Food operators cannot address an objective that states, for example, "there shall be no more than 20 domestic cases of a certain foodborne illness per 100,000 inhabitants per year in a country" (i.e., tolerable level of risk). While this may be a desirable goal, it requires the collective efforts of many parties. Food operators can only address factors over which they can exercise control. While all operators along the food chain must understand their role and manage their operation to satisfy an FSO, they cannot assume responsibility for the actions of all others along the food chain. It is important that each FSO clearly communicates the level of hazard that is considered tolerable in such a manner that food operators can effectively establish control measures to meet the objective.

FSOs are a relatively new concept offering considerable advantages when incorporated into a risk management framework. They are statements of the maximum frequency and/or concentration of a microbiological hazard in a food considered tolerable for consumer protection. Whenever possible, FSOs should be quantitative and verifiable.

FSOs can play an important role in modern food safety management systems by linking information from the risk assessment and risk management processes with measures to control the identified risk(s). Basic information is provided in Chapter 3 that can be used to establish scientifically based control measures.

FSOs should be flexible because information is often lacking on the relevant properties of the hazard, the factors that lead to adverse public health effects, the conditions necessary to control the hazard, and how control measures can be implemented effectively in the food chain. This is a common situation with newly recognized or emerging hazards, e.g., VT *E. coli*. As more information becomes available, risk assessments should be updated and FSOs adjusted accordingly.

For foods in international commerce, FSOs should be developed within the Codex framework. This is consistent with the concepts of the World Trade Organization and the SPS agreement and provides a framework for harmonization of acceptance criteria for foods in international trade. Food safety criteria developed in one country frequently differ from those of other countries. The principles presented in this text can lead to a scientific basis for comparing the relative level of protection afforded by different food safety systems. These principles are applicable to issues of equivalency, levels of protection, and non-tariff trade barriers. Their application should facilitate the harmonization of international trade where the practices of one country may differ from those of another, yet the practices of both countries provide safe products. Furthermore, they can be applied by control authorities and food operators for the establishment of equivalent criteria. FSOs are a preferred approach to food safety management because they focus on protection of human health while offering flexibility in achieving that goal.

FSOs differ from microbiological criteria by communicating the level of hazard that is considered tolerable for consumer protection. FSOs specify goals that can be incorporated into the design of control measures (e.g., GHP, HACCP) for the production and preparation of foods. FSOs can also provide the basis for assessing the adequacy and effectiveness of control systems adopted by industry and inspection systems adopted by regulatory authorities. FSOs are limited to food safety and do not address quality.

1.11 PERFORMANCE, PROCESS, PRODUCT, AND DEFAULT CRITERIA

When designing and controlling food operations, it is necessary to consider pathogen contamination, destruction, survival, growth, and possible recontamination. Consideration should also be given to subsequent conditions to which the food is likely to be exposed, including processing and potential abuse (time, temperature, cross-contamination) during storage, distribution, and preparation for use. The ability of those in control of foods at each stage in the food chain to prevent, eliminate, or reduce food safety hazards varies with the type of food and the effectiveness of available technology. Since GHP and HACCP are the primary tools available to help industry control microbial hazards in food operations, it is essential that the technical achievability of the FSO be confirmed.

A *performance criterion* is the required outcome of one or more control measures at a step or combination of steps that contribute to ensuring the safety of a food. When establishing performance criteria, account must be taken of the initial levels of the hazard and changes of the hazard during production, processing, distribution, storage, preparation, and use. An example of a performance criterion is a 6D kill of salmonellae when cooking ground beef or <10% of fresh or frozen broilers contaminated with salmonellae.

Process criteria are the control parameters (e.g., time, temperature, pH, a_w) at a step or combination of steps that can be applied to achieve a performance criterion. For example, the control parameter for milk pasteurization in the US is 71.7 °C for 15 s (FDA, 1997). This combination of temperature and time will ensure the destruction of *Coxiella burnetii*, as well as other non-spore-forming pathogens that are known to occur in raw milk.

Product criteria consist of parameters that ensure that the level of a hazard does not increase to unacceptable levels before preparation or consumption. They also can be used to assess the acceptability of a food. There is increasing recognition and acceptance that the microbial response in foods is dependent on the composition and environment in the food. Consequently, measurement of pH, water activity, temperature, and gas atmosphere affords a more rapid means of judging the safety of particular foods in which those factors are the main factors determining food safety. A food could be considered acceptable, for example, if it has been determined that a certain pH (e.g., pH 4.6 or below) or a_w (e.g., 0.86 or below) ensures that the food will meet an FSO for growth of a pathogen (e.g., *C. botulinum* or *S. aureus*).

Default criteria are conservative values established to ensure the safety of a process or a food. If insufficient resources are available to perform the research needed to arrive at sound process or product criteria, then default values can be applied. An example of a default value is heating for 10 min at 90 °C internal temperature to destroy nonproteolytic *C. botulinum* in extended shelf-life ready-to-eat chilled foods (ACMSF, 1992). Default values have most commonly been developed by control authorities or advisory groups. The values specify the minimum criteria that must be met to ensure the production of safe food.

1.12 ESTABLISHMENT OF CONTROL MEASURES

Control measures are actions and activities that prevent or eliminate a food safety hazard or reduce it to a tolerable level. One or more control measures may be necessary at each stage along the food chain to ensure that a food is safe when consumed.

From the information provided in an FSO, regulatory authorities and food operators can select appropriate control measures to achieve the intended results. In practice, FSOs are met through the establishment and implementation of performance, process, and product criteria, through the application of GHP and HACCP. To compensate for process variability and ensure that FSOs are consistently met, industry may implement more stringent performance criteria for their end products.

Because of increasing recognition of the importance of contamination from the environment in which ready-to-eat foods are produced, information is provided on methods to assess the effectiveness of the GHP control measures (Chapter 11).

1.13 ASSESSMENT OF CONTROL OF A PROCESS

Food management systems that incorporate the above principles will require some means of verification to ensure that the systems are being implemented as planned. Criteria, e.g., process, product, etc., may have been established to serve as a basis for meeting an FSO and can be used to assess whether a process is under control. A process is deemed to be under control when correct procedures are being followed and established criteria are being met. The procedures that are followed for assessing and adjusting control of a process will ideally be based upon the same principles applied in selecting the control measures. Statistical process control may be necessary when validating, monitoring, and assessing control of a process. Information on the use of statistical process control charts to monitor the performance of food operations is provided in Chapter 13.

1.14 ACCEPTANCE CRITERIA

Acceptance criteria are statements of conditions that differentiate acceptable from unacceptable food operations. Acceptance criteria may be sensory, chemical, physical, or microbiological and should specify ancillary information, such as the number of samples to be collected, how and where the samples are collected and held prior to analysis, the analytical unit, the method of analysis, and the range of values considered acceptable. The assessment can be performed by a control authority, a customer, or even by an independent auditor hired by the food operator, each for a different purpose. Acceptance criteria are also used to assess the acceptability of individual lots or consignments of food (Chapter 5).

1.15 MICROBIOLOGICAL CRITERIA

Under certain circumstances, microbiological criteria may be established to determine the acceptability of specific production lots of food, particularly when the conditions of production are not known. Microbiological criteria should specify the number of sample units to be collected, the analytical method, and the number of analytical units that should conform to the limits (Chapter 5). Criteria may be established for quality as well as safety concerns (CAC, 1997). The composition of a *lot* (batch) of food product is considered in Chapter 6. Only in well-mixed liquid foods does the distribution of microorganisms approximate to homogeneity. If the distribution of microorganisms differs greatly from log-normal, the sensitivity of attributes plans is affected (see Chapter 7). The number of microbiological tests applied routinely to a lot of food product at port-of-entry is rarely adequate, in a statistical sense, to detect low levels of defectives (e.g., salmonellae in dried milk or dried egg). Moreover, random sampling is often not possible, which influences the statistical interpretation of the results. Greater reliability can be attained by acquiring food from suppliers known to have HACCP and GHP in place and with a record of trouble-free production. If a lot is sampled, the stringency of the sampling plan should reflect risk to consumers (Chapter 8).

In the event samples are collected for microbiological analysis, the method of collecting and handling the samples is very important. Otherwise, the analytical results may have no bearing on the acceptability of the food. These factors are briefly summarized in Chapter 12.

Chapters 14 to 17 provide four examples to illustrate how the principles can be applied. Each example discusses a different aspect of control that may be necessary in a food safety management system.

1.16 MICROBIOLOGICAL TESTS

Microbiological tests are used for different purposes. It is very important to consider for which purposes microbiological testing is used. The purpose determines the type of test (indicator or pathogen), the method (rapidity, accuracy, repeatability, reproducibility, etc.), the sample (line-residue or end product), the interpretation of the result, and action to be taken (rejection of a lot, investigational sampling, readjustment of the process, etc.). Table 1–2 shows some of the many different aspects of microbiological testing that are discussed in the following chapters.

1.17 SUMMARY

The purpose of this book is to introduce the reader to a structured approach to managing food safety. The uses and limitations of testing for control of microbiological hazards are discussed. In addition, lot acceptance testing within such safety management systems is discussed with reference to its strengths and weaknesses. The text describes the use of existing Codex documents to develop stronger, more reliable food safety management systems.

Application of Codex documents in a logical sequence, together with FSOs, can provide the basis for addressing issues of equivalency, levels of protection, and non-tariff trade barriers.

Table 1-2 Examples of Microbiological Testing in Food Safety Management

Type of testing	Purpose	User	Sample type	Sampling plan	Microbes
Acceptance	Lot inspection	Government	End products	Attributes	Pathogens, indicators
Acceptance	Verification, lots (batches) of known history	Government	End products	Attributes	Pathogens, indicators
		Industry	Raw materials	Attributes	Pathogens, indicators
Monitoring, checking	CCPs, lines	Industry	Line samples	Variables, attributes	Indicators
Environmental sampling	Line, environments	Industry	Residues, dust, water	Targeted, to find source of contamination	Indicators
Verification	HACCP	Industry	End products	Attributes	Pathogens, indicators
Surveillance	Compliance	Governments, industry	Products in commerce	Attributes (usually $n=1$)	Pathogens
Investigation	Food chain	Governments, industry	All types of samples	Investigational, rarely statistically based	Pathogens

1.18 REFERENCES

ACMSF (Advisory Committee on Microbiological Safety of Food, UK) (1992). *Report on Vacuum Packaging and Associated Processes*. London: HMSO.

ACMSF (Advisory Committee on Microbiological Safety of Food, UK) (1993). *Report on* Salmonella *in Eggs*. London: HMSO.

ACMSF (Advisory Committee on Microbiological Safety of Food, UK) (1995). *Report on Foodborne Viral Infections*. London: HMSO.

ACMSF (Advisory Committee on Microbiological Safety of Food, UK) (1996). *Report on Poultry Meat*. London: HMSO.

ACMSF (Advisory Committee on Microbiological Safety of Food, UK) (2001). *Second Report on* Salmonella *in Eggs*. The Stationery Office, London.

Altekruse, S. F., Stern, N. J., Fields, P. I. & Swerdlow, D. L. (1999). *Campylobacter jejuni*—an emerging foodborne pathogen. *Emerg Infect Diseases* 5, 28–35.

Bean, N. H., Goulding, J. S., Daniels, M. T. & Angulo, F. J. (1997). Surveillance for foodborne disease outbreaks—United States, 1988–1992. *J Food Prot* 60, 1265–1286.

Bean, N. H., Griffin, P. M., Goulding, J. S. & Ivey, C. B. (1990). Foodborne disease outbreaks, 5 year summary 1983–1987. *J Food Prot* 53, 711–728.

Bell, B. P., Goldoft, M., Griffin, P. M. *et al.* (1994). A multistate outbreak of *Escherichia coli* O157:H7-associated bloody diarrhea and hemolytic syndrome from hamburgers: the Washington experience. *J Am Med Assoc* 272, 1349–1353.

Buzby, J. C. & Roberts, T. (1997). Economic costs and trade impacts of microbial foodborne illness. *World Health Stats Quarterly* 50, 57–66.

CAC (Codex Alimentarius Commission) (1997). *Joint FAO/WHO Food Standards Programme, Codex Committee on Food Hygiene. Food Hygiene, Supplement to Volume 1B-1997. Principles for the Establishment and Application of Microbiological Criteria for Foods.* CAC/GL 21-1997. Secretariat of the Joint FAO/WHO Food Standards Programme. Rome: Food and Agriculture Organization of the United Nations.

CAST (Council for Agricultural Science and Technology, USA) (1994). *Foodborne Pathogens: Risks and Consequences.* Task Force Report No. 122. Ames, IA: CAST.

CAST (Council for Agricultural Science and Technology, USA) (1998). *Foodborne Pathogens: Review of Recommendations.* Special Publication No. 22. Ames, IA: CAST.

Caul, E. O. (2000). Foodborne viruses. In *The Microbiological Safety and Quality of Food*, Vol. II, pp. 1457-1489. Edited by B. M. Lund, T. C. Baird-Parker & G. W. Gould. Gaithersburg, MD: Aspen Publishers, Inc.

Clark, J., Sharp, M. & Reilly, W. J. (2000). Surveillance of foodborne disease. In *The Microbiological Safety and Quality of Food*, pp. 975-1010. Edited by B. M. Lund, T. C. Baird-Parker & G. W. Gould. Gaithersburg, MD: Aspen Publishers, Inc.

de Roever, C. (1998). Microbiological safety evaluations and recommendations on fresh produce. *Food Control* 9, 321-347.

Doyle, M. P., Zhao, Z., Meng, J. & Zhao, S. (1997). *Escherichia coli* O157:H7. In *Food Microbiology: Fundamentals and Frontiers*, pp. 171-191. Edited by M. P. Doyle, L. R. Beuchat & T. J. Montville. Washington, DC: ASM Press.

FDA (Food and Drug Administration) (1997). Milk and cream, pasteurized (21 CFR 131.3b), *Code of Federal Regulations.* Washington, DC: U.S. Government Printing Office.

FSA (Food Standards Agency (UK)) (2000). *A Report of the Study of Infectious Intestinal Disease in England.* London: The Stationery Office.

Halliday, M. I., Kang, L. Y., Zhou, T. K. *et al.* (1991). An epidemic of hepatitis A attributable to the ingestion of raw clams in Shanghai, China. *J Infect Diseases* 164, 852-859.

Hathaway, S. C. (1997). Development of risk assessment guidelines for foods of animal origin in international trade. *J Food Prot* 60, 1432-1438.

Hathaway, S. C. & Cook, R. L. (1997). A regulatory perspective on the potential uses of microbial risk assessment in international trade. *Int J Food Microbiol* 36, 127-133.

ICMSF (International Commission on Microbiological Specifications for Foods) (1980a). *Microbial Ecology of Foods. Vol. 1. Factors Affecting Life and Death of Microorganisms.* New York: Academic Press.

ICMSF (International Commission on Microbiological Specifications for Foods) (1980b). *Microbial Ecology of Foods. Vol. 2. Food Commodities.* New York: Academic Press.

ICMSF (International Commission on Microbiological Specifications for Foods) (1986). *Microorganisms in Foods 2: Sampling for Microbiological Analysis: Principles and Specific Applications*, 2nd edn. Toronto: University of Toronto Press.

ICMSF (International Commission on Microbiological Specifications for Foods) (1988). *Microorganisms in Foods 4. Application of the Hazard Analysis Critical Control Point (HACCP) System to Ensure Microbiological Safety and Quality.* Oxford: Blackwell Scientific Publications Ltd.

ICMSF (International Commission on Microbiological Specifications for Foods) (1996). *Microorganisms in Foods 5: Characteristics of Microbial Pathogens.* Gaithersburg, MD: Aspen Publishers, Inc.

ICMSF (International Commission on Microbiological Specifications for Foods) (1998a). *Microorganisms in Foods 6: Microbial Ecology of Food Commodities.* Gaithersburg, MD: Aspen Publishers, Inc.

ICMSF (International Commission on Microbiological Specifications for Foods) (1998b). Principles for establishment of microbiological food safety objectives and related control measures. *Food Control* 9, 379-384.

Jouve, J. L. (1992). HACCP et Systèmes Qualité (ISO 9000). *Option Qualité*, 97, 11-15.

Jouve, J. L. (ed.) (1994, 1996). *La Qualité Microbiologique des Aliments: Maitrise et Critères.* 1st edn., 2nd edn., Paris: Centre National d'Etudes et de Recherches sur la Nutrition et l'Alimentation/Centre National de la Recherche Scientifique.

Liston, J. (2000). Fish and shellfish poisoning. In *The Microbiological Safety and Quality of Food*, Vol. II, pp. 1518-1544. Edited by B. M. Lund, T. C. Baird-Parker & G. W. Gould. Gaithersburg, MD: Aspen Publishers, Inc.

Mahon, B. E., Ponka, A., Hall, W. N. *et al.* (1997). An international outbreak of *Salmonella* infections caused by alfalfa sprouts grown from contaminated seeds. *J Infect Diseases* 175, 876-882.

Mead, P. S., Slutsker, L., Dietz, V. et al. (1999). Food-related illness and death in the United States. *Emerg Infect Diseases* 5, 607–625.

MMWR (Morbidity and Mortality Weekly Reports) (2000). Preliminary FoodNet data on the incidence of foodborne illnesses—selected sites, United States, 1999. *Morbidity Mortality Wkly Rpts* 49, 201–205.

Motarjemi, Y. & Käferstein, F. K. (1997). Global estimation of foodborne diseases. *World Health Stats Quarterly* 50, 5–11.

Notermans, S. & Hoogenboom-Verdegaal, A. (1992). Existing and emerging foodborne diseases. *Int J Food Microbiol* 15, 197–205.

Osewe, P., Addiss, D. G., Blair, K. A. et al. (1996). Cryptosporidiosis in Wisconsin: A case-control study of post-outbreak transmission. *Epidemiol & Infect* 117, 297–304.

POST (Parliamentary Office of Science and Technology, UK) (1997). *Safer Eating: Microbiological Food Poisoning and Its Prevention.* London: POST.

Public Health Laboratory Service, UK. http://www.phls.co.uk.

Speer, C. A. (1997). Protozoan parasites acquired from food and water. In *Food Microbiology: Fundamentals and Frontiers*, pp. 478–493. Edited by M. P. Doyle, L. R. Beuchat & T. J. Montville. Washington, DC: ASM Press.

Tauxe, R. V. (1997). Emerging foodborne diseases: an evolving public health challenge. *Emerg Infect Diseases* 3, 425–434.

Todd, E. C. D. (1996). Worldwide surveillance of foodborne disease: the need to improve. *J Food Prot* 59, 82–92.

Todd, E. C. D. (1997). Epidemiology of foodborne diseases: a worldwide review. *World Health Stats Quarterly* 50, 30–50.

Weltman, A. C., Bennett, N. M., Ackman, D. A. et al. (1996). An outbreak of hepatitis A associated with a bakery, New York, 1994: The 1968 "West Branch, Michigan" outbreak repeated. *Epidemiol & Infect* 117, 333–341.

Wheeler, J. G., Sethi, D., Cowden, J. M. et al. (1999). Study of infectious intestinal disease in England: rates in the community, presenting to general practice, and reported to national surveillance, The Infectious Intestinal Disease Executive. *Brit Med J* 318, 1046–1050.

Whittle, K. & Gallacher, S. (2000). Marine toxins in health and the food chain. Edited by D. I. Thurnham & T. A. Roberts. *Brit Med Bull* 56, 236–253.

WHO (World Health Organization) (1997). Foodborne Safety and Foodborne Diseases, *World Health Stats Quarterly* 50, 154 pp.

WTO (World Trade Organization) (1995). The WTO Agreement on the Application of Sanitary and Phytosanitary Measures (SPS Agreement). http://www.wto.org/english/tratop_e/sps_e/spsagr_e.htm.

Chapter 2

Evaluating Risks and Establishing Food Safety Objectives

i2.1	Introduction	2.7	Evaluation of Risk by Quantitative Risk Assessment
2.2	Tolerable Level of Consumer Protection	2.8	Establishment of an FSO Based on Quantitative Risk Assessment
2.3	Importance of Epidemiologic Data	2.9	Stringency of FSOs in Relation to Risk and Other Factors
2.4	Evaluation of Risk		
2.5	Food Safety Objectives (FSOs)	2.10	Summary
2.6	Establishment of an FSO Based on a Risk Evaluation by an Expert Panel	2.11	References

2.1 INTRODUCTION

Although food products should be enjoyable and healthy, they sometimes cause illness. Consequently, societies charge institutions or organizations with defining the level of protection that should be achieved to ensure the health and safety of the public. In the case of food safety, this responsibility usually resides with food control authorities given this mandate by national or local legislation. Companies are responsible for ensuring the safety of their products. Government and industry* function as risk managers and share the common goal of ensuring that consumers can enjoy safe and wholesome foods.

Control authorities may use different approaches when responding to an emerging food safety concern or when seeking to enhance current levels of food safety. Their choice of actions to minimize risk to consumers depends on the circumstances and urgency of the situation. This flexibility is necessary because the factors surrounding food safety concerns vary (e.g., nature of the hazard, population affected, severity of the disease, frequency of occurrence, and potential for wider dissemination of the disease agent). It is neither possible nor desirable to prescribe specific steps for control authorities to follow when responding to food safety concerns. However, some general guidelines can be given.

---

*The term "industry" will mean an organization, company, or group of individuals (cooks) working professionally in the food chain from primary agricultural production to the sale to, or preparation of food for, the consumer. The particular meaning in the text will depend on the context in which it is used.

In addressing a food safety concern, risk managers must evaluate whether the situation—either a current concern or an increasing risk—is under sufficient control or whether there are sound reasons for concern. In many situations, concerns are raised that, upon closer examination, are already adequately managed by existing control measures or that do not constitute a public health issue. In the latter instance, a rapid decision must be made to avoid wasting time and money on issues that have little impact on public health.

As new food safety concerns are recognized, some understanding of the nature and properties of the hazard and how it leads to foodborne illness are essential for control. Risk managers in government and industry have been obliged to consider the frequency or concentration of the hazard that would be acceptable in foods and not cause illness when the food is handled and prepared as expected. Food safety managers have depended on epidemiologic studies to identify problems and determine their cause. This information then forms the basis for control options that could be applied to prevent, minimize, or reduce the hazard.

A formal process has seldom been applied to determine what a society or country would consider an appropriate level of consumer protection in regard to a foodborne microbiological hazard. Yet risk managers have such goals in mind when developing and implementing policies and strategies for control of microbiological hazards. There is a long history of implicitly, or intuitively, selecting public health protection options that provide the basis for existing, robust food safety management systems.

The level of protection may not be explicitly expressed, but may be estimated from data on incidences of domestically occurring illnesses. Taking foodborne listeriosis as an example, the reported prevalence in a number of countries ranges from 0.1–1.3 cases per 100,000 (Table 2–1). Most case are considered to be sporadic and not associated with identified outbreaks. Although foods are recognized as the primary source of listeriosis, little is known about factors leading to sporadic cases or how they may be reduced. Most

Table 2–1 Reported Incidence of Listeriosis in Selected Countries

Nation	Incidence estimate (cases/100,000/year)	Period	Comment
Australia	0.18–0.39	1991–2000	
Canada	0.1–0.2	1990–1999	
	0.17–0.45	1987–1994	
Denmark	0.48 / 0.64	1991/1992	
	0.75–0.88	1996–1998	
Germany	0.34	Pre –1984	
	0.25	Late 1990s	
France	0.68 / 1.30	1991/1992	
	0.38 / 0.67	1995/1996	
Italy	0.35	1991/1992	
Netherlands	0.13–0.19	1996–1999	
New Zealand	0.4/0.61	1991/1992	
Sweden	0.42	1990s	
UK	0.14–0.23	1984–1996	excludes outbreaks 1987–1989
	0.40–0.46	1987–1989	includes outbreaks 1987–1989
US	0.46	1983–1992	active reporting
	0.14	1983–1992	passive reporting

countries have, therefore, albeit perhaps not deliberately, set protection at current levels and respond to unusual increases above the country's baseline. This does not mean that future reductions in the incidence of listeriosis cannot be a goal of a country's food safety enhancement program. As the factors causing sporadic cases become clearer, food safety policies can be modified to reduce the incidence of listeriosis and thereby achieve a higher level of protection.

Knowledge of the actual level of protection in a society depends on the disease surveillance system. Not all countries have detailed epidemiologic data describing the current situation for every foodborne pathogen; however, some level of surveillance is in place in most countries. Moreover, analysis of a particular food production system will allow identification of the hazards and factors either increasing or controlling a particular risk.

Understanding the level of protection requires evaluating the public health risk associated with a concentration or frequency of the hazard in a food. The evaluation may be done in different ways depending on the issue, the understanding of the scientific basis, and the extent of disagreement between the parties involved. In practice, the evaluation ranges from a simple qualitative estimation of risk to a quantitative risk assessment (see sections 2.4 and 2.7).

Existing or enhanced food safety goals may be expressed in terms of risk. The risk that society regards as tolerable in the context of, and in comparison with, other relevant risks in everyday life is called the *tolerable level of risk* (TLR). The TLR is established following consideration of public health impact, technological feasibility, economic implications, etc. The TLR can be expressed as the number of illnesses occurring per annum due to a certain microbial hazard per 100,000 inhabitants of a country. A hypothetical example could be 0.5 case of listeriosis per 100,000 population per year. Although deciding on a TLR is a societal issue, sound scientific principles should underline the evaluation of risk.

The World Trade Organization Agreement on the Application of Sanitary and Phytosanitary Measures (WTO/SPS agreement) (WTO, 1995) defines the *appropriate level of sanitary or phytosanitary protection* (ALOP) as "the level of protection deemed appropriate by the Member [country] establishing a sanitary or phytosanitary measure to protect human, animal or plant life or health within its territory." The agreement recognizes that "many members refer to this concept as the acceptable level of risk." The International Commission on Microbiological Specifications for Foods (ICMSF) prefers the term *tolerable level of risk* (TLR) instead of *acceptable level of risk,* because risks related to the consumption of food are seldom accepted, but at best tolerated.

However, food operators cannot use a public health goal or level of protection to control the conditions of food processing to eliminate, prevent, or reduce microbiological hazards that may contribute to the incidence of a disease. In this book, the concept of a *food safety objective* (FSO) is introduced to allow risk managers to communicate effectively to industry and trade partners precise food safety goals. FSOs are stated as the maximum frequency and/or concentration of a microbiological hazard in a food at the time of consumption that provides the appropriate level of protection. FSOs are typically expressions of concentrations of microorganisms or toxins at the moment of consumption; the concentrations at earlier stages of the food chain are considered to be *performance criteria*. As described in Chapter 5 (Table 5–1), FSOs differ from microbiological criteria in being suitable to help design control of food operations, but FSOs are not intended for determination of lot acceptance. FSOs are useful when comparing the safety goals of different countries or trade partners and can assist in determining equivalence of seemingly different control measures used for health protection that may, in fact, be equivalent.

Depending on the urgency of the situation, the complexity of the hazard, and the level of disagreement, the FSO may be derived by advice from a few specialists, by larger expert panels, or by conducting a quantitative risk assessment. The FSO may be based on a realistic estimate of the risk, but can also, when time and/or knowledge are short, be based on a detailed examination of the frequency or concentration of a hazard that is expected to keep the situation under control.

Below, the concept of FSO is introduced as a tool to express and communicate in practical terms the desired level of consumer protection. The following sections also provide information on some of the tools that have been used to characterize the public health situation.

2.2 TOLERABLE LEVEL OF CONSUMER PROTECTION

The food chain—from primary production through harvesting, processing, marketing, distribution, and preparation for consumption—is complex. Hazards may enter along that entire chain, starting with the source of the food and ending with its final preparation. In particular, effective control measures do not exist for the many pathogens that occur on raw agricultural commodities and seafood. At best, it is possible to reduce, but not prevent or eliminate, their presence in these foods and still provide them to consumers in a raw, unprocessed state. Eliminating foodborne hazards is further complicated by the fact that controlling one hazard may increase the potential for other hazards or other adverse consequences. Thus, food control programs are oriented toward ensuring that foods are as free as practicably possible from hazards through appropriate hazard management.

For a number of foodborne diseases, the TLR is effective absence of disease (e.g., < 0.1 case per 100,000). This typically occurs when Good Hygienic Practices (GHP) and Hazard Analysis Critical Control Point (HACCP) procedures and other sanitary measures completely control the pathogen and effectively eradicate it from the food chain, as has occurred with foodborne brucellosis in certain regions.

Food safety management is similar to managing other risks in human life. For a range of hazards, society balances the risks and benefits and, although rarely stated publicly, accepts that a certain risk has to be tolerated. For example, the risk of injury while driving a car can be reduced by designing safer vehicles, regulating traffic, setting speed controls, wearing seat belts, and driving defensively. However, despite all such efforts, society has come to accept that a certain number of accidents will occur. However, many consumers expect food to be safe and have little or no tolerance for purchasing food that may cause illness.

Managing tolerable risk implies balancing public health considerations with other factors such as economic costs, public acceptability, etc. Based on the epidemiologic data and evaluation of risk, managers should decide whether the risk is so small that no further action need be taken or that the risk only needs to be contained (to keep it at the present level because systems already in place are adequate). Alternatively, managers may decide that the risk needs to be reduced. When deciding to reduce microbiological risks, managers should consider control measures that reduce the level of a hazard in a food. However, it must be realized that there will be a point at which further reductions in risks associated with specific foods may have additional "costs" that society is not willing to bear. Therefore, there is a need to balance the benefits of risk reduction with the costs incurred.

The most obvious cost is economic impact. For example, assume a proposal has been made to reduce the maximum allowable concentration of aflatoxin in peanuts from 15 parts per million (ppm) to 5 ppm, and it has been estimated that this change would reduce the lifetime risk of liver cancer per person from 10^{-11} to 10^{-12}. If implementation of this new standard increased the average cost of peanuts by $0.01 per pound, society might consider the additional cost warranted and reduce the lifetime risk of liver cancer due to aflatoxin in peanuts to 10^{-12}. However, if the change increased the cost of peanuts by $2.00 per pound, society might consider this to be an unacceptable additional cost and the lifetime risk level tolerated would remain 10^{-11}. Various tools for cost-benefit analysis have been developed to estimate the economic impact of decisions of this nature.

A key component of any estimate of economic costs is the impact that foodborne disease has in terms of medical costs, lost productivity, loss of consumer confidence, etc. In recent years, attempts have been made to estimate the economic cost associated with foodborne illness. As an example, one study estimated that between 3.3 and 12.3 million cases of foodborne illness occurred each year in the US due to six bacteria (*Campylobacter jejuni/coli*, *Clostridium perfringens*, *Escherichia coli* O157:H7, *Listeria monocytogenes*, salmonellae, and *Staphylococcus aureus*) and one parasite (*Toxoplasma gondii*). The medical costs and losses of productivity, in 1995 US dollars, associated with the seven pathogens were estimated to be between $5.6–9.4 billion (Buzby & Roberts, 1997).

Costs are not limited to economic concerns. As another hypothetical example, suppose that changing a traditional, religiously based food handling practice would reduce the risk of foodborne illness. While modifying these practices might provide a small reduction in risk, the cost associated with the loss of choice based on religious beliefs may be considered unacceptable, and the affected population may prefer to tolerate the slightly higher risk.

Similarly, the impact that food control requirements have on the freedom of consumers to make decisions on the foods they eat must be recognized. However, there are several instances where the potential public health consequences of specific foods or food handling practices are so severe that consumers' choices are restricted; that is, society will not communally tolerate the risks that some consumers as individuals are willing to take. For example, a number of countries prohibit the commercial sale of non-pasteurized milk for direct consumption. Other examples of societally enforced risk reductions are the prohibition in a number of countries on the sale of uneviscerated salted fish due to the potential risk of growth and toxin production by *C. botulinum* or on the import of fish from the order *Tetraodontiae* (e.g., puffer fish) due to potential risk of tetrodotoxin poisoning.

In these examples, the products inherently exceed the risk tolerated and, other than prohibition, there are no viable control measures. The foods have a long history of public health problems and risk managers in some governments have concluded that the risks were not tolerable within the context of their society. Nevertheless, it is equally possible that another society values these foods highly and would be willing to tolerate the associated public health consequences.

Another cost that must be considered is when reducing the risk of one hazard is likely to increase the risk associated with another hazard or produce other adverse effects, e.g., on nutritional properties. A pertinent microbiological example is the use of chlorine for the treatment of water for drinking and food processing. While chlorination reduces risks associated with waterborne microbial diseases, there is a concentration-dependent risk that results from formation of organo-chlorine compounds. These situations require risk management decisions that balance two competing hazards such that the benefits outweigh the risks to the affected population and environment.

Most societies consider consumer protection a moral responsibility deserving high priority. With increased reporting of foodborne illness by the media, consumers have become more aware of its frequency and public health impact. This increased awareness has led to greater pressure on governments and industry to make changes that will further reduce risk. This could be reflected in the future levels of risk that will be tolerated for various foodborne microbiological hazards.

A decision not to take an action also has its costs in relation to all of the categories identified above.

The net result of balancing the various risks and benefits is a decision on which actions have to be taken. To implement the actions, an objective must be defined. This objective may be expressed in terms of a level of a hazard in a food or as a tolerable level of risk. Expressing the objective in terms of risk (i.e., cases per 100,000 per annum) does not provide the guidance required by producers, manufacturers, or control authorities. The most effective means to ensure that the actions taken will be effective is to express the objective in terms of the level of hazard. This is the basis of the FSO concept.

2.3 IMPORTANCE OF EPIDEMIOLOGIC DATA

Tolerable levels of risk are generally articulated in relation to the mortality or morbidity of a disease, expressed as a number of cases or deaths per 100,000 population per year. For many reasons, the *true* incidence of foodborne disease is not known. At best, reasonable estimates can be developed for particular diseases because the impact on the consumer is profound and the characteristics of the disease are sufficiently evident.

This information is very important to understand the burden on a population and to assess the effect of policies instituted to thwart foodborne disease. Several approaches are currently used to monitor and report the incidence of foodborne diseases:

- passive notification systems
- active surveillance systems
- case-control studies
- outbreak investigations
- sentinel studies

None of these systems yield all the data necessary for a quantitative risk assessment, and some (e.g., passive notification systems) often fail to identify food as a source. Passive notification systems follow trends in disease and can be useful for measuring the impact of changes in technology, preventive measures, and regulatory policies. For example, Figure 2–1 shows the reported incidence of salmonellosis and shigellosis in the US from 1939–1998. The data were developed through reports from local sources in each state and then submitted from the state to the Centers for Disease Control and Prevention. In addition, physicians have been required to report particular *notifiable* diseases. This mandatory requirement can strengthen the accuracy of the data, but many cases remain unreported. Similarly, reports from laboratories can identify trends in *Salmonella* serovars and identify for public health officials changing risks to the population (Figure 2–2).

Another approach to collecting data on the incidence of disease is through active surveillance systems such as EnterNet or FoodNet. EnterNet (http://www.enter-net.org.uk) was established to determine more accurately the incidence of salmonellosis and infections caused by *E. coli* O157 in Europe. Another goal of EnterNet is to identify out-

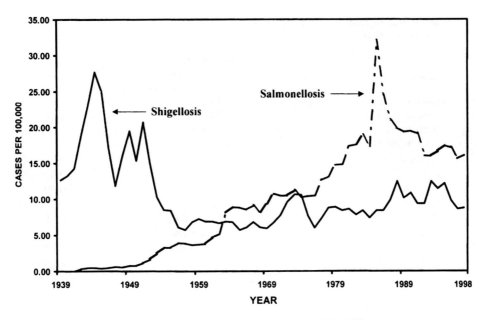

Figure 2–1 Trends in salmonellosis and shigellosis in the US, 1939–1998.

breaks in Europe from a common food source. The US surveillance program, FoodNet (http://www.cdc.gov/foodnet/), is an active, sentinel site program that collects weekly updates from clinicians in certain regions of the country for gastroenteritis and listeriosis. Isolates of selected pathogens are compared for commonality to identify outbreaks due to a common food source. These systems are relatively new and the data are just becoming available. Table 2–2 summarizes the data from 1996–1999 for FoodNet. Data from the older passive system (Morbidity and Mortality Weekly Report-MMWR) are provided for comparison, including several diseases not covered by FoodNet.

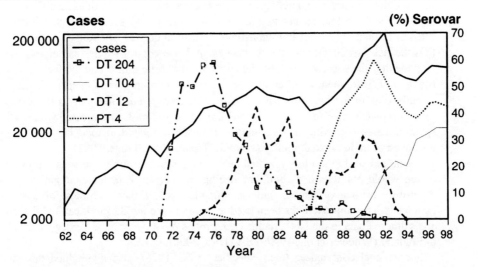

Figure 2–2 Trends in *Salmonella* serovars in Germany, 1962–1998. DT serovars *S. typhimurium*. PT serovars *S. enteritidis*.

Table 2-2 Annual Incidence of Selected Diseases in the US Following Two Different Reporting Systems

Disease	Cases / 100,000 / year MMWR 1986–1998	FoodNet 1996–1999
Cholera	0.00–0.04	
Botulism	0.01–0.02	
Trichinosis	0.01–0.05	
Brucellosis	0.03–0.05	
Typhoid fever	0.14–0.22	
Listeriosis		0.50–0.60
Yersiniosis		0.80–1.00
E. coli	0.82–1.18	2.10–2.80
Shigellosis	7.11–12.48	5.00–8.90
Hepatitis A	8.59–12.64	
Salmonellosis, non-typhoid	15.70–20.73	12.40–14.80
Campylobacteriosis		17.30–25.20

Data on foodborne disease are also collected through case-control studies by interviewing patients to learn their food consumption history and to identify problem food sources. In parallel, a number of individuals are selected to serve as controls. This methodology has been used to identify not only the foods that may be involved, but also risk factors that the patients may share and that may explain increased susceptibility to the disease. Case-control studies are useful for identifying pathogen-food combinations where it has been difficult to isolate the causative organism from the food source or the role of foods in diseases with long incubation times before onset of symptoms (e.g., listeriosis).

The identity of the problem food source and conditions leading to foodborne illness may also be determined through epidemiologic investigations of outbreaks. Unfortunately, not all outbreaks are adequately investigated or described fully in the scientific literature, particularly those that do not provide new information. Consequently, that literature is often of limited usefulness in establishing the true frequency of outbreak occurrence and thus the risk associated with the disease agent. Case-control studies can also be used to help identify the source(s) of sporadic cases of foodborne illness and the factors that contribute to their frequency. Different sources may be more important in sporadic cases than in outbreaks. In the US, outbreaks of *Campylobacter jejuni* infections in the spring and fall are typically caused by drinking raw unpasteurized milk or untreated water, whereas sporadic cases occurring in the summer are usually related to touching or consuming uncooked or undercooked poultry (Potter & Tauxe, 1997; Tauxe, 1992).

A sentinel study monitors selected health events in a group of persons representative of the whole population. Laboratory testing may be limited, for example, to patients reporting diarrhea or may include examination of all fecal samples for a range of pathogens. This approach has been used to estimate the incidence of campylobacteriosis and salmonellosis in the Netherlands (Notermans & Hoogenboom-Verdegaal, 1992) and in England (Wheeler *et al.*, 1999; FSA (UK), 2000).

International conferences (e.g., Noeckler *et al.*, 1998) have highlighted the value of national reporting systems for foodborne disease and provide a snapshot of national disease patterns. Their longer-term goal is to establish a multinational system of cooperation that would facilitate the transfer of information and expertise. With such a system in place, the possibility of establishing TLRs and/or FSOs on a regional basis would be greatly enhanced.

2.4 EVALUATION OF RISK

2.4.1 Introduction

When a food safety problem occurs, or an improvement in food safety is desired, a risk evaluation can determine the magnitude of the problem so that a decision can be made whether to take action and which actions should be taken. An evaluation of risk may be via one or more experts, an expert panel, or a more in-depth quantitative risk assessment.

Effective management of microbial food safety hazards requires identification of the hazards, assessment of the risks associated with those hazards (i.e., their significance, their severity, their likelihood of occurrence), and estimation of the effectiveness of potential control measures. The stringency of the control system should be proportional to the risk to public health (risk = the severity of an illness and the likelihood of its occurrence). This principle also underlies modern food safety systems, such as HACCP, that are based on preventive strategies.

Historically, food control agencies and industry have relied on the judgment of one or more experts to estimate the risk and the corresponding level of control needed to manage it. Although this approach has often been successful, it can be biased or inconsistent. Furthermore, it can be difficult to communicate adequately the underlying scientific basis and rationale for the decisions to interested parties because such evaluations were rarely documented.

An improved approach is the use of structured safety assessments, typically including broader scientific expertise and more formal consideration of available data and information. This approach has been widely used in recent years where expert panels have been called to address various food safety issues, e.g., psychrotrophic *C. botulinum* in refrigerated foods with extended shelf-life (ACMSF, 1992) or *Listeria* in ready-to-eat foods (ICMSF, 1994).

In recent years, quantitative microbial risk assessment has been initiated to evaluate more systematically the impact of various factors, such as the host, the pathogen, and the type of food (liquid, solid, or fat), that contribute to the risk associated with a foodborne microbiological hazard.

Risk assessment is mentioned as a tool in the WTO/SPS agreement to ensure that international trade in food is not hampered by unjustifiable safety requirement demands. This has led to an international harmonization of the concept of risk assessment and its practical implications by the Codex Alimentarius Commission (CAC, 1999), although there is, as yet, no general (or international) agreement on when a quantitative risk assessment is necessary or which statistical/mathematical approaches are appropriate.

The first step in any evaluation of risk is the identification of a food safety problem from one or more sources, e.g., accumulated epidemiologic data, regulators, public health sectors, the food industry, expert opinion/scientists, or consumers. Background information is generally assembled by a risk manager, who then provides a summary of the food safety problem and its context to the risk assessor. Risk managers must decide whether to direct risk assessors to initiate a qualitative or quantitative evaluation to obtain the information necessary for risk management decisions. It is important that there is a common understanding of the problem, including the approach to be taken (qualitative or quantitative assessment, deterministic or probabilistic), the available resources, time constraints, and the desired form of the output, e.g., a risk estimate (probability of disease per exposure or per year to the population or a sub-population) and/or suggested control measures. Circumstances may lead a risk manager to seek the information needed for a short-term fix through a qualitative evaluation (e.g., expert panel) and then rely on a more detailed quantitative assessment to assemble the information that may lead to longer-term solutions. Both qualitative and quantitative risk assessments will be discussed below. The use of expert panels as a form of qualitative risk assessment will be discussed first.

2.4.2 Use of Expert Panels

Control authorities, and others, have found expert panels to be an effective means to assemble information, interpret its content, and develop recommendations in a relatively short period of time. Expert panels have been used extensively by governments and international bodies (e.g., FAO/WHO consultations) to address concerns about the safety of a particular hazard-food combination. In addition, expert panels (e.g., Nickelson et al., 1996) have been used extensively by industry to consider the factors leading to foodborne disease and to develop recommendations for their control. Assessments by expert panels are often appropriate, especially if a decision is needed quickly to address a newly recognized concern, or if resources and/or data for a quantitative risk assessment are limited, or perhaps where there are few management options. Such panels are called upon when epidemiologic evidence indicates that a hazard is not under control and there is need for increased consumer protection. Concerns may be raised following change in food habits, food processing technologies, and food packaging or distribution systems. Such concerns must be addressed and, if reasonable and supported by scientific evidence, an evaluation of the risk must be undertaken.

In practice, risk managers will call upon people with expertise on the particular pathogen and/or food in question. The panels established usually consist of epidemiologists, public health specialists, food microbiologists, and food technologists with knowledge about actual food processing operations important for the evaluation. The panels typically go through a range of steps very similar to conducting a hazard analysis when developing a HACCP plan to address the concerns presented by risk managers. In some instances, a rough estimation of the risks associated with different likely scenarios is sufficient. One approach is to assign relative probability and impact rankings, such as negligible, low, medium, or high, to the factors used to determine likelihood of exposure and likelihood of an adverse outcome. If such a system is used, definitions and rationale for assigned rankings must be clearly described and justified to avoid misinterpretations of the information by users. An example appears in Chapter 8, where different hazards are ranked in categories, depending on the severity of the disease.

The examples in Chapters 14–17 illustrate the tasks of expert panels. The panel will be asked to provide the best information available at that given point in time considering several aspects. Due to the more simple nature, the procedure is faster than a quantitative risk assessment, described in section 2.7 below. Although the complexity of the risk evaluation may vary, any risk assessment or risk evaluation comprises four steps: hazard identification, exposure assessment, hazard characterization, and risk characterization (CAC, 1999). In addition, the expert panel should provide information on the conditions that lead to hazardous food. The outcome from an expert panel may be to recommend to risk managers one or more measures to control a hazard or, if necessary, ban the product or process. Where appropriate, an expert panel may recommend the establishment of an FSO where this can be an effective means to enhance the safety of the food under consideration.

2.5 FOOD SAFETY OBJECTIVES (FSOs)

Governments and food industries have a significant influence on the incidence of foodborne disease by controlling the frequency and extent of contamination of foodstuffs and other conditions that minimize or control foodborne diseases. Consequently, public health goals must be converted into parameters that can be controlled by food producers and monitored by government agencies. A food safety objective provides such a conversion. While FSOs can obviously be set for any food hazard (e.g., carcinogens, pesticides, toxins, microorganisms), in the context of this book, only hazards of microbial origin are considered. Therefore, it is implicit that the FSOs dealt with are microbiological food safety objectives.

2.5.1 Basic Concept

A microbiological FSO is the maximum frequency and/or concentration of a microbial hazard in a food considered tolerable for consumer protection. It is important to reemphasize that the primary purpose of an FSO is to translate a public health goal (i.e., a desired level of consumer protection) to measurable attributes that allow industry to set control measures for processes and that allow comparison between countries. Examples* of FSOs are:

- the amount of staphylococcal enterotoxin in cheese must not exceed 1 µg 100 g^{-1}.
- the concentration of aflatoxin in peanuts must not exceed 15 µg kg^{-1}.
- the level of *L. monocytogenes* in ready-to-eat foods must not exceed 100 cfu g^{-1} at the time they are consumed.
- the concentration of salmonellae must be less than 1 cfu 100 kg^{-1} of milk powder.

While the FSOs at first glance seem similar to microbiological criteria, they differ in several ways (see Chapter 5). FSOs are not applied to individual lots or consignments, and they do not specify sampling plans, number of analytical units, etc. FSOs define the level of control that is expected for a food operation and can be met through the implementation of GHP and HACCP systems and application of performance criteria, process/product criteria and/or acceptance criteria (see Chapter 3).

Whenever possible, FSOs should be quantitative and verifiable. However, this does not mean that they must be verifiable by microbiological testing. For example, an FSO for low-acid canned foods might be established in terms of the probability of a viable spore of *C. botulinum* being present as being less than 0.000000000001 per can. It would be impossible to verify this by end-product testing, but it would be verifiable by measurement of time/temperature protocols that are based on a performance criterion (see Chapter 3).

The FSO allows control authorities to communicate clearly to industry what is expected of foods produced in properly managed operations. The FSO establishes the stringency

---

*These examples are hypothetical and are intended only to illustrate the concept of FSOs.

under which food control systems must operate by specifying the frequency or concentration of a microbiological hazard that should not be exceeded at the moment of consumption. It thereby forms the basis by which control authorities can establish standards or guidelines and assess whether an operation is in compliance and is producing safe foods, i.e., the foods will under normal conditions of commercialization and use meet the established FSO. Thus, the FSO has very practical value and can be commonly understood and applied by industry and regulators alike. Since an FSO does not specify how an operator achieves compliance, the concept offers considerable flexibility of operation. This would enable one operator to use formulations, equipment, and procedures that differ from other operators, so long as the FSO is met. Furthermore, there can be a high level of confidence in the acceptability of food being produced by operations that have been designed and validated to meet the relevant FSOs. The foods from such operations need seldom be tested for the relevant pathogen(s) to verify compliance. Instead, verification can be achieved through record review and observation of GHP and HACCP (see Chapter 4).

2.5.2 Use of FSOs

In summary, the concept of an FSO offers many advantages for both control authorities and industry because they can be used to:

- translate a public health goal into a measurable level of control upon which food processes can be designed so the resulting food will be acceptable;
- validate food processing operations to ensure they will meet the expected level of control;
- assess the acceptability of a food operation by control authorities or other auditors;
- highlight food safety concerns, separate from quality and other concerns;
- force change in a food commodity and improve its safety; and
- serve as the basis for establishing microbiological criteria for individual lots or consignments of food when the source or conditions of manufacture are uncertain.

It is not necessary to establish an FSO for all foods or known hazard-food combinations. In some cases, the potential microbiological hazards associated with a food represent so little risk that an FSO is not needed (e.g., granulated sugar, sweetened condensed milk, most breads, pineapple, carbonated beverages). In other cases, the sources of a pathogen are so variable that identifying the foods for which FSOs should be set is not possible. An example of the latter is shigellosis, which can be transmitted by many routes, most of which are more important than food (e.g., water, person-to-person), and it is unpredictable which specific food may next be implicated.

Investigation of foodborne disease continues to identify new pathogens and new pathogen-food combinations. The emergence of listeriosis as a foodborne disease during the 1980s as a result of outbreaks traced to coleslaw and Mexican-style cheese is an example of a recently recognized foodborne pathogen. The finding that nonpasteurized juices and raw vegetables can be vehicles for *E. coli* O157:H7 is an example of a new pathogen-food combination. In such situations, a quick decision may be necessary to prevent more cases or outbreaks. Establishment of an interim FSO could be an initial step to communicate to the food industry, or exporting countries, the maximum level of a hazard considered acceptable. As further knowledge about the hazard, the food, and conditions leading to illness become available and effective control measures can be determined, the interim FSO can be adjusted.

FSOs also can be used to force change in an industry and enhance the safety of certain products. Many examples could be cited where epidemiologic data indicated that certain foods were linked to foodborne illness. In response to this information, governments used various mechanisms at their disposal to bring about the changes necessary to reduce or eliminate the risk of disease. In some cases, modifications in commercial practices were necessary, including the adoption of new or more reliable technologies. The establishment of an FSO could be used by risk managers in government to communicate to industry the level of control expected and thereby force the change. The FSO may require some operators to modify their operation, implement more effective technologies, adopt tighter control systems, or even cease operation.

The WTO/SPS agreement recognizes that governments have the right to reject imported foods when the health of the population may be endangered. The criteria used to determine whether a food is considered to be safe or unsafe should, however, be clearly conveyed to the exporting country (transparency) and should be scientifically sound. Integral to the treaties is the concept of *reasonableness*, a requirement that is inherent to the establishment of realistic FSOs. An exporting country can contest an FSO that does not reflect conditions existing in the importing country and argue that the FSO is an unjustified trade barrier. However, because an FSO also reflects commercialization conditions, eating habits, preparation, and use practices, FSOs may vary considerably between countries. Nevertheless, a country cannot demand that imported foods are "safer" than similar domestically produced foods. For example, if the tolerance for aflatoxin in domestically grown and processed peanuts is 15 $\mu g\ kg^{-1}$, then imported peanut products cannot be rejected if contaminated to the same or a lesser concentration. FSOs provide a means for implementing the concept of equivalence in Article 4.1 of the WTO/SPS agreement: "Members (countries) shall accept the SPS measures of other Members as equivalent, . . . , if the exporting Member objectively demonstrates . . . that its measures achieve the importing Member's appropriate level of sanitary or phytosanitary protection."

FSOs provide a maximum frequency or concentration for a hazard that is judged as tolerable for consumer protection, without prescribing how an exporting country or company should achieve this level of protection. The FSO also becomes useful when the safety of new products is evaluated. To put new products or novel foods on the market, their safety should be substantially equivalent to existing similar products.

2.6 ESTABLISHMENT OF AN FSO BASED ON A RISK EVALUATION BY AN EXPERT PANEL

Some risk evaluation must be conducted prior to the establishment of an FSO. Due to the time and money required to do quantitative risk assessments, FSOs will often be based on the advice of an expert panel. For a large number of hazard-food combinations, there is no need to do a quantitative risk assessment, as there is already agreement on the factors determining risk.

FSOs contain three elements, namely, the hazard, the food, and the frequency or concentration of the hazard that is considered tolerable. Some basic knowledge is necessary before an FSO can be established. For this reason panel members should be selected based on their knowledge, experience, and access to information that can ensure these basic needs are met. At a minimum, the panel must have knowledge about the microbiological hazard (e.g., infectious agent, toxic metabolite), its source, and the conditions along the food chain that lead to foodborne illness. The relationship between the microbiological hazard, the food, and the disease may be elucidated through a combination of passive and active epidemiologic programs, case-control studies, and other pertinent public health studies as described above. Investigations of foodborne illness should also provide information about whether certain populations are at risk for greater frequency and/or severity of the disease. This knowledge should be supplemented with data derived from laboratory research and from steps in the food chain that may be relevant to the disease. Records of foods processed for safety may provide useful data concerning the level of consumer protection normally achieved. This knowledge can form a solid basis for a risk evaluation and determination of an FSO.

If it is known, for example, that the source of a pathogen (e.g., *S. aureus*) is humans and animals, and that growth on cooked ham during storage at room temperature to a high concentration (e.g., 10^6 cfu g^{-1}) is necessary for toxin production, then this information can be used in establishing an FSO. Additional information about the concentration of toxin (i.e., dose) required to cause illness would be needed before a meaningful FSO could be developed. In this case, it would be necessary to determine whether to base the FSO on the concentration of *S. aureus* or the concentration of staphylococcal enterotoxin in the food.

As another example, the incidence of listeriosis in a number of countries has been between 0.1–1.3 per 100,000 people per annum (Table 2–1) (Ross *et al.*, 2000). When the vehicles were identified, high numbers of the organism were present in the food (McLauchlin, 1995, 1996; Anonymous, 1999). As *L. monocytogenes* is ubiquitous and low numbers are prevalent and ingested daily by consumers with no adverse effect, ICMSF, acting in the capacity of an expert panel, suggested that such low numbers are unlikely to pose a risk to healthy consumers and, therefore, proposed an FSO of 100 cfu g^{-1} at time of consumption (ICMSF, 1994). Based on data from Germany for the incidence of listeriosis, as well as prevalence and levels of the organism in foods, a hazard characterization was developed (Figure 2–3) (Buchanan *et al.*, 1997).

2.7 EVALUATION OF RISK BY QUANTITATIVE RISK ASSESSMENT

2.7.1 Quantitative Risk Assessment

Quantitative risk assessment typically involves persons similar to those of an expert panel, plus others with mathematical/computing skills. The process takes longer than a qualitative assessment. Quantitative assessments are normally undertaken for complex interactive situations, when there may be uncertainty over where control can best be exercised or the effectiveness of various control options, and/or when there is substantial disagreement among stakeholders concerning the level of control needed to achieve a tolerable level of consumer protection. The purpose of a quantitative risk assessment is the same as that discussed for expert panels, i.e., to provide scientific advice to the risk managers who will use the information to decide upon the risk management option(s) that will be implemented to achieve the desired level of consumer protection.

Risk assessment comprises four steps: hazard identification, exposure assessment, hazard characterization, and risk characterization (CAC, 1999). Each involves a systematic process for collecting, assembling, and providing the necessary knowledge to evaluate the public health significance of a microbial hazard in food. The final outcome of the four steps is a risk estimate, i.e., a measure of the magnitude of risk to a population attributable to the food, with the attendant uncertainties. It is impossible to predict the risk to an individual. The estimates are derived mathematically by calculating the

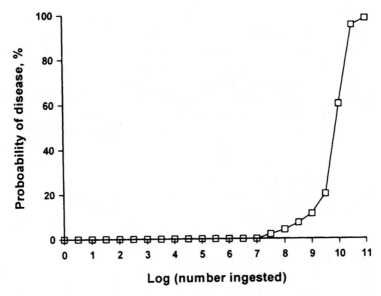

Figure 2-3 Hazard characterization of *L. monocytogenes*. Probability of disease as a function of ingested numbers.

likely frequencies and/or concentrations of the hazard in food at the time of consumption, combined with an estimate of the probability that disease will occur after the food is consumed.

Of necessity, assumptions will be made during the assessment when data or other information are missing. Data used and assumptions made should be clearly documented and their effect on the final risk estimate clearly stated. It is also important that risk assessors identify, describe, and, if possible, quantify sources of variability and uncertainty that affect the validity of the risk estimate.

2.7.2 Hazard Identification

The first step of risk assessment, *hazard identification*, assembles the knowledge about the pathogen and/or food in question and its association with adverse health effects. Sometimes epidemiologic data clearly indicate that foodborne transmission plays a role and which foods are implicated. Conversely, if a particular food is suspected, epidemiological and microbiological data may indicate which pathogens have been, or potentially could be, associated with the product. Epidemiologic data from disease monitoring programs or investigations of foodborne outbreaks are often the first indication of a food safety problem; in such a case, adverse effects associated with the pathogen are relatively well documented. Information may also come from animal disease monitoring when the pathogen is a zoonosis.

2.7.3 Exposure Assessment

Exposure assessment estimates the prevalence and levels of microbial contamination of the food product at the time of consumption and the amount of the product consumed at each meal by different categories of consumers. Programs for nutrition and consumption habits are often available nationally to gauge food intake; these can be used to estimate exposure. The exposure assessment may be limited to measurements of pathogen levels at the time of consumption. However, models are available that estimate how certain factors, such as prevalence of pathogens in raw ingredients, the potential growth of the pathogen in the food, and impact of handling and preparation practices, affect the frequency and levels of pathogens consumed (Product/Pathogen/Pathway (PPP) ("Farm to Fork") analysis). Data from baseline surveys of pathogens in foods and predictive microbial modeling techniques have proven to be valuable sources for deriving probable exposure estimates for pathogenic bacteria (ICMSF, 1998). Substantial amounts of information on microbial levels have been accumulated in food inspection data in many countries and could provide an additional source of information on the microbiological status of foods just before consumption.

The sensitivity, specificity, and validity of sampling and testing methods used to collect empirical information should be considered to assure that results from different studies are comparable. Some apparent differences in pathogen prevalence in the food chain may be attributable to underreporting or methods employed, however, there may be real variation due to ecological situations or differing food safety control measures and animal health control programs. For example, food distribution systems vary from country to country with respect to temperature control. Exposure assessments should also consider differences in the cultural, social, economic, or demographic structures of societies that may influence consumption patterns and practices.

2.7.4 Hazard Characterization

Hazard characterization describes the severity and duration of adverse health effects that may result from the ingestion of a microorganism or its toxin in food. A dose-response assessment provides an estimate of the probability that disease/illness will occur in a certain category of consumers after exposure to a certain number of a pathogenic microorganism and/or its metabolites/toxin (i.e., dose). The consequences of being exposed to a microbial pathogen or microbial toxin in a food will vary, ranging from no discernible effect to infection (colonization and growth in the intestinal tract) without symptoms of illness, to acute illness (usually gastroenteritis, but sometimes septicemia and meningitis), to long-term effects or sequelae (chronic illness such as reactive arthritis, Guillain-Barré syndrome, or hemolytic uremic syndrome), to death. The likelihood that exposure to a particular dose (i.e., number of cells) of a specific pathogen may have any one of these consequences is dependent on three factors:

1. characteristics of the microorganism itself (e.g., mechanism(s) of pathogenesis, virulence factors, ability to resist the host's defenses) that vary among strains and may be altered by prior conditions;
2. the susceptibility of the host (e.g., immune status, predisposing conditions, age); and
3. characteristics of the food in which the pathogen is carried (e.g., fat content, acidity, or other factors that affect the microorganism's capacity to resist acidity of the stomach, competing bacteria in the food, etc.).

In practice, estimates of the numbers of pathogen that may cause illness and the severity of illness relative to dose are derived from experimental studies with humans, animal models, and epidemiological data (and accumulated knowledge and experience) (see Chapter 8).

To perform the final risk characterization (see below), the relation between frequency of exposure of the population (or subpopulation) with various numbers of the pathogen in the food at the moment of consumption and the number of illnesses (diarrhea, death) per annum has to be established. Figure 2–4 depicts such a relationship. It is a hypothetical curve, not based on real data for any food-pathogen combination.

2.7.5 Risk Characterization

Risk characterization combines the information generated in hazard identification, exposure assessment, and hazard characterization to produce a complete picture of risk. The result is a risk estimate that is an indication of the level of disease (e.g., number of cases per 100,000 per year) resulting from the given exposure.

Whenever possible, the resulting risk estimate should be compared with epidemiological data or other reference information to assess the validity of the risk assessment's models, data, and assumptions. The risk estimate should reflect a distribution of risk that represents the range of contamination of a food product, factors that might affect growth or inactivation of the pathogen, and the variability of the human response to the microbial pathogen.

Risk characterizations should also provide insights about the nature of the risk that are not captured by a simple qualitative or quantitative statement of risk, e.g., identifying the most important factors contributing to the average risk, the uncertainty and variability of the risk estimate, and gaps in data and knowledge. The consequences of any default assumptions provided to the risk assessment team should be documented. The risk assessor may also compare the effectiveness of alternative methods of risk reduction, enabling the risk manager to consider risk management options.

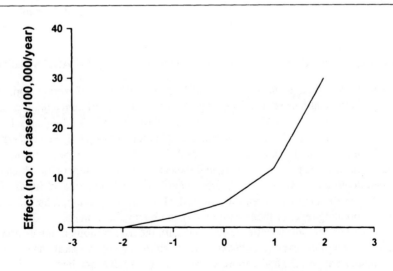

Figure 2–4 Hypothetical hazard characterization curve for a pathogen that causes gastroenteric infection.

The resulting risk estimate must be compared to the tolerable level of risk decided upon, and if the risk estimate is higher than that which can be tolerated, obviously actions must be taken to reduce the risk. Establishing a food safety objective (see section 2.5) is recommended.

2.7.6 Mathematical Approach to QRA

Quantitative risk assessments use mathematical models to estimate risk as a function of one or more inputs. Point estimates, or single values, such as the mean or maximum value of variable data sets, have been used historically to generate a single numerical value for the risk estimate.

Until recently, the most common approach was to use either the mean or worst-case (95th percentiles) estimate calculated from the available data for each step of the assessment. These values were then used to calculate an overall mean or worst-case single value estimate (e.g., 1 per 100,000 exposures will result in illness; 100 cases/100,000 population). Such risk estimates were termed *deterministic* or *point estimate* risk assessments. A major shortcoming of these approaches is that the variability of diverse and dynamic biological phenomena is ignored and consideration is not allowed for how much uncertainty there may be about the data and how it may influence the risk estimate.

Probabilistic assessments represent all the information available for each parameter (i.e., information or data sets about a factor that is important in determining risk) described as a distribution of possible values. A mathematical description of the production and consumption of a food using probability distributions is very difficult to calculate analytically. While some analysis is practical on very small and simple models, a compound model of food production involving pathogen growth, destruction, and infection is too complex to interpret without computational tools. Probabilistic risk assessments for food safety are feasible using commercial software. Monte Carlo simulation is a computational tool that aids in the analysis of models involving probability distributions.

2.8 ESTABLISHMENT OF AN FSO BASED ON QUANTITATIVE RISK ASSESSMENT

Risk assessments can help to identify how the frequency and/or concentration of a microbiological hazard in a food or group of foods can influence the incidence of a disease. There is a relationship between the level of a hazard in a food and the incidence of the disease it causes in a given population. This relationship may be represented by a hazard characterization curve (e.g., Figures 2–3 and 2–4). The slope of this curve is specific to the hazard, the food, the illness, and the category of consumers for which the curve has been determined. If such curves are available for the incidence of disease for a specific pathogen-food combination, the selected TLR can be positioned on the y-axis, and the corresponding level of the hazard (FSO) can be obtained on the x-axis.

However, the uncertainties associated with the assumptions in the estimations used in elaboration of the exposure-disease relationship may oblige risk managers to take a more cautionary approach, and a more stringent (lower) FSO may be adopted. In many situations, the public health goal is to have nearly zero cases from a given pathogen-food combination (e.g., botulism from commercially canned shelf-stable low-acid foods), and consequently a very conservative FSO may be adopted. However, the costs to realize this goal may be higher than a society will accept (see section 2.2). It must be remembered that the FSO will reflect the balance between the costs associated with reducing risk and the costs associated with accepting risk.

When interpreting graphs such as those depicted in Figs. 2–3 and 2–4, it is important to note that for infectious agents seldom, if ever, will the relationship between risk and the level of the hazard reach zero unless the hazard has been eradicated. Instead, as the predicted incidence falls below unity, the population size and the timeframe need to be adjusted. Even when the risk to a population falls to the level where the predicted value is less than 1 case per year, the risk is not zero. For example, a predicted incidence of 0.2 cases per year would be more appropriately expressed as 2 cases per 10 years.

Ideally, an FSO would be based on the frequency or concentration of a pathogen in a food that would not produce disease. This would be equivalent to finding a no-effect dose, the value that is used for setting tolerable levels of daily exposure for acutely toxic chemicals. It is clear that certain foodborne pathogens have clearly definable threshold levels below which they pose no risk to the consumer. In general, it is assumed for certain toxigenic foodborne pathogens (i.e., microorganisms that cause disease through the production of a toxin) that there is a threshold concentration of cells below which the microorganism does not produce sufficient toxin to cause an adverse effect. For example, it is generally accepted that low levels (i.e., $< 10^4$ cfu g^{-1}) of $S.$ $aureus$ in a food do not represent a direct risk to humans, whereas higher levels ($> 10^5$ cfu g^{-1}) do so through the production of enterotoxin to levels that will probably cause staphylo-enterotoxicosis.

For infectious pathogens, most current models are based on the assumption that there is a chance, however remote, that one single cell may cause disease. For $L.$ $monocytogenes$, current dose-response data indicate that the chance of illness varies from 1 in 10^{-5} (of contaminated butter causing listeriosis in a vulnerable population) to 8 in 10^{-15} (US Food and Drug Administration (FDA) model for listeriosis in the elderly population) (FAO/WHO, 2000). Establishing microbial FSOs is substantially more complex when a no-effect level cannot be assumed; however, the basic process for establishing an FSO remains the same.

2.9 STRINGENCY OF FSOs IN RELATION TO RISK AND OTHER FACTORS

Examples of hypothetical FSOs are provided throughout this book to describe the FSO concept and how it can be used as a food safety management tool. Some of the hypothetical FSOs are more restrictive for a food-pathogen combination, even though another food-pathogen combination may be of greater risk. Ideally, as risk increases, the corresponding FSO would become more restrictive. Thus, there would be a quantitative relationship, reflecting risk, in the allowable frequency and/or concentration of the various hazards.

The ICMSF has not endorsed the FSOs provided as examples in this book, except for the FSO for $L.$ $monocytogenes$ (see section 2.5.1) (ICMSF, 1994). To illustrate the use of the concept, FSOs are provided for $Salmonella$ in milk powder (< 1 cfu 100 kg^{-1}), $E.$ $coli$ O157:H7 in raw ground beef (≤ 1 cfu per 250 g or not to exceed -2.4 log$_{10}$ g^{-1}), $L.$ $monocytogenes$ in ready-to-eat foods (≤ 100 cfu g^{-1}), staphylococcal enterotoxin in cheese (≤ 1 μg 100 g^{-1}), and aflatoxin in peanuts (15 μg kg^{-1}).

Relative risk is not reflected in the stringency of the FSOs. For example, the FSO for ground beef (recognizing that as many as 15% of the US population consumes undercooked ground beef) is much less stringent than the FSO for milk powder. In addition, the risk associated with $Salmonella$ is lower than $E.$ $coli$ O157:H7. In the case of milk powder, there is about 30 years of commercial experience in identifying and implementing the necessary controls for *Salmonella*. The FSO is commercially achievable with current technology and equipment, and processes can be validated as meeting an FSO of this stringency. The available control measures are much less effective for *E. coli* O157:H7 from slaughter through grinding (mincing). In the absence of an effective kill step (e.g., irradiation, high pressure), a more stringent FSO is not achievable. Due to the many factors influencing the decisions of risk managers, it is uncertain whether an internationally recognized system can be developed for FSOs that would be based on relative risk ranking.

2.10 SUMMARY

Establishing an FSO for hazard-food combinations provides an element that is missing from current food safety management systems based on GHP and HACCP. Hitherto, rarely has there been any expression of the expected level of control in a food operation. Instead, reliance has been placed on using process or product criteria as specified in regulations, or in end-product testing to ensure compliance with microbiological criteria. While those criteria will continue to be used, food processors that use the FSO concept will have the capacity/facility to design management systems to achieve specific food safety goals based on an expected level of consumer protection. This approach will further strengthen the scientific basis of existing systems.

Having established and communicated an FSO to enhance the safety of a food, industry must verify that the FSO is achievable, by a process that must necessarily be iterative. Information must be exchanged between those proposing an FSO (and possible control measures) and the affected industry. If the FSO is not achievable and/or the possible control measures are not possible, for example, with existing equipment, then some adjustments may be necessary. The adjustments may involve modifying the FSO and/or changing equipment or industry practices.

The process of verifying that an FSO is achievable and then establishing and validating the effectiveness of control measures is discussed in the next chapter.

2.11 REFERENCES

ACMSF (Advisory Committee on Microbiological Safety of Food, UK) (1992). *Report on Vacuum Packaging and Associated Processes*. London: HMSO.

Anonymous (1999). *Management of* Listeria monocytogenes *in Foods*. Draft document prepared by a Codex working group, hosted by Germany, for the 32nd session of the Codex Committee on Food Hygiene.

Buchanan, R. L., Damert, W. G., Whiting, R. C., & van Schothorst, M. (1997). The use of epidemiologic and food survey data to estimate a purposefully conservative dose-response relationship for *Listeria monocytogenes* levels and incidence of listeriosis. *J Food Prot* 60, 918–922.

Buzby, J. C. & Roberts, T. (1997). Economic costs and trade impacts of microbial foodborne illness. *World Health Stats Quarterly* 50, 57–66.

CAC (Codex Alimentarius Commission) (1999). *Joint FAO/WHO Food Standards Programme, Codex Committee on Food Hygiene. Principles and Guidelines for the Conduct of Microbiological Risk Assessment.* CAC/GL 30–1999. Secretariat of the Joint FAO/WHO Food Standards Programme. Rome: Food and Agriculture Organization of the United Nations.

CDC (Centers for Disease Control and Protection) (2000). Preliminary FoodNet data on the incidence of foodborne illnesses—selected sites, United States, 1999. *MMWR Morb Mortal Wkly Rep* 49, 201–205.

FAO/WHO (Food and Agriculture Organization/World Health Organization) (2000). *Joint FAO/WHO Expert Consultation on Risk Assessment of Microbiological Hazards in Foods*. FAO Food and Nutrition Paper 71. Rome: FAO.

FSA (Food Standards Agency, UK) (2000). *A Report of the Study of Infectious Intestinal Diseases in England*. London: The Stationery Office.

ICMSF (International Commission on Microbiological Specifications for Foods) (1994). Choice of sampling plan and criteria for *Listeria monocytogenes. Int J Food Microbiol* 22, 89–96.

ICMSF (International Commission on Microbiological Specifications for Foods) (1998). Potential application of risk assessment techniques to microbiological issues related to international trade in food and food products. *J Food Prot* 61, 1075–1086.

McLauchlin, J. (1995). What is the infective dose for human listeriosis? In *Proceedings of the XII International Symposium on Problems of Listeriosis*, Perth, Australia, pp. 365–370. Promaco Conventions Ltd., Canning Bridge, Australia.

McLauchlin, J. (1996). The relationship between *Listeria* and listeriosis. *Food Control* 7, 187–193.

Nickelson, R., Luchansky, J., Kaspar, C. & Johnson, E. (1996). *Update on Dry Fermented Sausage Escherichia coli O157:H7 Validation Research*. Research Report No. 11-316. Chicago: National Cattleman's Beef Association.

Noeckler, K., Teufel, P., Schmidt, K. & Weise, E. (1998). *Proceedings of the 4th World Congress on Foodborne Infections and Intoxications*, Vol. I, Vol. II. Federal Institute for Health Protection of Consumers and Veterinary Medicine, Berlin.

Notermans, S. & Hoogenboom-Verdegaal, A. (1992). Existing and emerging foodborne diseases. *Int J Food Microbiol* 15, 197–205.

Potter, M. E. & Tauxe, R. V. (1997). Epidemiology of foodborne diseases: tools and applications. *World Health Stats Quarterly* 50, 24–29.

Ross, T., Todd, E. & Smith, M. (2000). *Exposure Assessment of* Listeria monocytogenes *in Ready-to-Eat Foods. Preliminary Report for Joint FAO/WHO Expert Consultation on Risk Assessment of Microbiological Hazards in Foods*. Rome: FAO.

Tauxe, R. V. (1992). Epidemiology of *Campylobacter jejuni* infections in the United States and other industrialized nations. In Campylobacter jejuni: *Current Status and Future Trends*, pp. 9–19. Edited by I. Nachamkin, M. J. Blaser & L. S. Tompkins. Washington, DC: American Society for Microbiology.

Wheeler, J. G., Sethi, D., Cowden, J. M. *et al.* (1999). Study of infectious intestinal disease in England: rates in the community, presenting to general practice, and reported to national surveillance, The Infectious Intestinal Disease Executive. *Brit Med J* 318, 1046–1050.

WTO (World Trade Organization) (1995). The WTO Agreement on the Application of Sanitary and Phytosanitary Measures (SPS Agreement). http://www.wto.org/english/tratop_e/sps_e/spsagr_e.htm

Chapter 3

Meeting the FSO through Control Measures

3.1 Introduction
3.2 Control Measures
3.3 Confirm That the FSO Is Technically Achievable
3.4 Importance of Control Measures
3.5 Performance Criteria
3.6 Process and Product Criteria
3.7 The Use of Microbiological Sampling and Performance Criteria
3.8 Default Criteria
3.9 Process Validation
3.10 Monitoring and Verifying Control Measures
3.11 Examples of Control Options
3.12 Assessing Equivalency of Food Safety Management Systems
3.13 References

3.1 INTRODUCTION

Effective food safety management systems are designed to meet:
- FSOs established by control authorities through the risk management processes or
- objectives established by food processors during the development of a HACCP plan.

These objectives are achieved through the application of control measures intended to prevent, eliminate, or reduce microbiological hazards such as those discussed in this chapter.

3.2 CONTROL MEASURES

While microbiological criteria have played an important role in defining what has been considered acceptable, their use in testing of food has seldom proven to be effective for control of microbial hazards. The food manufacturers controlled their processes to ensure that their products would meet the criteria if sampled. Greater emphasis was placed on lot acceptance testing than the conditions of operation. Despite the weakness of relying on end-product testing, advances have been made in preventing or reducing foodborne illness. In ICMSF Book 4 (ICMSF, 1988) it was recognized, however, that more emphasis should be placed on the use of selected, targeted control measures and less on microbiological testing of food.

Historically, the major advances in consumer protection have resulted from the development and implementation of selected, targeted control measures at one or more steps along the food chain, from the farm up to the consumer. These advances have followed periods of extensive investigation to gain the information necessary to understand the pathogens (e.g., sources, life cycle, parameters influencing growth, survival, death or metabolite production).

Control measures are the actions and activities used to prevent, eliminate, or reduce a food safety hazard to a tolerable level. They generally fall into three categories:

Controlling Initial Levels

- avoiding foods with a history of contamination or toxicity (e.g., raw milk, raw molluscan shellfish harvested under certain conditions).
- selecting ingredients (e.g., pasteurized liquid eggs or milk).
- using microbiological testing and criteria to reject unacceptable ingredients or products.

Preventing Increase of Levels

- preventing contamination (e.g., adopting GHPs that minimize contamination during slaughter, separating raw from cooked ready-to-eat foods, implementing employee practices that minimize contamination, using aseptic filling techniques).
- preventing growth of pathogens (e.g., proper chilling and holding temperatures, pH, a_w, preservatives).

Reducing Levels

- destroying pathogens (e.g., freezing to kill certain parasites; disinfectants, pasteurization, irradiation).
- removing pathogens (e.g., washing, ultra-filtration, centrifugation).

One or more of the above activities may be necessary to control a hazard. In addition, one or more may be applied at different steps along the food chain to eliminate, prevent, or reduce a hazard to an acceptable level. Each participant along the food chain has a responsibility to apply those control measures that contribute to providing safe foods.

These control measures also fall into one of two programs applied by food manufacturers, the Good Hygienic Practices (GHP) and the Hazard Analysis Critical Control Point (HACCP) systems discussed below.

The first program, GHP, can be viewed as the basic sanitary conditions and practices that must be maintained to produce safe foods. It also includes certain support activities such as raw material selection, product labeling, and coding or recall procedures. Effective application of GHP provides the foundation upon which the second program, HACCP, is developed and implemented. The development of an effective HACCP system involves, therefore, a systematic approach to the identification, evaluation, and control of food safety hazards in a food operation.

HACCP is not implemented in lieu of GHP, and failure to maintain and implement GHP can invalidate a HACCP system and result in production of unsafe food. It is necessary to consider the hazards that are most likely to occur in each particular food operation and to pay particular attention to those elements of GHP and HACCP that will contribute most to controlling those hazards.

3.2.1 Good Hygienic Practices (GHP)

The General Principles of Food Hygiene (CAC, 1997b) describe the major components of GHP as:

- design and facilities (location, premises and rooms, equipment facilities)
- control of operation (control of food hazards, key aspects of food hygiene control, incoming material requirements, packaging, water, management and supervision, documentation and records)
- maintenance and cleaning (maintenance and cleaning programs, pest control systems, waste management, monitoring effectiveness)
- personal hygiene (health status, illness and injuries, personal cleanliness and behavior, visitors)
- transportation (general requirements, use, and maintenance)
- product information and consumer awareness (lot identification, product information, labeling, consumer education, handling/storage instructions)
- training (awareness and responsibilities, training programs, instruction and supervision, refresher training)

As previously stated, effective application of GHP provides the foundation upon which HACCP systems are developed and implemented. Failure to maintain and implement GHP can invalidate a HACCP system and result in production of unsafe food.

Effective control of a hazard in a food necessitates consideration of the components of GHP likely to have significant impact in controlling the hazard. For example, incoming material requirements would be very important to control the risks of certain hazards in seafood (e.g., paralytic shellfish poisoning, ciguatera toxin, scombroid poisoning). Incoming material requirements would be of lesser importance for a food that will be cooked sufficiently to eliminate enteric pathogens (e.g., salmonellae in raw meat or poultry) that may be present. Thus, the various components of GHP do not carry equal weight in all food operations. It is necessary to consider the hazards that are most likely to occur and then apply those GHPs that will be effective for controlling the hazards. This does not mean that the other components of GHP, such as equipment maintenance or calibration of temperature probes, are ignored. Some may be very important to ensure that a food meets established quality requirements.

In certain situations, selected components of GHP may carry particular significance and should be incorporated into the HACCP plan. For example, equipment maintenance and calibration are important for large, continuous ovens used in cooking meat products. The procedure and frequency (e.g., monthly, quarterly) for conducting checks on heat distribution during cooking could be incorporated into the HACCP plan as a verification procedure. In addition, it is normally necessary to verify the accuracy of the thermometers used for monitoring oven temperatures during cooking.

Information on hygienic design of facilities and equipment, cleaning and disinfection, health and hygiene of personnel, and education and training are discussed in ICMSF Book 4 (ICMSF, 1988).

3.2.2 Hazard Analysis and Critical Control Point (HACCP)

The Codex document on the Hazard Analysis and Critical Control Points (HACCP) System and Guidelines for its Application (CAC, 1997a) lists the following seven principles of the HACCP system:

1. Conduct a hazard analysis.
2. Determine the critical control points.
3. Establish critical limits.
4. Establish monitoring procedures.
5. Establish corrective actions.
6. Establish verification procedures.
7. Establish recordkeeping and documentation procedures.

The development of an effective HACCP system involves a systematic approach to the identification, evaluation, and control of food safety hazards in a food operation. HACCP plans specify the actions to be taken in a food operation to control food safety hazards. HACCP plans also specify records to be generated during the operation for use in verification that critical limits have been met at critical control points in the operation. In the event that a deviation occurs at a critical control point, the deviation should be detected in time to ensure that corrective actions will prevent unsafe food from reaching consumers. This may necessitate collecting and analyzing samples from across the questionable quantity of food. The principles described in this text for sampling food can be applied to help assess the safety of a suspect lot and lead to appropriate disposition of the food (see Chapter 11, Tightened Inspection).

The production of safe food requires food operators selectively to apply GHP and the principles of HACCP to develop and implement a total food safety management system that will control the significant hazards in the food that is being produced.

The Codex Alimentarius Commission has defined a critical control point as "a step at which control can be applied and is essential to prevent or eliminate a food safety hazard or reduce it to an acceptable level" (CAC, 1997a, p.19). The interpretation of what is considered an acceptable level has been left to the judgment of the HACCP team and respective control authority. The FSO concept can be used to communicate the level of control necessary for a hazard to be reduced to "an acceptable level."

3.3 CONFIRM THAT THE FSO IS TECHNICALLY ACHIEVABLE

Following the development of an initial FSO for a particular food or food category, risk managers in industry and government must confirm that it is technically achievable through GHP and HACCP (see Figure 1–1). This decision should be based on preliminary process and product requirements and exchanges of information between industry and government.

If the proposed FSO is (or seems to be) technically achievable, then final requirements for product and process can be developed to ensure that this objective is met. Individual food manufacturers can then implement the necessary control measure(s) to fulfill the overall performance criteria.

However, if the proposed FSO is not technically achievable, then modifications of the product, the process, if technically possible, and/or the FSO may be necessary. Modifications should result from discussions between industry and government aiming at achieving optimal public health protection based on a tolerable level of risk. If no technically achievable solution can be found, then it may be necessary to ban the product and/or the process. In addition, as new information regarding a particular hazard or product emerges, FSOs may be modified.

3.4 IMPORTANCE OF CONTROL MEASURES

Historically, the major advances in consumer protection have resulted from the development and implementation of selected, targeted control measures at one or more steps along the food chain. Appendix 3–A summarizes a number of foodborne diseases and the measures that have been found to be most effective for their control. Certain biological hazards are influenced by environmental conditions that are beyond the control of a food operator (e.g., ciguatera toxin, scombroid poisoning, and shellfish intoxication, mycotoxin formation in crops in the field). For these hazards, it has been essential to understand the effect of climatic conditions in the affected regions of the world. This information is used to determine when to screen for the hazard during periods of higher risk. Thus, to avoid these hazards, raw material selection is an important control measure. In addition, dehydration of certain crops can be used following harvest to prevent mycotoxin development during subsequent storage and distribution. For certain pathogens associated with livestock, selected on-farm practices have led to reductions in human disease. For example, brucellosis and trichinosis in the US and Europe have been reduced by specific control measures (Figures 3–1 and 3–2).

3.4.1 Examples of the Effectiveness of Control Measures

Historically, foods have been preserved by salting, pickling, curing, heating, drying, etc., to save the foods for those periods when they would not normally be available. These

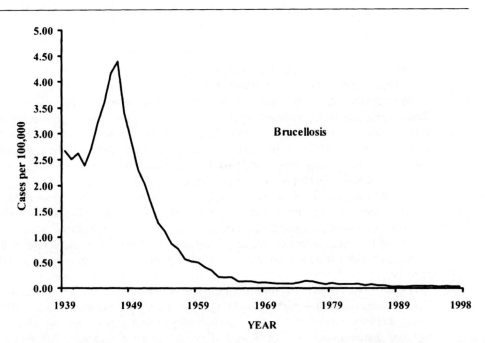

Figure 3–1 Cases of human brucellosis per 100,000 population in the US, 1939–1998.

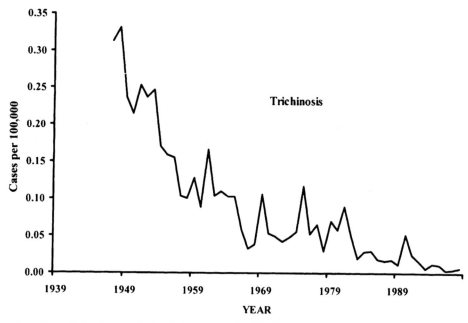

Figure 3–2 Reduction of trichinosis in the US after World War II.

basic technologies have led to considerable diversity in the world's food supply. They also can be viewed as control measures that have evolved through experience and, when applied correctly, have been found to yield safe food for later use. Table 3–1 summarizes the control measures that have been used to produce a wide variety of shelf-stable foods (Tompkin & Kueper, 1982). In some cases, a single important control measure has been adequate. In most, however, combinations of factors have been used to achieve shelf stability.

Another means of demonstrating the importance of control measures is to examine trends of foodborne disease. For example, Figure 3–1 shows the number of cases of human brucellosis per 100,000 population in the US from 1939–1998. A marked increase occurred during World War II that was attributed to greater movement of livestock among farms in response to the increased demand for meat and higher livestock prices and diversion of veterinarians to the war effort (Simms, 1950). Similar disruptions in on-farm pathogen control practices would have been commonplace elsewhere in the world during that time period. During the period 1940–1947, human brucellosis was strongly associated with animal contact. For example, in Iowa, the rate per 100,000 population during 1940–1947 was 562 for veterinarians, 276 for packinghouse employees, 59 for male farm workers, and 2 for the "no contact" group (Jordan, 1950). Following the war, efforts resumed for control of brucellosis through a combination of on-farm practices and culling of livestock. Increased use of pasteurization of milk for manufacture of dairy products provided an additional important control measure that led to reduced consumer exposure.

Prior to 1940, the incidence of human trichinosis in the US also decreased due to specific on-farm control measures, processing requirements in commercial meat establishments, and education of consumers of the necessity to cook pork adequately. After investigations demonstrated the significance of household kitchen garbage containing raw

Table 3–1 Processes That Have Been Used for Making Commercially Prepared Foods Shelf-stable

Food	Process	Food	Process
Milk	D, H	Vegetable juices	B, D
Eggs	H	Vegetables, pickled	A
Natural cheese	J	Peanut butter	I
Processed cheese and cheese sauce	C	Ham, bacon	A, E
		Beef	B, E
Cake, bread	K, L	Pork sausage	E, G
Cookies, crackers	H	Dry and semi-dry sausage	E, F, G
Fruit	A, C, J, H	Franks, viennas, meat spreads	B
Fruit juices	C, D	Luncheon meats, cured	A
Jams, jellies	C	Fish, smoked salmon, pickled herring, salt cod	B, E, H
Vegetables	A, B, H		

<u>Process Codes</u>:
A. Mild thermal process ($F_0 = < 2.78$) in hermetically sealed container.
B. High thermal process ($F_0 = \geq 2.78$) in hermetically sealed container.
C. Thermal process, hot fill, seal and hold before cooling.
D. Thermally process, chill, then aseptically package.
E. Salted, perhaps cured, at low temperature then dried during heating or at ambient temperatures. Smoke commonly applied.
F. Fermented at 20–40 °C, perhaps heated to $\geq$ 46 °C, then dried at cool temperature (e.g., 13 °C).
G. Fermented and/or cooked, dried, then sealed in lard.
H. Thermal process and dehydration.
I. Roasted, ground, and filled at moderate temperature.
J. Dehydration.
K. Filled into container, sealed, then heated sufficiently to bake product.
L. Baked in container, then sealed while hot.

pork as a source of infection for pigs, regulations were implemented that required cooking garbage before feeding to pigs (Leighty, 1983). Figure 3–2 shows the reduction of trichinosis in the US after World War II.

Trichinosis is an interesting example illustrating the differences in control measures adopted by the US and Europe to achieve similar levels of protection. While in the US control measures have been focused on destruction, e.g., by cooking, in Europe, microscopic and later immunologic methods have been used to detect and eliminate carcasses with *T. spiralis* (Pozio, 1998). The different approaches have led to differences in regulatory requirements, quality characteristics of pork products, and culinary habits. In this particular example, however, the European approach was much more effective and achieved more rapid reductions in the prevalence of *T. spiralis* in the hog population and human trichinosis than in the US. It is only in recent years that the prevalence of human trichinosis in the US has begun to approach the levels found in Europe before 1940. In the European Union (EU), 74.4% of domestic pigs are raised on industrialized farms where *Trichinella* infection has not been diagnosed. The reservoir for *Trichinella* in the EU consists of pigs raised on small family farms or grazing in wild areas, as well as wild boars and horses (Pozio,

1998). The prevalence of trichinellae in pigs in Germany has been three or fewer infected pigs among about 40 million pigs slaughtered each year (Rehmet et al., 1999). In a review of trichinellosis in the Netherlands, it was stated that such infections had not occurred in humans or animals during the previous 40 years (Ruitenberg & Sluiters, 1974). For comparison, farm-raised swine in the US, which comprised 98.5% of the swine production, had prevalence rates of about 0.125% (Zimmerman, 1974).

The third example is the sharp decline in typhoid fever in the US from 1939–1998 (Figure 3–3). The source of *Salmonella typhi* is the intestinal tract of humans. Growth outside the human host has not been an important factor. The primary routes of infection have been fecal-oral routes involving untreated water and infected food handlers or health care providers. Thus, improvements in waste water management; availability of safe, potable water for drinking and food processing; and education on the importance of hand washing have led to reductions in the prevalence of typhoid fever. This enabled a more directed approach toward identifying and treating individuals who harbored the pathogen due to infection or as asymptomatic carriers. Finally, vaccination of individuals traveling to regions where typhoid fever was endemic reduced the likelihood of reintroducing the pathogen into the population.

In Figure 3–4, the trends for shigellae and nontyphoid salmonellae are provided. While progress was initially made after World War II toward reducing the rate of shigellosis, there has been no improvement over the past four decades. Shigellae are similar to *S. typhi* in that humans are the source, and transmission involves contaminated water and food. An additional significant route involves person-to-person transmission via the fecal-oral route, particularly in institutional settings (e.g., child care centers). Can effective control measures be implemented that would decrease the prevalence of shigellosis below the level of 5–8.9 cases per 100,000 (CDC, 2000a)?

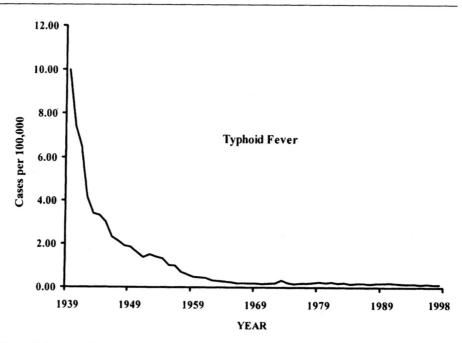

Figure 3–3 Reduction in typhoid fever in US from 1939–1998.

The data for nontyphoid salmonellae show an increase of salmonellosis until about 1990, followed by a decrease (Figure 3–4). The decrease may be attributed to reductions in egg-borne transmission of *S. enteritidis* through improved on-farm controls (CDC, 2000b). In addition, some reduction may eventually prove to be due to improved controls implemented by the meat and poultry industry following the establishment of performance standards by the US Department of Agriculture (USDA)-Food Safety and Inspection Service in 1996 (USDA, 1996a). The performance standards in Table 3–2 specify the maximum frequencies for salmonellae on carcasses and in ground meat. In this case, the performance standards were incorporated into a regulation and assumed the nature of a regulatory standard, but not a microbiological standard, as described in Chapter 5.

In anticipation of the regulation and other requirements, the meat and poultry industry in the US invested extensively to modify its processing equipment, plant layouts, and processing procedures. This included the use of antimicrobial treatments, such as improved control of chlorinated processing water for poultry and the use of steam or hot water pasteurization for beef carcasses. Some modifications were implemented well in advance of the effective dates for the regulatory requirements.

All categories of raw meat and poultry for which performance standards were established have shown reductions in the prevalence of salmonellae in large plants from 1996–2000. The prevalence rate for salmonellae in broilers in large plants, for example, declined to approximately 10% (USDA, 2000). Thus, it would appear that the performance standards have led to reductions in the prevalence of salmonellae while allowing the industry to decide how to modify its processing conditions. The question remains whether the reductions in salmonellae in raw meat and poultry will lead to a measurable reduction in human salmonellosis. In addition, it is uncertain whether further reductions will be possible in the absence of new technology or improved control at the farm level.

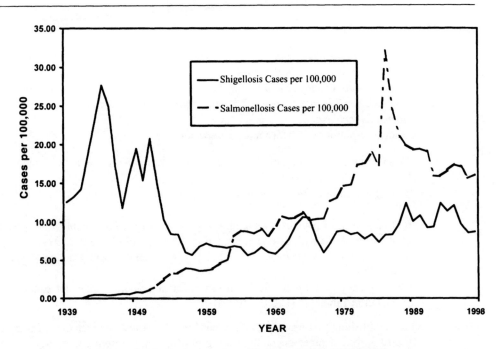

Figure 3–4 Incidence of shigellosis and salmonellosis in the US, 1939–1998.

Table 3–2 Maximum Allowable Frequencies for Salmonellae on Carcasses and Ground Meat and Poultry

	Allowable percent positive samples	Number of samples	Maximum allowable number of positives
Carcasses:			
Steer/heifer	1.0	82	1
Cows/bulls	2.7	58	2
Hogs	8.7	55	6
Broilers	20.0	51	12
Ground products:			
Beef	7.5	53	5
Chicken	44.6	53	26
Turkey	49.9	53	29

In Finland, Norway, and Sweden, greater emphasis has been placed on testing and control at the farm level to achieve on-farm prevalence rates of less than 1% for salmonellae in livestock and poultry. These control measures are intended to meet a performance standard of less than 1% prevalence rate for salmonellae in foods of animal origin (Kruse, 1999).

A final example of an effective control measure has been the apparent 40% reduction in human yersiniosis in Norway following a rather simple but effective change in the slaughtering procedure for hogs. Data indicate that the reduction occurred after about 90% of the hog slaughterhouses implemented a procedure of sealing off the rectum with a plastic bag immediately after the rectum was cut free. A similar decline (about 30%) reportedly occurred following adoption of the same practice in Sweden (Nesbakken, 2000).

3.5 PERFORMANCE CRITERIA

To achieve the defined FSO, it is necessary to implement one or more control measures at one or more steps in the food chain. At these steps, hazards can either be prevented, eliminated, or reduced. If appropriate control measures are not applied at these steps, then the hazards may increase. The outcomes of these control measures are defined as *performance criteria*.

Examples of published performance criteria include:

- 12 D reduction of proteolytic *Clostridium botulinum* in low-acid canned foods (Stumbo, 1973; Brown, 1997).
- 6 D reduction of *Listeria monocytogenes* in ready-to-eat chilled foods (Lund et al., 1989).
- 6 D reduction of psychrotrophic strains of *Clostridium botulinum* in pre-prepared chill-stored foods with extended shelf-life (ACMSF, 1992; Gould, 1999).
- 5 D reduction of *Escherichia coli* O157:H7 for fermented meat products (Nickelson et al., 1996).
- the frequency of salmonellae in/on raw meat and poultry shall not exceed those specified in Table 3–2 (USDA, 1996a).

When establishing performance criteria, consideration must be given to the initial level of a hazard and to changes occurring during production, distribution, storage, preparation, and use of a product. A performance criterion is preferably less than but at least equal to the FSO and can be expressed by the following equation (1):

$$H_0 - \Sigma R + \Sigma I \leq FSO \tag{1}$$

Where: FSO = Food safety objective

H_0 = Initial level of the hazard

ΣR = Total (cumulative) reduction of the hazard

ΣI = Total (cumulative) increase of the hazard

FSO, H_0, R and I are expressed in $\log_{10}$ units.

These criteria are usually not established for control measures designed to avoid certain foods, although they may be applied to ensure that the initial level of hazards in ingredients is not excessive. Microbiological testing may thus be used to select ingredients or to obtain information on the initial level of a hazard.

3.5.1 Examples of Performance Criteria

For examples 1 to 3 below, a hypothetical dose-response curve for a certain infectious pathogen has been included at the top of each figure (Figures 3–5, 3–6, and 3–7). In these examples, the estimated number of cases per 100,000 population increases as the pathogen concentration in the food exceeds the minimum infectious dose of 1 cfu g^{-1}. In all three examples, the FSO has been established at 100-fold less than this dose, i.e., < 1 cfu 100 g^{-1} at the time of consumption.

These examples could be representative for *E. coli* O157:H7 in products submitted to heat treatment and/or other processing steps.

3.5.1.1 Example 1

In this example, the initial population (H_0) in the raw material is estimated to be as high as 10^3 cfu g^{-1}, but growth (ΣI) can be prevented. The required performance criterion would be expressed as:

$$H_0 - \Sigma R + \Sigma I \leq FSO$$
$$3 - \Sigma R + 0 \leq -2$$
$$\Sigma R = 5$$

Based on equation (1), the process must result in an overall reduction of 5 $\log_{10}$ to meet the FSO. This corresponds to a performance criterion of a 5 D reduction of the pathogen and could be achieved by one, or a combination of, control measures (Figure 3–5).

3.5.1.2 Example 2

In this example, the initial population (H_0) in the raw material can be controlled to ≤ 100 cfu g^{-1} and growth (ΣI) can be prevented. Here, a sampling plan allowing for raw material selection can be used. The required performance criterion would be expressed as:

$$H_0 - \Sigma R + \Sigma I \leq FSO$$
$$2 - \Sigma R + 0 \leq -2$$
$$\Sigma R = 4$$

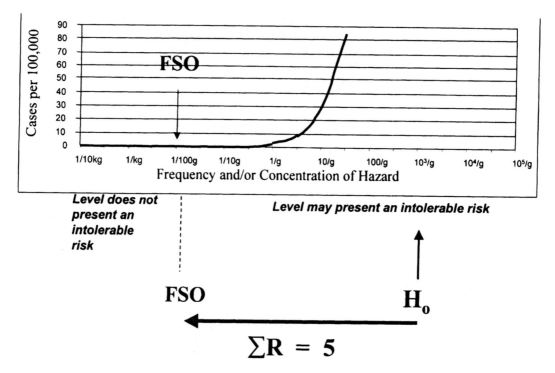

Figure 3–5 Example 1.

Based on equation (1), the process must result in an overall reduction of 4 $\log_{10}$ units to meet the FSO. This corresponds to a performance criterion of a 4 D reduction of the pathogen and can be achieved using selected raw materials and other control measures (Figure 3–6).

3.5.1.3 Example 3

In this example, the initial population (H_0) in the raw material is 10^3 cfu g^{-1}, and recontamination or growth (ΣI) leading to not more than a 100-fold increase is possible. The required performance criterion would be expressed as:

$H_0 - \Sigma R + \Sigma I \leq FSO$

$3 \; - \Sigma R + 2 \; \leq -2$

$\Sigma R = 7$

Based on equation (1), the process must result in an overall reduction of 7 $\log_{10}$ units (i.e., from 10^5 cfu g^{-1} to 1 cfu 100 g^{-1}) to meet the FSO. This corresponds to a performance criterion of a 7 D reduction of the pathogen and can be achieved by various control measures (Figure 3–7).

For examples 4 and 5, a hypothetical dose-response curve is included at the top of each figure (Figures 3–8 & 3–9). In these examples, the estimated number of cases per 100,000 population increases as the pathogen concentration in the food exceeds the minimum infectious dose of 10^3 cfu g^{-1}. In the two examples, the FSO has been established at 10-fold less than this dose, i.e., < 100 cfu g^{-1} at the time of consumption.

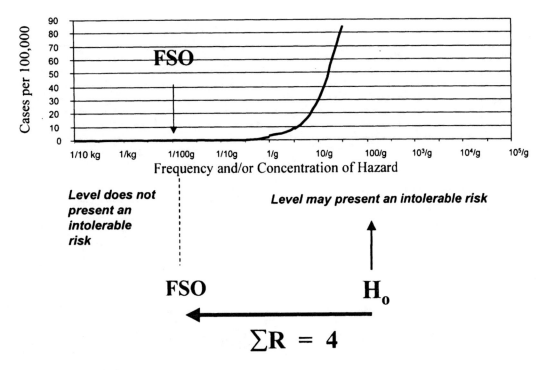

Figure 3-6 Example 2.

These examples could be representative of *Listeria monocytogenes* and of ready-to-eat product not submitted to any killing effect ($\Sigma R = 0$) before the food is consumed.

3.5.1.4 Example 4

In this example, the initial level (H_0) is unknown but there is no recontamination or growth (ΣI) up to the time of consumption. The required performance criterion would be expressed as:

$$H_0 - \Sigma R + \Sigma I \leq FSO$$
$$H_0 - 0 + 0 \leq 2$$
$$H_0 \leq 2$$

Based on equation (1) and to ensure that the FSO is met, the pathogen level must not exceed 100 cfu g^{-1}. This could be achieved through the application of microbiological criteria. In this case, the performance criteria must be below the FSO (Figure 3-8).

3.5.1.5 Example 5

In this example, the initial level (H_0) of the pathogen can be controlled to ≤ 10 cfu g^{-1} and growth (ΣI) can be prevented. Here, a sampling plan allowing for raw material selection can be used. The required performance criterion would be expressed as:

$$H_0 - \Sigma R + \Sigma I \leq FSO$$
$$1 - 0 + \Sigma I \leq 2$$
$$\Sigma I \leq 1$$

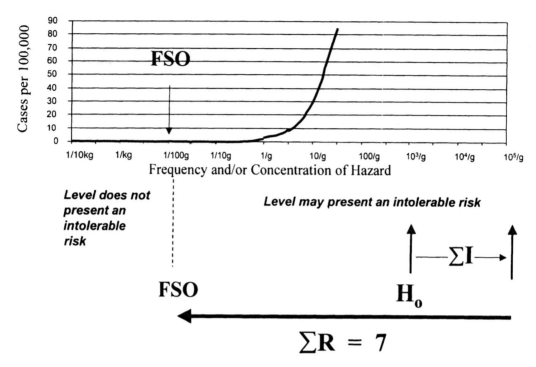

Figure 3–7 Example 3.

Based on equation (1) and to ensure that the FSO is met, the pathogen level must not increase more than 10-fold. This could be achieved through the establishment of product criteria such as a_w or pH, alone or in combination (Figure 3–9).

It should be recognized that the parameters that may be used in equation (1) are point estimates, whereas in practice, they will have a distribution of values associated with them. If data exist for the variance associated with the different parameters, then the underlying probability distributions may be established using an approach similar to that in risk assessment.

Certain aspects must be considered when determining the elements of equation (1). For the initial level of the hazard (H_0), there is a temptation to use the highest level reported in the literature. However, care should be used in this approach as this can lead to an unnecessarily restrictive (conservative) performance criterion. H_0 may vary from one situation to another and should be considered as the initial upperlevel that can be expected under normal conditions at a specific operation. The level of hazard detected in one situation may have no relevance to another. Thus, the value for H_0 in one food operation may differ from other operations. H_0 may range from zero (undetectable) to its upper level in the absence of any effort for its control. It may be necessary to establish a maximum value to assure the production of safe food, as shown in Example 4.

In the case of ΣR, any type of reduction needs to be considered, such as killing by thermal treatments, washing, or filtration.

In the above examples, the increase in a hazard (ΣI) was through pathogen growth. While growth is likely to be the most significant factor in many foods, an increase in a hazard through recontamination also must be considered.

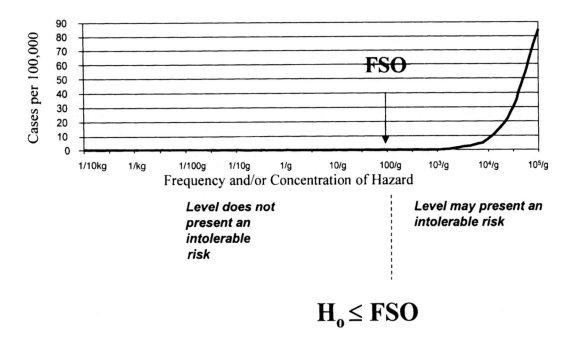

Figure 3–8 Example 4.

3.6 PROCESS AND PRODUCT CRITERIA

A performance criterion is met by implementing process criteria, such as time and temperature of a heat treatment, and/or product criteria, such as the water activity of the product, alone or in combination, to achieve control over a specific hazard. For example, the process criterion for milk pasteurization is 71.7 °C for 15 s; this criterion has been adopted as a regulatory requirement in numerous countries. It defines the processing conditions considered necessary to ensure inactivation of *Coxiella burnetti*. This process criterion will also ensure elimination of other non-spore-forming pathogenic enteric bacteria known to occur in raw milk.

An example of a product criterion is pH ≤ 4.6 in canned shelf-stable acidified foods, which is the parameter known to inhibit growth of *C. botulinum*.

In the case of minimally processed refrigerated foods, combinations of process and product criteria have been proposed to ensure the safety of products. In these cases, heat-treatments milder than 90 °C for 10 min in combination with other factors, such as reduced pH, water activity, or shelf-life (AMCSF, 1992; Gould, 1999), are recommended.

3.7 THE USE OF MICROBIOLOGICAL SAMPLING AND PERFORMANCE CRITERIA

Increasingly, it has been realized that food safety management systems based on preventing hazards through GHP and HACCP are much more effective in ensuring safe foods than is end-product testing.

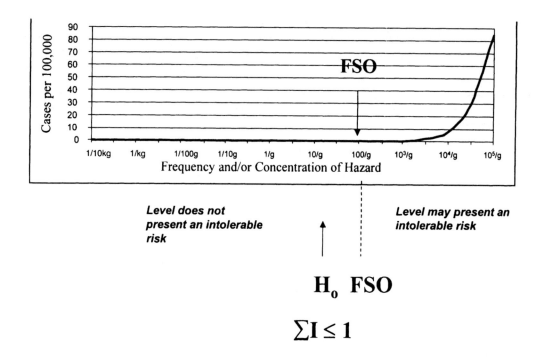

Figure 3–9 Example 5.

In the ICMSF scheme for managing microbiological risks (Figure 1–1), two uses of microbiological criteria are identified:

1. to validate that control measures meet the performance criteria.
2. to determine acceptability of a food when no more effective means of providing such assurance of safety is available, i.e., in the absence of knowledge that GHP and HACCP have been properly applied (see Chapter 4 for more details).

Different control measures or options can be applied to manufacture safe foods to meet the FSO. However, the equivalence of these measures, in comparison to the established performance criterion, needs to be established. For a number of processes and products, this can be expressed in terms of frequency or concentration of a microbiological hazard in the food.

Traditionally, the performance of sampling plans and microbiological criteria have been expressed in terms of their ability to detect defective sample units within a lot and, thereby, determine acceptance or rejection of the lot. The proportion of defective sample units can be used to estimate the concentration of the hazard within the food (Foster, 1971; Legan et al., 2001); however, homogeneous distribution of the defective units throughout the lot and random sampling (Chapter 7) must be assumed.

Figure 3–10 shows the number of 25 g samples that must test negative to be able to conclude (with a confidence of 95%) that the concentration of organisms is at a given level or below. For example:

- 13×25 g negative corresponds to < 1 cell 125 g^{-1}
- 29×25 g negative corresponds to < 1 cell 250 g^{-1}
- 60×25 g negative corresponds to < 1 cell 500 g^{-1}

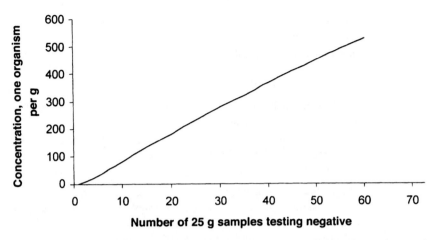

Figure 3–10 Number of 25 g samples testing negative to have 95% confidence in meeting standard.

Using this approach, microbiological testing can be used to ensure that the initial concentration of the hazard (H_0) in an ingredient does not exceed a given concentration (assuming a homogeneous distribution or random sampling, as per Chapter 7). In this way, a microbiological criterion can be used as a control measure to help meet a performance criterion and, thereby, an FSO as in Example 5.

It must, however, be realized that for a number of foods, the application of effective control measures will allow the achievement of much lower defect rates and that under such circumstances, testing remains questionable.

3.8 DEFAULT CRITERIA

In the absence of an FSO, it may be appropriate to establish default criteria for certain control measures. These fail-safe criteria, developed by advisory bodies of experts or control authorities, are intended to control hazards under "worst-case" situations and will, of necessity, be less flexible. They may assume a higher than normal initial level of a hazard (H_0) and/or increase (ΣI) during processing or prior to consumption.

An example of a default criterion is cooking roast beef to an internal temperature of 62.8 °C to destroy enteric pathogens, such as salmonellae and pathogenic *E. coli* (USDA, 1999). Another example is in-pack thermal treatment for 10 min at an internal temperature of 90 °C for ready-to-eat chilled foods with extended shelf-life to destroy nonproteolytic *Clostridium botulinum* (ACMSF, 1992; Gould, 1999).

Default criteria provide a safe harbor for food operators lacking either the resources or the desire to develop information necessary for alternative criteria that may be more appropriate for their specific operation or product.

3.9 PROCESS VALIDATION

Control of food operations depends upon operator knowledge and the conditions that influence the production of safe and unsafe food. A considerable amount of information is available in the literature and other sources. For novel processes it may be necessary to develop information to verify the efficacy of the control measures. Some operations may be so unique and different from other operations producing similar foods that control is less certain. In other situations, an operator may wish to use minimal processing techniques for improved product quality or reduced cost. In such instances, it may be necessary to validate the efficacy of the adopted control measures.

Validation can involve:

- developing data through challenge tests in the laboratory that are intended to mimic the conditions of operation.
- collecting data during normal processing in the food operation.
- comparison with similar processes/products.
- other expert knowledge.

Each method has its strengths and weaknesses, and in certain cases, two or more methods may be desirable. Data developed through laboratory challenge tests can involve the food, culture media, or other material that may be appropriate. Challenge studies in a food processing environment can provide a higher degree of assurance concerning the ability to meet performance criteria; however, this requires the use of surrogate test microorganisms (see below). Pathogenic microorganisms should never be introduced into the food production or processing environment for the purpose of process validation. In some cases, it may be possible to follow changes in the population of naturally occurring pathogens throughout a process. Such studies, for example, could be conducted during the preparation and processing of raw agricultural commodities into ready-to-eat foods. Ideally, validation could involve laboratory challenge tests with pathogens in the laboratory and then revalidation after the control measures have been implemented. This, however, may be impractical in situations where the prevalence of a pathogen is very low and large numbers of samples are necessary to develop meaningful data.

3.9.1 Laboratory Challenge Tests

When conducting laboratory challenge studies, factors such as the following should be considered:

- intrinsic resistance of the pathogen;
- use of nonpathogens for validation;
- composition of the food; and
- conditions of storage, distribution, and preparation for use.

3.9.1.1 Intrinsic Resistance of the Pathogen

Studies to evaluate the resistance of a pathogen to different parameters (e.g., heat, cold, acid) that may be incorporated into a control measure should be performed using several strains (e.g., five or more), including outbreak-associated isolates from the food in question. Resistance of the strains used for testing is a key factor when establishing effective control parameters. The inocula should be prepared under conditions that yield resistance of the pathogen appropriate to the process. Vegetative cells of salmonellae and pathogenic *E. coli*, for example, exhibit a maximum resistance to heat and acidic conditions when in the stationary phase after having been grown at elevated temperatures. Sufficient numbers of the pathogen (e.g., cells, spores, viral particles, oocysts) should be used to eliminate biovariability effects.

Strains to be tested should not include isolates with unrealistically extreme resistances or growth characteristics when these are not associated with public health concerns. For example, *Salmonella senftenberg* 775W should not be used as the basis for thermal processing requirements for liquid egg products because of its unusually high heat resistance and rare occurrence, but is appropriate to evaluate survival in chocolate and similar products (ICMSF, 1998).

3.9.1.2 Use of Nonpathogens for Validation

Validation of control measures in a food operation can be accomplished through the use of nonpathogenic microorganisms if they have been shown to have the same growth pattern or resistance as the pathogen of concern.

3.9.1.3 Composition of the Food

The composition of the food can affect inactivation, survival, and/or growth of pathogens and therefore must be known and taken into account. Factors such as pH, a_w, Eh, humectants, acidulants, solutes, antimicrobials, substrates, and competing microflora can affect the chemical and physical properties of the food and subsequently the pathogen of concern. Normal variation in the concentration and distribution of food constituents and microorganisms also must be known and understood.

3.9.1.4 Conditions of Storage, Distribution, and Preparation for Use

Factors affecting the safety of a food during storage, distribution, and preparation for use must be identified and controlled. Information on the intended use and an estimate of likely misuse of the product may be necessary. Examples of parameters that often have a significant effect include time and temperature, the potential for contamination, and faulty preparation before consumption.

3.9.2 Data Collected from Food Operations

A considerable amount of data can be collected from a food operation to better understand the potential microbial hazards. The data can consist of a variety of chemical, physical, and microbiological measurements. For example, if the chemical composition of a food is known as it undergoes processing, estimates can be made of the potential for certain pathogens to survive or multiply in it. Similarly, measurements of processing times and temperatures must be understood if the potential for survival and growth during processing is to be estimated. While generalizations often can be made from published data, the source and type of raw materials may differ among food operators. The best means to establish H_0 is to analyze raw material samples from an operation. Opportunities also may exist for collecting microbial data from samples collected as a food is being processed. Such in-plant data can be used to validate a process or to verify results obtained in the laboratory. For a variety of reasons, however, it may be necessary to measure changes in the population of a nonpathogen that has similar or greater resistance than the pathogen. This may be necessary, for example, when the numbers or prevalence of the pathogen are too low to develop meaningful data. The variability in a pathogen population can be influenced, for example, by season, geographic location of the operation, source and type of raw materials, and processing conditions. These and other factors should be considered when collecting data for use in process validation.

3.9.2.1 Process Variability

The variability that occurs in a food operation must be considered when establishing the critical limits associated with control measures. Examples of factors that can influence variability of a process include equipment performance and reliability, integrity of container seals, processing times and temperatures, pH, humidity, flow rates, and turbulence.

It is essential that the variability of process parameters and product formulation be considered when setting critical limits. In general terms, the critical limits at a CCP for a process operating under a high degree of control (low variability) can be closer to the conditions necessary for control of a hazard, as discussed above. Conversely, the critical limits for a less controlled process (high variability) must be more conservative and more restrictive. In other words, critical limits must be based on the capability of the process to achieve a given criterion under normal operating conditions taking into account variability. Monitoring and verification procedures specified in a HACCP plan should be designed to determine when the process is operating outside this normal variability so appropriate corrective actions can be taken.

These principles are illustrated in Figure 3–11. Three different process/product capabilities are illustrated, each of which must meet a product criterion of pH < 4.6 to ensure the safety of a high-acid product with respect to *C. botulinum*. In the first example, there is poor control of final product pH and a high variation (distribution a), hence the operating target pH (mean) or *set point* must be at pH 3.8 to be sure that pH < 4.6 will always be met. In the second example, there is better control of the process and resulting final product pH (distribution b); hence, the set point for the process is pH 4.0 and closer to the required product criterion. In the final example, there is excellent control of the process (distribution c), and the set point can be at pH 4.3.

An effective process control system is a key element in the management of food safety and can also provide economic benefits. Processes under control are less likely to yield foods that will cause harm to consumers. Food processors who understand the factors that can cause variability in their operations have established monitoring systems to detect and prevent unacceptable loss through inefficiencies, reduced yield, or poor quality. Similarly, by incorporating the elements of GHP and HACCP into their process control systems, food operators can ensure the production of safe foods. Whether for economic gain or food safety, criteria are established at selected points in the operation to enable the operator to assess control. The operation is considered under control while established criteria are being met. If they are not, adjustments must be made to bring the

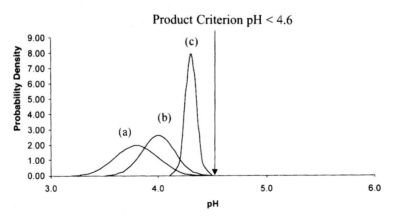

Figure 3–11 Set point depends on the variability in the process.

process back under control. A number of statistical tools can be used to aid in the evaluation of process control and trend analysis both for microbiological testing and for the physical and chemical parameters (see Chapter 13). Through knowledge of the process and use of the data, operators can plan for and achieve continuous improvement, thereby further reducing variability and achieving greater control.

Process control systems can involve two types of measurement—real time and delayed time. In the former, data are collected and used to adjust processes during the operation. Examples include measurements for pH, temperature, and humidity. Ideally, there is continuous feedback to provide automatic adjustment as the operation proceeds. Delayed time measurements do not yield data that permit adjustment to an ongoing operation. Examples include measurements using conventional microbiological methods and certain chemical analyses. Due to the time lapse between when samples are collected and results are obtained, these methodologies yield historical data and document what has happened rather than what is happening. While of less value for current production, the data can be used to detect trends, and with proper adjustments, reduce the likelihood that future lots will be unacceptable. These concepts will be discussed more fully in Chapter 13.

3.10 MONITORING AND VERIFYING CONTROL MEASURES

After effective control measures have been established, it is necessary to establish procedures to monitor each CCP in HACCP plans and verify that the control measures are being implemented as planned. Monitoring and verification can consist of a variety of measurements, such as:
- sensory assessments based on visual inspection, aroma, taste, touch, and sound
- chemical measurements, such as for sodium chloride, acetic acid, or water content
- physical measurements, such as pH, a_w, humidity, temperature
- time measurements
- packaging, such as integrity of the container closure (e.g., hermetical seal)
- records for incoming raw materials (e.g., records showing an ingredient is from an approved source or region)
- microbial tests, including tests for toxic metabolites
- environmental sampling (see Chapter 10)

It is important to note that the microbiological safety of most foods can be assessed by a variety of methods besides microbiological testing. These determinations can be used for monitoring or verifying that the specified control measures in a food operation are or have been implemented correctly. For example, measurements of pH and/or acid content are commonly used for monitoring fermentation processes and foods that are acidulated.

Some of these methods also are satisfactory for assessing the acceptability of foods at port-of-entry or upon receipt for use in a food operation. For example, the acceptability of a high-acid, canned, shelf-stable food can be easily determined by performing a pH measurement. Many foods of relatively low moisture content that are intended to be distributed and sold at ambient temperature can be assessed by a measurement of a_w or moisture content (see Chapter 4). Food operators become very familiar with the normal appearance, aroma, and feel of the foods with which they are involved. Foods that appear atypical should raise doubts in the operator's mind and indicate a need for additional evaluation.

3.11 EXAMPLES OF CONTROL OPTIONS

Two examples will be used to demonstrate the relationship between performance criteria, product/process criteria, and microbiological criteria (Baird-Parker & Tompkin, 2000). The first is a recommendation to industry by the UK Advisory Committee on the Microbiological Safety of Foods (ACMSF, 1992) about four options to control the risk of psychrotrophic *C. botulinum* in refrigerated cooked foods having an extended shelf-life of more than 10 days. The recommendations were:

1. A heat treatment of 90 °C for 10 min, or equivalent lethality.
2. A pH of 5 or less throughout the food and throughout all components of complex foods.
3. An a_w of 0.97 or less throughout the food and throughout all components of complex foods.
4. A combination of heat and preservative factors that can be shown consistently to prevent growth and toxin production by *C. botulinum*.

Option one is directed toward killing vegetative cells and spores of psychrotrophic strains of *C. botulinum* that may be present in the raw materials used in the production of the foodstuff. Options two and three are intended to prevent the growth of the organism and hence prevent toxin production. Option four could involve thermal destruction/thermal injury of spores and/or inhibitory factors to prevent the outgrowth of surviving *C. botulinum* spores.

Underlying each control option is an unstated performance criterion. Thus, the performance criterion for option one could be stated as a 6 D reduction of spores of psychrotrophic strains of *C. botulinum*, as this is the intended result of a heat treatment of 10 min at 90 °C. For options two and three, the performance criterion could be stated as less than a 1 log increase of *C. botulinum* within the use-by-date when stored at the recommended storage temperature. The ACMSF report provides considerable background information on the likely occurrence of the hazard and factors that can be used for control.

The second example concerns the risk of *E. coli* O157:H7 and similar enteric foodborne pathogens in fermented sausages. In December 1994, an outbreak of foodborne illness caused by *E. coli* O157:H7 in a fermented sausage product occurred on the West Coast of the US. In response, the U.S. Department of Agriculture established a requirement that all manufacturers use processes that control the risk of illness from *E. coli* O157:H7. In this case, the agency proposed a performance criterion (i.e., 5 D kill of *E. coli* O157:H7) and left it to industry to decide how to satisfy the criterion and still produce products of acceptable quality. The agency's proposal of a 5 D kill was based on very limited evidence suggesting that up to 1000 g^{-1} of *E. coli* O157:H7 could occur in the raw meat used for processing. Industry-sponsored research led to five options that were accepted by the agency (Nickelson *et al.*, 1996):

1. Apply an existing, approved heat treatment as specified in former USDA Regulation 9CFR 318.17 (i.e., heating to an internal temperature of 62.8 °C for 4 min or to a lower temperature for such time to obtain an equivalent level of safety (USDA, 1996b)).

2. Apply a process that is validated by research to cause a 5 D kill of *E. coli* O157:H7 before the product is released for distribution.
3. Combine raw material testing with a process that is validated by research to cause a 2 D kill of *E. coli* O157:H7 before the product is released for shipment. The sample procedure must ensure that the level of *E. coli* O157:H7 in the raw sausage blend does not exceed 1 cell per g. One such sampling procedure could consist of analyzing 15 samples (25 g each) collected at the time of stuffing the meat blend into casings.
4. Apply a program of holding the product until the test results become available indicating the lot has met the criteria of acceptance (i.e., a hold and test program). Products intended to be heated before serving (e.g., pepperoni for pizza) would be sampled at a rate of 15 samples per lot. Products normally consumed without heating before serving (e.g., salami) would be sampled at a rate of 30 samples per lot. The analytical unit for each sample tested would consist of 25 g.
5. To allow for new technology or ideas, this option permits the use of alternate processes that provide the equivalent of a 5 D reduction.

All of the options are intended to ensure that the level of *E. coli* O157:H7 is 1 cell in 100 g or less when the products are released for distribution. This was considered by the agency at the time to provide an acceptable level of consumer protection for this class of product. The five options include process criteria, performance criteria, and microbiological criteria. Options one and two assume an initial level of 1,000 *E. coli* O157:H7 per g in the raw sausage blend. The process criterion for option one (heating to an internal temperature of 62.8 °C and holding for 4 min) is derived from an existing regulation for roast beef and is based on research data demonstrating a 5 D kill of salmonellae and *E. coli* O157:H7 in beef. The 4-min hold time was an added requirement since the roast beef regulation does not require a hold time at 62.8 °C. Processors choosing this option would not take advantage of the faster rate of kill that would occur with the reduced pH of a fermented product. The performance criterion in option two specifies a 5 D kill of *E. coli* O157:H7. To satisfy this option, the processing plant must have on file, and available for review, research data that validate that the process being used will achieve a 5 D kill. The validation research must have been developed with a protocol approved by the USDA.

Option three incorporates both a reduction step and an elimination step based on microbiological criteria. It involves a performance criterion of a 2 D kill, in combination with microbiological testing to verify that the level of *E. coli* O157:H7 in each production lot does not exceed 1 cell per g in the raw sausage blend. Collectively, the net result is equivalent to the 5 D kill of option two, which assumed an initial level of 1,000 per g in the raw sausage blend. Subsequent industry-sponsored research has demonstrated that fifteen 25 g samples of raw sausage blend can be composited for analysis with no significant loss in sensitivity of detection. If, for example, a processor elects to collect 15 samples from across a lot during stuffing and have each tested using a 25 g analytical unit (375 g total), then this would provide 95% confidence that the level of *E. coli* O157:H7 in the blend is no more than 1 cell in 125 g if a negative result is obtained (Foster, 1971). While this may be a prudent sampling plan for some operations, this level of sampling exceeds by approximately 100-fold the detection level required in option three.

Option four establishes microbiological criteria for the finished product and assumes no prior knowledge of the level of *E. coli* O157:H7 in the raw sausage blend or the lethality of the process. Reliance is placed solely on the use of a sampling plan that may detect *E. coli* O157:H7 if it is present in the finished product. The sampling plans are based upon information in ICMSF Book 2 (ICMSF, 1986) and an assignment of *E. coli* O157:H7 to cases 13 and 14 for pepperoni and salami, respectively (see Chapter 8). Case 13 involves n = 15 and c = 0. Case 14 involves n = 30 and c = 0. A negative result with fifteen 25 g samples (375 g total) provides a 95% confidence of no more than 1 cell in 125 g. A negative result with thirty 25 g samples (750 g total) provides a 95% confidence of no more than 1 cell in 250 g (Foster, 1971).

3.12 ASSESSING EQUIVALENCY OF FOOD SAFETY MANAGEMENT SYSTEMS

Food safety management systems based upon an FSO and the use of performance criteria provide greater flexibility in how food operators can control hazards. An assessment for equivalency of systems that use performance criteria may require a holistic approach when the food operation involves multiple steps. In another situation, an assessment for equivalency may be limited to a single step in a process (e. g., pasteurization). The assessment should involve a review of data that demonstrate the scientific basis for the alternative process and processing records that demonstrate the control measures are being implemented as planned. Finally, the product must be able to meet any acceptance criteria that may have been established.

It should be evident that the TLR, FSO, and measures subsequently established by a country can serve as one basis for assessing the equivalency of a country's food control system.

3.13 REFERENCES

ACMSF (Advisory Committee on Microbiological Safety of Foods) (1992). *Report on Vacuum Packaging and Associated Processes.* London: HMSO.

Baird-Parker, A. C. & Tompkin, R. B. (2000). Risk and microbiological criteria. In *The Microbiological Safety and Quality of Food,* pp. 1852–1885. Edited by B. M. Lund, A. C. Baird-Parker & G. W. Gould. Gaithersburg, MD: Aspen Publishers, Inc.

Brown, B. E. (1997). Thermal processes—development, validation, adjustment and control. In *Food Canning Technology,* pp. 451–488. Edited by J. Larousse. New York: Wiley-VCH.

CAC (Codex Alimentarius Commission) (1997a). *Joint FAO/WHO Food Standards Programme. Codex Committee on Food Hygiene. Food Hygiene, Supplement to Volume 1B-1997. Hazard Analysis and Critical Control Point (HACCP) System and Guidelines for Its Application.* CAC/RCP 1-1969, Rev. 3 (1997).

CAC (Codex Alimentarius Commission) (1997b). *Joint FAO/WHO Food Standards Programme, Codex Committee on Food Hygiene. Food Hygiene, Supplement to Volume 1B-1997. Recommended International Code of Practice, General Principles of Food Hygiene.* CAC/RCP 1-1969, Rev. 3.

CDC (Centers for Disease Control and Prevention) (2000a). Preliminary FoodNet data on the incidence of foodborne illnesses—selected sites, United States, 1999. *MMWR Morb Mortal Wkly Rep* 49, 201–205.

CDC (Centers for Disease Control and Prevention) (2000b). Outbreaks of Salmonella serotype Enteritidis infection associated with eating raw or undercooked shell eggs—United States, 1996–1998. *MMWR Morb Mortal Wkly Rep* 49, 73–79.

Foster, E. M. (1971). The control of salmonellae in processed foods: A classification system and sampling plan. *J Assoc Off Anal Chem* 54, 259–266.

Gould, G. W. (1999). Sous vide foods: conclusions of an ECFF Botulinum Working Party. *Food Control* 10, 47–51.

ICMSF (International Commission on Microbiological Specifications for Foods) (1986). *Microorganisms in Foods 2. Sampling for Microbiological Analysis: Principles and Specific Applications*, 2nd edn. Toronto: University of Toronto Press.

ICMSF (International Commission on Microbiological Specifications for Foods) (1988). *Microorganisms in Foods 4. Application of the Hazard Analysis Critical Control Point (HACCP) System to Ensure Microbiological Safety and Quality*. Oxford: Blackwell Scientific Publications Ltd.

ICMSF (International Commission on Microbiological Specifications for Foods) (1998). *Microorganisms in Foods 6: Microbial Ecology of Food Commodities*. London: Blackie Academic and Professional.

Jordan, C. F. (1950).The epidemiology of brucellosis. In *Brucellosis. A Symposium under the Joint Auspices of National Institutes of Health of the Public Health Service, Federal Security Agency, United States Department of Agriculture, National Research Council*, pp. 98–115. Washington, DC: American Association for the Advancement of Science.

Kruse, H. (1999). Globalization of the food supply—food safety implications. Special regional requirements: future concerns. *Food Control* 10, 315–320.

Legan, J. D., Vandeven, M. H., Dahms, S. & Cole, M. B. (2001). Determining the concentration of microorganisms controlled by attributes sampling plans. *Food Control* 12, 137–147.

Leighty, J. C. (1983). Public-health aspects (with special reference to the United States). In *Trichinella and Trichinosis*, pp. 501–513. Edited by W. C. Campbell. New York: Plenum Press.

Lund, B. M., Knox, M. R. & Cole, M. B. (1989). Destruction of *Listeria monocytogenes* during microwave cooking. *Lancet* (Jan. 28), 218.

Nesbakken, T. (2000). *Yersinia* species. In *The Microbiological Safety and Quality of Food*, pp. 1363–1393. Edited by B. M. Lund, A. C. Baird-Parker & G. W. Gould. Gaithersburg, MD: Aspen Publishers, Inc.

Nickelson, R., Luchansky, J., Kaspar, C. & Johnson, E. (1996). *Update on Dry Fermented Sausage* Escherichia coli *O157:H7 Validation Research*. Research Report No. 11–316. Chicago: National Cattleman's Beef Association.

Pozio, E. (1998). New strategies for reducing the cost of the control of trichinellosis in the European Union. In *Proceedings of the 4th World Congress Foodborne Infections and Intoxications*, Vol. 1, pp. 526–530. Edited by K. Noeckler, P. Teufel, K. Schmidt, & E. Weise. Berlin: Federal Institute for Health Protection of Consumers and Veterinary Medicine.

Rehmet, S., Sinn, G., Robstad, O. *et al.* (1999). Two outbreaks of trichinellosis in the state of Northrine-Westfalia, Germany, 1998. *Eurosurveillance* 4, 78–81.

Ruitenberg, E. J. & Sluiters, J. F. (1974). *Trichinella spiralis* infections in the Netherlands. In *Trichinellosis. Proceedings of the Third International Conference on Trichinellosis*, pp. 539–548. Edited by C. W. Kim. New York: Intext Educational Publishers.

Simms, B. T. (1950). Federal aspects of the control of brucellosis. In *Brucellosis. A Symposium under the Joint Auspices of National Institutes of Health of the Public Health Service, Federal Security Agency, United States Department of Agriculture, National Research Council*, pp. 241–246. Washington, DC: American Association for the Advancement of Science.

Stumbo, C. R. (1973). *Thermobacteriology in Food Processing*, 2nd edn. New York: Academic Press.

Tompkin, R. B. & Kueper, T. V. (1982). Microbiological considerations in developing new foods. How factors other than temperature can be used to prevent microbiological problems. In *Microbiological Safety of Foods in Feeding Systems*. ABMPS Report No. 125, pp. 100–122. Washington, DC: Advisory Board on Military Personnel Supplies, National Research Council, National Academy Press.

USDA (United States Department of Agriculture) (1996a). Pathogen reduction; hazard analysis and critical control point (HACCP) systems; final rule. *Federal Register* 61, 38806–38989.

USDA (United States Department of Agriculture) (1996b). Requirements for the production of cooked beef, roast beef, and cooked corned beef. *Federal Register* 61, 19564–19578.

USDA (United States Department of Agriculture) (1999). *Compliance Guidelines for Meeting Lethality Performance Standards for Certain Meat and Poultry Products*. Appendix A to Compliance Guidelines, January 1999, updated June 1999. Washington, DC: US Department of Agriculture, Food Safety and Inspection Service.

USDA (United States Department of Agriculture) (2000). *HACCP Implementation: First Year* Salmonella *Test Results, January 26, 1998 to January 25, 1999.* Washington, DC: US Department of Agriculture, Food Safety and Inspection Service.

Zimmerman, W. J. (1974). The current status of trichinellosis in the United States. In *Trichinellosis. Proceedings of the Third International Conference on Trichinellosis,* pp. 603–609. Edited by C. W. Kim. New York: Intext Educational Publishers.

Appendix 3–A

Control Measures Commonly Applied to Foodborne Diseases

PRIMARY SOURCE: HUMAN

Group	Agent	Common vehicles	Control measures: Controlling initial level	Control measures: Reducing level	Control measures: Preventing increase in level
Bacteria	V. cholerae, S. typhi	Untreated water, foods contaminated by infected food handlers	Treat waste water, provide potable water for drinking and food preparation; educate food handlers in proper personal hygiene	Cook	Pathogen increase in the implicated food is not an important factor
	Shigellae	Foods contaminated with untreated waste water or by infected food handlers	Educate food handlers in proper personal hygiene; exclude infected persons from high risk positions (e.g., food preparation, day care centers)	Cook	Pathogen increase in the implicated food is not an important factor
Virus	Hepatitis A, SRSV, Norwalk virus	Ready-to-eat foods, bivalve molluscs	Treat waste water, provide potable water for drinking and food preparation; educate food handlers in proper personal hygiene; avoid bivalves from contaminated waters	Cook	Viruses cannot multiply in food

PRIMARY SOURCE: SEAFOOD

Group	Agent	Common vehicles	Control measures: Controlling initial level	Control measures: Reducing level	Control measures: Preventing increase in level
Bacteria	C. botulinum type E V. parahaemolyticus	Seafood, meat from marine mammals, fresh water fish	Avoid certain traditional fermented foods (e.g., muktuk) and fish dried with intestines intact; consumer education	Apply thermal process sufficient to kill spores (e.g., canning of seafood)	Safety depends on preventing growth to levels sufficient for toxin production; apply combinations of temperature, a_w, pH and time that inhibit growth (e.g., temperature < 3 °C; a_w < 0.97; 3.5% NaCl on water at 5 °C; or pH below 5.0); control fermentation conditions to prevent growth and toxin production

Meeting the FSO through Control Measures 73

	Hazard	Food	Control measure	Effect
	V. vulnificus	Raw seafood, recontaminated cooked seafood	Avoid uncooked fish where pathogen is prevalent	Cook	Avoid recontamination, maintain low storage temperatures
		Oysters	Avoid bivalves from regions of higher risk, avoid wounds to hands when handling crustaceans	Cook	Increase of V. vulnificus in oysters post-harvest is not a factor
Toxin	Scombrotoxin	Tuna, mackerel, mahi mahi	Select freshly harvested and properly chilled fish; specification for histamine: mean of 100 ppm, 2 of 9 samples tested may be > 100 but < 200 ppm, no sample > 200 ppm	Histamine level cannot be reduced after harvest	Cool fish rapidly and maintain at low temperature (< 8 °C; preferably lower) to prevent histamine formation
	Tetrodotoxin	Puffer fish	Avoid fish from the Tetraodontidae family	Toxin cannot be reduced after harvest; toxin is stable	Toxin cannot increase
	Ciguatera	Certain tropical reef fish	Avoid tropical reef fish from regions and seasons of high risk	Toxin cannot be reduced after harvest; toxin is stable	Toxin cannot increase
	Shellfish toxins (PSP, DSP, ASP, NSP)	Bivalves	Monitor shellfish harvesting areas for toxic phytoplankton, suspend harvesting if necessary, specification PSP 80 µg 100 g^{-1}, ASP 20 µg 100 g^{-1}	Toxin cannot be reduced after harvest; toxin is stable	Toxins cannot increase
Parasites	Anisakis	Herring, cod	Inspect and discard infected fish	Freeze (−18 °C for 24 hrs); hold in combination of NaCl and acetic acid; cook	Parasite cannot increase
PRIMARY SOURCE:					
	Brucellae				

continues

LIVESTOCK, POULTRY
Bacteria

Group	Agent	Common vehicles	Control measures		
			Controlling initial level	Reducing level	Preventing increase in level
	M. bovis	Unpasteurized milk and cheeses made therefrom, contact with infected livestock (e.g., cattle, goats)	Eradicate brucellosis from domesticated animals; avoid cheeses made from unpasteurized milk in regions where brucellae are endemic	Pasteurize milk	Brucellae do not increase during cheese manufacture or subsequent storage
	C. jejuni/coli	Unpasteurized milk	Eradicate bovine TB	Pasteurize milk	M. bovis cannot increase during storage of milk or milk-containing foods
		Raw or undercooked poultry, unpasteurized milk	Effective on-farm and slaughtering control measures for raw poultry and meat are uncertain; education of food handlers	Cook; pasteurize milk	C. jejuni/coli are not likely to multiply in food due to the high temperature (≥ 32 °C) and other requirements for growth; cross contamination to ready-to-eat foods is a significant concern
	Salmonellae, non-typhoid	Wide variety of foods	Select foods and ingredients from suppliers with effective control systems; understanding the limitations of current control measures to prevent or eliminate salmonellae from raw agricultural commodities	Cook; pasteurize; acidify; other combinations (e.g., a_w, pH, temperature, preservatives); irradiate	Use of inhibitory factors (preservatives, a_w, acidifiers, low temperature storage) alone, or in combination, to inhibit growth; minimize cross contamination
	Pathogenic E. coli	Undercooked ground beef, raw produce, unpasteurized juice, sprouts, untreated water, contact with ruminants	Effective on-farm control measures have not been identified; control measures during slaughter can reduce but not eliminate these pathogens; effective GAPs (good agricultural practices) for crops have yet to be	Cook; pasteurize; wash and disinfect (e.g., fruits, vegetables)	Multiplication can occur in many foods, but this has not been a significant factor; survival and low infective dose among the more sensitive populations are much more important

Y. enterocolitica	Raw or undercooked pork, untreated water, milk	demonstrated; selecting raw agricultural commodities from suppliers with a system of comprehensive control measures can minimize but not prevent the presence of these pathogens; avoid untreated water and contact with livestock. Improve slaughter hygiene (i.e., sealing off rectum in plastic bag)	Cook	Minimize contact of ready-to-eat foods with raw pork
S. aureus	Cooked foods contaminated with *S. aureus* and held at conditions permitting growth and enterotoxin production; *S. aureus* does not compete well against the normal flora associated with raw agricultural commodities and, thus, these foods are rarely involved	Selecting foods and ingredients that are likely to have lower initial numbers; preventing food handlers with boils, infected cuts from handling food	Cook, pasteurize, acidify, etc. to reduce the number of *S. aureus* in food; staphyloccocal enterotoxin is very stable and cannot be reduced by normal methods of food preparation	Control temperature and time of holding cooked perishable foods to prevent sufficient growth for enterotoxin production; use inhibitory factors (e.g., preservatives, a_w, acidifiers, smoke) to inhibit growth; control conditions for manufacture of certain cheeses and dry sausages where growth during fermentation and, in the case of cheese, pressing is important
Parasites				
Taenia	Raw or undercooked pork and beef	Break epidemiologic chain at final host (man) or intermediate host (cattle, pigs); animal husbandry; treat sewage and sludge; meat inspection		
T. spiralis			Cook to >60 °C; freeze to −10 °C 6 days	This parasite cannot increase during processing or storage

continues

Group	Agent	Common vehicles	Control measures		
			Controlling initial level	Reducing level	Preventing increase in level
		Raw or undercooked pork, wild game, horse meat	Select pigs from sources (e.g., industrialized farms) and regions of little or no risk; eradicate infested pigs and manage the risk of consumer exposure to infested wild game; avoid feeding uncooked kitchen waste to pigs; adopt meat inspection procedures to eliminate infested carcasses	Freeze −25 °C for 10–20 days; cook to 58 °C minimum	*T. spiralis* cannot increase in food
PRIMARY SOURCE: SOIL Bacteria	*L. monocytogenes* (Sources include the soil, environment of food establishments and homes, raw agricultural commodities, animals, including humans) *B. cereus*	Perishable ready-to-eat foods in which growth can occur (e.g., certain soft cheeses, milk, cooked meats and poultry, smoked seafood)	Select foods subjected to a kill step, where possible; control the ready-to-eat environment to minimize contamination; educate consumers about foods to avoid	Cook; pasteurize milk; pasteurize in the container in which a food will be sold; formulate foods to create listericidal conditions during subsequent processing, storage and distribution (e.g., acidification, dehydration); wash and disinfect	Formulate foods with additives or create conditions that can prevent multiplication; store perishable foods at low temperatures that can retard or prevent growth; manage the environmental conditions of the operation to minimize the likelihood of contamination from raw foods/ingredients or contaminated surfaces; use controlled fermentations that prevent growth

Meeting the FSO through Control Measures   77

Organism	Foods			
C. botulinum type B (non-proteolytic)	Rice, cereal products, certain dairy products, sauces, cooked pasta, vegetable or meat dishes	Control measures cannot ensure the absence of this pathogen in the foods commonly involved	Cooking can kill the vegetative cells, but spores may survive	Cool rapidly from 50 °C to 10 °C after heating; store below 10 °C or above 60 °C
	Dry, cured ham and other salt cured meats and seafood	Control measures cannot ensure the absence of this pathogen in the foods commonly involved	Apply thermal process sufficient to kill spores (e.g., canning)	Safety depends on preventing growth to levels sufficient for toxin production; apply combinations of temperature, a_w, pH and time that inhibit growth (e.g., temperature < 3 °C; a_w < 0.97; 3.5% NaCl on water at 5 °C; or pH below 5.0). Control fermentation conditions to prevent growth and toxin production
C. botulinum type A, B (proteolytic)	Low-acid foods, meat, home-canned vegetables and low-acid fruits	Control measures cannot ensure the absence of this pathogen in the foods commonly involved; consumer education to avoid tasting suspect food	Apply thermal process sufficient to kill spores (e.g., canning)	Safety depends on preventing growth to levels sufficient for toxin production; apply combinations of temperature, a_w, pH and time, etc. that inhibit growth (e.g., temperature < 10 °C; a_w < 0.94; 10% NaCl on water; pH below 4.6); sodium nitrite in cured meats
C. perfringens	Cooked, uncured meat or poultry, meat pie, stew, gravy	Control measures cannot ensure the absence of this pathogen in the foods commonly involved	Cook to kill vegetative cells, retort to kill spores; storage at low temperatures can cause significant decrease in vegetative cells	Cool rapidly from 51 °C to 15 °C after cooking; store below 10–12 °C or above 51 °C, preferably at or above 60 °C
A. flavus, A. parasiticus				

continues

Group	Agent	Common vehicles	Control measures		
			Controlling initial level	Reducing level	Preventing increase in level
Fungi	A. ochraceus	Corn, peanuts	Monitor weather and test crops when conditions indicate higher risk; minimize insect infestation; reject moldy products	The mycotoxins are very stable	Rapid drying and dry storage to prevent growth of fungi and toxin production
	A. versicolor	Beans, dried fruit, nuts, cereals, biltong, dried fish	Monitor weather and test crops when conditions indicate higher risk; reject moldy products	Toxins are stable	Rapid drying and dry storage to prevent growth of fungi and toxin production
	Fusarium graminearum	Wheat, corn, flour, rice, nuts, spices, fermented meats, cured meats, biltong, cheese	Prompt harvest when ready; reject or test suspect foods for the toxin	Toxins are stable	Rapid drying and dry storage to prevent growth of fungi and toxin production
	F. moniliforme	Wheat	Prompt harvest when ready; reject or test suspect lots for the toxin	Toxins are stable	Rapid drying and dry storage to prevent growth of fungi and toxin production
	F. sporotrichoides	Rice, sorghum, sugar cane, corn	Prompt harvest when ready; reject or test suspect foods for the toxin	Toxins are stable	Rapid drying and dry storage to prevent growth of fungi and toxin production
	Penicillium citrinum,	Cereals, peanuts, soya beans	Prompt harvest when ready; reject or test suspect foods for the toxin	Toxins are stable	Rapid drying and dry storage to prevent growth of fungi and toxin production
	P. crustosum, P. islandicum, P. verrucosum	Ubiquitous, especially rice, wheat and corn	Prompt harvest when ready; reject or test suspect foods for the toxin	Toxins are stable	Rapid drying and dry storage to prevent growth of fungi and toxin production

Chapter 4

Selection and Use of Acceptance Criteria

4.1 Introduction
4.2 Equivalence
4.3 Establishment of Acceptance Criteria
4.4 Application of Acceptance Criteria
4.5 Determining Acceptance by Approval of Supplier
4.6 Examples To Demonstrate the Process of Lot Acceptance
4.7 Auditing Food Operations for Supplier Acceptance
4.8 References

4.1 INTRODUCTION

Available statistics for the volume of food crossing borders between countries, or being shipped from a seller to a buyer, are estimates that may represent only the tip of the iceberg. For example, the number of formal customs entries for foods imported into the US is about 1.5 million each year (Schultz, 1997). All these products should be safe and of suitable quality. However, the safety of a product is seldom verifiable by tasting or other organoleptic examination, although appearance and odor can suggest a hazardous product under certain circumstances. Foods of questionable safety should never be tasted. Thus, buyers and consumers must rely on the supplier to deliver ingredients or food with the expected level of safety or on government control for domestic and imported foods. This expectation occurs all along the food chain. In many instances, the buyer (a food manufacturer, retailer, etc.) or an inspector, for example, at port-of-entry has no reliable means to check the safety of incoming raw materials or finished products. There is, however, a difference in the relationship between commercial buyers and consumers. Usually, there is a business deal between commercial partners, and product liability is an important issue. Also, industry buyers can impose specifications for the food being purchased and a variety of options to verify compliance.

To harmonize food control systems and facilitate trade, World Trade Organization (WTO) members are encouraged to adopt international food standards. Governments also can impose standards or guidelines for imported food. Governments often have a working relationship with the exporting country's food control authority that can lead to various forms of verification. Thus, governments may intervene in the transfer process between buyers and suppliers at the port-of-entry. Also, governments can impose criteria for acceptance on local food products, such as those produced at the retail level.

This chapter will discuss the use of criteria to determine the acceptability of individual lots or consignments of food and operations where they are produced. Criteria for assessing the acceptability of a lot of food can take several forms based on sensory, chemical, physical, or microbiological parameters. Acceptance criteria should include additional information, for example, such as the number of samples to be collected, how and where the samples are collected and held prior to analysis, the analytical unit, the method of analysis, and what is considered acceptable. The acceptability of a food operation can be assessed through inspections by control authorities or through audits by buyers. The inspection or auditing processes are very similar in nature but are for different reasons, i.e., to assess compliance with regulations vs. approval of a supplier. The results of an inspection or audit can influence whether the food(s) or ingredients being produced should be sampled.

4.2 EQUIVALENCE

For foods received at a port-of-entry, the origin (i.e., country or region) of the food is an important factor influencing acceptance. It is impossible for the control authorities of an importing country to inspect and approve foods while they are being produced, harvested, processed, and readied for shipment. Reliance must be placed on the system of inspection in the exporting country or region. Thus, to facilitate the movement of foods across international borders, it is necessary to develop a mechanism for acceptance based on mutual agreement between the authorities in the participating countries. While it is not possible for historical and cultural reasons to have identical regulations and codes of practice, it is necessary to assess whether they result in an equivalent level of consumer protection. Equivalence is the capability of different inspection and certification systems to meet the same objectives (CAC, 2000).

The equivalency concept was introduced in the WTO Agreement on Sanitary and Phytosanitary (SPS) Measures. According to the agreement, each member of the WTO must accept as equivalent to its own the food regulatory systems of other members as long as they provide the same level of public health protection. Thus, equivalent regulatory systems need not be identical. The outcome, however, of the exporting country's food control system must meet tolerances established for a particular hazard or hazards for the same foods within the importing country. The burden of demonstrating equivalence is on the exporting country. The importing country has the right to decide whether the evidence provided to demonstrate equivalence is adequate.

The following information derived from a proposed Codex Alimentarius Commission draft (CAC, 2000) to describe a framework for determining equivalence suggests how the material in this book can be used when food safety issues of microbiological origin are the focus of a determination. A determination of equivalence requires establishment of a relationship between two "building blocks": the appropriate level of protection (ALOP) and the sanitary measures. It is the sovereign right of an importing country to set any level they deem appropriate in relation to their food supply, while having regard to Codex standards as international benchmarks. An ALOP can be expressed in qualitative or quantitative terms. In determining equivalence, the ALOP of the importing country must be met by the exporting country. For the purposes of determining equivalence, the sanitary measures that comprise a food control system can be broadly categorized as follows:

- infrastructure, including the legislative base (e.g., food and enforcement law), and administrative systems (e.g., organization of national and regional authorities);
- program design/implementation, including documentation of systems, performance, decision criteria and action, laboratory capability, and provisions for certification and audit; and

- specific requirements, including individual facilities (e.g., premises design), equipment (e.g., design of food contact machinery), processes (e.g., retorting, Hazard Analysis Critical Control Point (HACCP) plans), procedures (e.g., post mortem inspection) and tests (e.g., tests for microbiological and chemical hazards).

Sanitary measures are often narrowly focused and specific while the ALOP is more flexible. It may be expressed in specific terms (e.g., number of cases of disease per year in a defined population) or far less specific terms (e.g., broad goals relating to food safety). Irrespective of the expression of the ALOP, an objective basis for comparison of sanitary measures must be established.

Presentation of an objective basis for comparison may include the following elements:

1. the reason/purpose for the sanitary measure;
2. the relationship to the ALOP, with quantitative expression wherever possible;
3. where risk assessment information is available, expression of the risk;
4. the level of a hazard in a food that is tolerable in relation to an ALOP (i.e., a food safety objective (FSO), as described in Chapter 2 and elsewhere).

Any sanitary measure or combination of sanitary measures can be identified for determination of equivalence.

Determination of equivalence presumes that the exporting country has already reviewed all applicable importing country requirements for the food involved and has identified those it will meet and those for which it seeks a determination of equivalence. The determination of equivalence is facilitated by both exporting and importing countries following a sequence of steps such as those illustrated in Figure 4–1.

The exporting country should present a submission for equivalence that facilitates the judgment process applied by the exporting country. Importing and exporting countries should utilize an agreed process for exchange of information. This information should be limited to that which is necessary to facilitate the determination of equivalence and minimize administrative burden. When achievement of equivalence is agreed upon by the importing country, the importing and exporting countries may enter into a formal agreement to the effect of that decision.

Further development of the concept of equivalence and the establishment of international standards for food are activities of the Codex Alimentarius Commission. Since these concepts and their application continue to evolve, the most recent documents of the Commission should be reviewed for more information and for their current status.

4.3 ESTABLISHMENT OF ACCEPTANCE CRITERIA

4.3.1 Role of FSO in Lot Acceptance

As described in the previous chapters, an FSO specifies the outcome expected of food operations that have implemented a system of effective control measures. The outcome may be expressed as a frequency or maximum concentration of a microbiological hazard in the food. Furthermore, the FSO is based on what is considered acceptable for consumer protection. Thus, an FSO is a goal that food operators use when designing and implementing their food safety management systems. Operators who can demonstrate their ability consistently to meet an FSO will be more reliable as suppliers. Shipments of

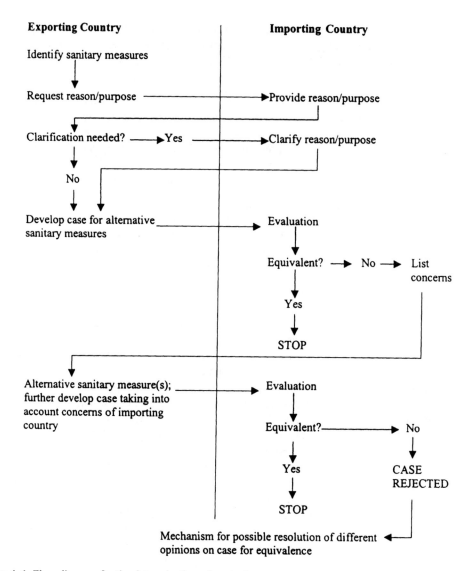

Figure 4–1 Flow diagram for the determination of equivalence.

food from these suppliers can be sampled at a much lower frequency. FSOs can therefore be used as the basis for establishing lot acceptance criteria.

4.3.2 Types of Acceptance Criteria

Governments are responsible for establishing food regulations and policies that ensure the safety of the foods for which they have regulatory responsibility. Thus, food control authorities serve as risk managers and establish FSOs for hazards in foods. This information is then used to develop standards (e.g., performance, process, product, default criteria) or guidelines that can be used by both industry and government as the basis for assessing the acceptability of a food operation or food. These criteria are applied to domestic and imported foods.

Industry must produce foods that meet regulatory requirements. This includes regulations that define the conditions of operation as well as standards that may have been established for the food. To ensure compliance with regulatory requirements and the operator's own requirements for quality, food operators frequently establish purchase specifications for food ingredients and other materials. Under certain circumstances, an operator may specify the conditions of operation in the form of an operating instruction or best practices guideline, particularly when the supplier needs to assure that a critical control point (CCP) is controlled for the buyer.

Acceptance criteria can involve a variety of parameters (sensory, physical, chemical, biological) and generally fall into three categories, namely:

- **Standard**—a mandatory criterion that is part of a law or ordinance.
- **Guideline**—an advisory criterion issued by a control authority, industry association, or food producer to indicate what might be expected when best practices are applied.
- **Specification**—Part of a purchasing agreement between a buyer and supplier of a food; such criteria may be mandatory or advisory according to use.

The Codex document on the "Principles for the Application of Microbiological Criteria for Foods" (CAC, 1997a) recognizes only one category, i.e., a *microbiological criterion* (see Chapter 5).

4.3.2.1 Standards

Standards may be established for a wide variety of reasons, but are best applied when risk to consumers is sufficiently high and compliance with the standard is essential for consumer protection. Standards are established by governments and define the parameters that processes or foods must meet to be in compliance with regulatory policies or regulations. In the case of a process (e.g., pasteurization of eggs or milk), the conditions of operation, including design and performance of the equipment, may be defined. Processes that do not meet the standard conditions of operation must be brought into compliance or may be subject to rejection. A standard for a food defines the acceptance criteria (e.g., pH, a_w, coliform or *Escherichia coli* concentration) that must be met. Foods that do not meet the standard are in noncompliance and would be subject to removal from the market. Depending on the circumstances (e.g., severity of the noncompliance), the food may be reprocessed, have restrictions imposed on its use, or be destroyed. Standards that define criteria of importance to the microbiological safety or quality of a food should be established following the principles outlined by Codex (CAC, 1997a) and as further elaborated in this book. Thus, standards could be developed following a qualitative or quantitative risk assessment and establishment of an FSO.

4.3.2.2 Guidelines

Guidelines may be established for a process or product. In the case of a process (e.g., production of cheese, fermented sausage, sprouts), the conditions that should be controlled are described so the resulting food will be safe. A guideline for a product normally provides the product characteristics (e.g., pH, a_w) and microbiological criteria (e.g., *E. coli*), that the operator should strive for to produce an acceptable product. Thus, the criteria are advisory in nature and do not necessarily lead to rejection of a food. Such criteria describe the characteristics of a food when best practices are applied. Guidelines may be the preferred means to inform food operators that change is needed within a segment of the industry. In other situations, guidelines may be used to describe the conditions believed to correct a newly recognized microbiological problem but for which insufficient data prevent the establishment of a standard. Thus, guidelines can serve as an interim measure providing guidance to food operators until sufficient information and technology is available to establish standards. A guideline alone may be adequate to bring about the changes necessary to improve food safety and quality, thereby negating the need to establish standards.

4.3.2.3 Specifications

Implicit in the transfer process between buyer and supplier is the necessity for the buyer to make a decision whether to accept or reject ingredients, food, and other materials. This decision involves a variety of factors, most important of which is the buyer's experience with the material being purchased. As buyers become more sophisticated and familiar with the risk(s) associated with certain foods, they recognize the utility of having preestablished acceptance criteria that are agreed upon with their supplier(s). The purpose of such purchase specifications is to reduce the likelihood of accepting an unacceptable ingredient or food. Purchase specifications define the expected characteristics of an ingredient, for example, so that when it is added, the final product will meet all requirements for safety, quality, wholesomeness, and other requirements of concern to the buyer (e.g., cost, nutrition, compatibility with ingredient listing on label). It is now common for buyers along the food chain to establish their own specifications for the materials they purchase, particularly when the safety of a raw material or ingredient is a CCP in the buyer's HACCP plan.

4.4 APPLICATION OF ACCEPTANCE CRITERIA

A variety of criteria traditionally have been used to decide whether to accept individual lots of food. These criteria define what is considered acceptable, preferably taking into account how the food will be used. As will be discussed later, there may be no need to test each lot of food produced by a supplier. In other situations, adherence to the specifications may be essential for safety or quality concerns. In extreme situations, this may even involve product testing by the supplier prior to delivery and providing a certificate of analysis with the shipment. In other situations, the buyer may analyze the material after it is received. Tests that are typically performed may involve a variety of parameters, such as product pH, a_w, and moisture content. Other tests may involve microbiological analysis for indicator microorganisms or pathogens.

Regardless of the criterion, consideration must be given to the factors influencing the analytical results (e.g., method of sampling, transportation and storage of samples, analytical procedure, etc.; see Chapter 12). These important factors determine the utility and reliability of criteria to differentiate acceptable from unacceptable lots. Considering the limitations of microbiological testing to ensure the safety of individual lots food (see Chapters 6, 7, and 8), greater emphasis is now being placed on the conditions under which foods are produced–the application of Good Hygienic Practices (GHP) and HACCP.

Control authorities are responsible for verifying that the established regulations and policies are being met. This can involve inspections of food operations and, if deemed necessary, sampling and testing the food. Operations that are in compliance are typically granted a certificate or permit that is required for continued operation. Failure to comply with the regulatory requirements may result in the certificate or permit being revoked and, thereby, closing the operation. Ideally, governments organize the results of their assessments to determine if trends indicate a need for change in the regulations or enforcement actions. These data can influence the control authority's stance on sampling and testing certain categories of food for compliance with established criteria.

Today, food operators recognize the necessity for assessing suppliers of sensitive ingredients or foods to consistently control the food safety hazards. Assurance that suppliers control the hazards is best accomplished through a system of auditing and approving a supplier's total system for the management of food safety, such as described in Chapter 3. Suppliers that are considered unacceptable are removed from the list of approved suppliers. In certain cases, this may still lead to sampling and testing (verifying) incoming lots of purchased food. Lots that fail to meet the established specifications can be rejected. In general, however, this approach shifts the decision to accept or reject individual lots of food to supplier approval.

4.5 DETERMINING ACCEPTANCE BY APPROVAL OF SUPPLIER

4.5.1 Role of FSO in Supplier Approval

If an FSO has been established for a commodity, the first step is to determine whether the FSO is being met. If there is agreement on the achievement, further detailed information can be shared on the control measures that have been adopted to meet the FSO. Specifically, this would involve evaluating whether the adopted control measures (i.e., GHP, HACCP) can be expected to meet the FSO. Thus, auditors can incorporate the information in Chapter 3 into their auditing procedures as the basis for evaluating whether the operation being audited has established effective controls that can be justified. For example, will the system meet established performance criteria? Can the critical limits at CCPs (i.e., process criteria) be expected to prevent, eliminate, or reduce the hazard(s) to acceptable levels? If default criteria are selected as a basis for control, are they being correctly applied?

During the auditing process, the FSO provides a common yardstick for evaluation. The supplier and buyer of an ingredient or food share the same values with regard to safety, and the supplier considers the buyer's risk equal to his own. The buyer is aware of the intended use and performance criteria that may be necessary to ensure that safety can be mutually agreed upon. This will influence the HACCP plan and GHP procedures that are applied. Ideally, this information can be used for auditing by a mixed team of experts from both companies. If laboratories are involved, they would be expected to use the same or comparable (validated) methods and, perhaps, even participate in the same proficiency-testing program.

Assessment of incoming raw materials and ingredients in such a situation by the buyer may be limited to verifying accuracy and completeness of data concerning the lot and inspection of the shipment upon receipt. For certain foods, very critical parameters (e.g., temperature, moisture content) may be checked, but microbiological testing would seldom be used as an acceptance procedure from approved suppliers.

It is not always possible to have such an integrated quality assurance system between two business partners. This integrated approach requires a step-by-step approach over time until a level of mutual trust and acceptance is achieved. Preferably, such partnerships should be applied along the entire food chain. Very effective partnerships of this nature have been established among suppliers of certain major retail and fast food operators who rely upon a limited core of approved suppliers.

In situations where an FSO has not been established, conventional systems must be used to determine whether to accept a supplier. These include inspection and auditing of GHP and HACCP systems and licensing of the production site. Where necessary, such an approach may be supported by testing representative lots of production from the supplier to determine consistency of compliance with purchase specifications. Chemicophysical and/or microbiological tests may be used for this purpose. The number of lots that should be tested should reflect hazards and risks and the consequences of noncompliance with the criteria. When it has been determined that the supplier can reliably meet purchase specifications, then product testing can be discontinued or sharply curtailed.

In contrast to the FSO, microbiological criteria are of limited value in the approval process for a potential supplier. This is because microbiological criteria can be applied to specific lots of product to determine their acceptability, but this information provides only a point-in-time snapshot and provides little or no confidence for what may be produced over an extended period of time. By continued testing of incoming raw material from a supplier, it would be possible to establish a history of compliance with established criteria. However, when evaluating a new supplier, microbiological testing of current production lots would not be adequate to assess variability and, in particular, the presence of a pathogen that may occur intermittently and at low frequency. Thus, confidence in a supplier's ability to meet an FSO has greater value than reliance upon microbiological testing of incoming lots.

4.5.2 Approval Procedures

At each transfer point in the food chain, there is normally a buyer and a supplier. Ultimately, the final buyer will be a consumer. Buyers and suppliers can be in government as well as in industry. Suppliers may be domestic or in another country.

The preferred approach to managing the safety of food is to select suppliers who can be relied upon consistently to provide ingredients or foods that meet food safety requirements. Food safety systems based on prevention are much more effective than attempting to differentiate safe from unsafe lots by microbiological testing. While there may be a role for testing certain ingredients or foods, microbiological testing should be applied cautiously and used as a supplement to other information, particularly the conditions under which the material is produced. A listing of parameters that can be used to approve suppliers appears in Table 4–1. Many of these parameters can be verified by auditing experts that the conditions under which the ingredient or food is produced are acceptable. This could lead to licensing or certification.

The net result of these activities is to develop a base of suppliers that can be consistently relied upon to provide ingredients and foods that will be safe when used as intended. This approach requires that all parties be knowledgeable of the significant hazards that may be associated with the ingredient or food. The FSO concept can be an effective means to communicate the significant hazards that must be controlled to ensure consumer protection.

Certificates may be used for foods in international trade in the form of an export certificate. An example of a certificate appears in Exhibit 4–1 (CAC, 1997b). In addition to the expected information about the source of the food, the mode of transportation, and

Table 4–1 Parameters That Can Be Used To Assess the Acceptability of a Supplier

Component of the Food Control System	Expectation
Good Hygienic Practice	In place and designed to help control known hazards
HACCP plan	In place and designed to control significant hazards
FSO	Process is designed and validated to meet an FSO
Performance criteria	Validated process(es)
Process criteria	Process criteria incorporated into HACCP plan as critical limits
Product criteria: Organoleptic, chemical, physical, and biological specifications	Meets specifications
Records	Records are complete, accurate, and facilitate validation and verification

the quantity involved, the sanitary or phytosanitary status of the food are attested to by an issuing authority. The certificate is a legal document that specifies that the lot is in conformity or meets:

1. the specified product standards required by the importing country;
2. provisions of bilateral or multilateral agreements between the importing and exporting countries; and
3. in the absence of such provisions, the standards and requirements as agreed upon, with emphasis on the use of standards and codes of practice of the Codex Alimentarius Commission.

4.6 EXAMPLES TO DEMONSTRATE THE PROCESS OF LOT ACCEPTANCE

The decision to accept or reject foods upon receipt at port-of-entry or at a buyer's operation is influenced by a number of factors. Among the most common is prior experience with the supplier for compliance with all established criteria. Another is the likely impact if an incorrect decision is made to accept a defective lot. The extent of an adverse impact will depend on the likely presence and severity of a hazard in the food and whether its intended use can result in a decrease, no change, or an increase in the hazard prior to consumption as described in detail in Chapter 8. The following examples describe the decision making process for food being received at port-of-entry or at a food operation when there are different degrees of knowledge about the source.

4.6.1 Ideal Situation—Extensive Knowledge

- Food is produced applying control measures as described in Chapter 3.
- Supplier is on approved list.
- Supplier has a history of consistent compliance with all established criteria during operation and for finished product.
- There is a favorable history of audit results.

- All pertinent records are complete, accurate, and facilitate verification.
- Supplier is from a country judged to have an equivalent inspection system and to provide the same level of protection and FSO for the food in question.
- Product and packaging looks normal. No evidence of spoilage or damage to the product during shipment (e.g., excessive humidity and mold growth, collapsed cartons indicating thawing of frozen product).
- Product is received within time-temperature limitations that have been specified for perishable items.

The extensive knowledge available in this example leads to a high level of confidence that the ingredient or food will meet the FSO and provide the expected level of protection. In this case, it would be redundant and nonproductive to sample and test the incoming material.

Exhibit 4–1 Example of a Certificate for the Export of Food and Food Products

Exporter/Consignor		Certificate No.	
Consignee		**TITLE**	
		Name and address of issuing authority	
	Port of loading	Country of origin of goods	
Vessel/ Aircraft	Date of departure		
Port of discharge	Final destination (if on carriage)		
Identification, Shipping marks	No. and kind of packages	Description of goods	Quantity
Container number, Seal number			
Details of producing establishments			
Details of treatment			
Attestation			
DECLARATION			
Dated at _____ (place) On _____ (date)			
Signature of signing officer		Printed name	

4.6.2 Less Than Ideal—Some Knowledge

- It is uncertain whether the food is produced applying the control measures described in Chapter 3.
- Supplier does not appear on the list of approved suppliers.
- Supplier has a history of compliance with established criteria for finished product.
- Supplier is infrequently audited.
- All pertinent records accompanying the lot are complete and accurate.
- Supplier is from a country judged to have an equivalent inspection system and to provide the same level of protection and FSO for the food in question.
- Product and packaging looks normal with no evidence of spoilage or damage to the product during shipment (e.g., excessive humidity and mold growth, collapsed cartons indicating thawing of frozen product).
- Product is received within time-temperature limitations that have been specified for perishable items.

Considering the limited knowledge available, the decision to accept the lot may be determined by the consequences of an incorrect decision. It may be prudent to test the incoming material if the chance of a food safety problem is sufficiently high and testing can be expected to provide useful information. The tests that may be performed would depend on the material and the significant hazards that may be expected to occur.

4.6.3 Uncertain Situation—No Knowledge

- No knowledge of how the food is produced.
- Supplier is not on approved list.
- A history of compliance with established criteria is not available.
- There is no history of auditing results.
- Available records accompanying the lot meet minimum requirements.
- Supplier is from a country that has yet to be judged on whether it has an equivalent inspection system.
- Product and packaging looks normal. No evidence of spoilage or stress to the product during shipment (e.g., excessive humidity and mold growth, collapsed cartons indicating thawing of frozen product).
- Product is received within time-temperature limitations expected of perishable items

In the absence of knowledge that can be used to decide whether to accept the lot, it is necessary to develop information for the lot. The tests that should be conducted should be based on the hazards that may be expected to occur. There is little choice but to place greater reliance on the use of a sampling plan that reflects potential risk as described in Chapters 5 and 8.

4.6.4 Factors Influencing Whether a Lot Should Be Tested

Criteria of various types are commonly established for foods at different stages along the food chain. In many cases, the criteria serve as purchase specifications and are intended to inform suppliers of the expected safety and/or quality characteristics for the food. In a similar manner, a standard may be developed by governments that food operators are required to meet. Despite these criteria, foods are not sampled at each step to verify compliance. In reality, the majority of foods are not sampled. The decision to sample a lot of food depends on a wide variety of factors, as outlined in Exhibit 4–2. The specific use of sampling and testing for compliance with microbiological criteria is dealt with in Chapter 5. Several aspects of Exhibit 4–2 are also incorporated into Figure 8–1.

To respond differently to the questions in Exhibit 4–2 would decrease the likelihood that a lot of food would be sampled and tested. Thus, foods believed to be of low risk to consumers would be sampled infrequently or not at all.

4.6.5 Situations Warranting Reduced/Tightened Inspection or Testing

Foods being received from suppliers that meet the characteristics described in section 4.6.1 would be of low risk and would warrant a reduced level of inspection and testing. Changes in the characteristics that lead to uncertainty about the safety of the food (e.g., an unfavorable audit report) would indicate a need to increase the level of inspection and/or testing. As described above, the degree of inspection and/or testing would depend on the risk associated with an incorrect decision. The factors outlined in Exhibit 4–2 would influence the decision whether to sample and test the food. Further discussion on tightened inspection is found in Chapter 9.

4.7 AUDITING FOOD OPERATIONS FOR SUPPLIER ACCEPTANCE

4.7.1 Basic Stages of Auditing To Determine Acceptability of a Supplier

Regardless of the availability and application of an FSO, the method of auditing is principally the same. This chapter will not discuss in detail how audits should be organized and executed, because text books and international standards are available on this subject. However, some aspects considered to be important will be mentioned. Part of the following text is adapted from the report of an FAO/WHO Consultation (WHO, 1998).

The auditing process is a formal procedure that is agreed upon by both parties and conducted in accordance with a format that is specific to the circumstances. Auditors should ensure that they plan the process properly, and that:

- sufficient time is allocated;
- the required skills are available within the audit team (in accordance with resources available);
- arrangements are agreed upon with the management of the site being audited.

There are typically two stages to auditing a production, processing, or preparation business. The first stage consists of an initial review of documentation that may be carried out on- or off-site. Although it is possible to carry out an audit without a preassessment or document review, experience has shown that a review of pertinent documents before visiting the site leads to a more focused, thorough and informed assessment. The second stage is the on-site determination of whether the company applies Good Hygienic Practices and HACCP as documented, and whether the company can consistently deliver the product as specified and when expected for delivery.

It is helpful to prepare an agenda for the audit program to ensure that relevant personnel are available during the assessment and can participate in the on-site discussion.

Exhibit 4–2 Factors that Can Influence the Decision Whether to Sample and Test a Lot of Food

Factor influencing decision to sample and test	Condition that would increase the likelihood of sampling the lot
What is the expected outcome if the food is not sampled and is accepted?	Illness among consumers is probable.
Has the food commonly been involved in foodborne illness?	The food has a recent history of causing illness.
What is the severity of the hazard(s)?	The expected hazard is of high severity.
Is there reason to suspect the food will not meet established criteria?	Certain lots occasionally fail established criteria.
Is the food intended primarily for a sensitive population?	The food is designed and intended for high risk populations.
Is the food from a country or region with endemic disease of importance to food safety?	Recent reports indicate that endemic disease has resulted in foodborne illness among consumers of an importing country.
Is the country or supplier known to exercise control over the food?	The exporting country's inspection program is considered inadequate for the expected hazard(s) in the food.
What is known about distribution of the hazard (e.g., homogeneous, heterogeneous, stratified)?	The hazard is expected to be homogeneously distributed throughout the lot, thereby increasing the likelihood of detecting an unacceptable lot.
Can a sampling plan be used to detect unacceptable lots, particularly when a small number of defective units are expected?	When defective lots occur, the number of defective units is sufficiently high to be detected.
Can the sampling plan detect a low prevalence of a pathogen of concern?	When present, the pathogen occurs in sufficiently high prevalence that it can be detected.
What is the expected complexity, accuracy, sensitivity and time for the result?	The laboratory method is easily performed, accurate, sensitive, and the time to obtain a result will not lead to a decrease in the quality of the product.
Does the laboratory have the equipment and expertise to analyze the samples?	Yes.
Where is the food located; can the lot be easily sampled?	The lot is located nearby and can be easily sampled.
Can samples be transported to the laboratory easily?	The laboratory is situated locally.
What is the cost of the product that will be sampled?	The product samples need not be purchased.
Are there sufficient funds, personnel, and laboratory support to collect and analyze the samples?	These resources are available.
Are there any outside influences to consider (e.g., political, supplier/buyer relationship)?	There are no outside influences.

4.7.1.1 Preassessment Document Review (Stage 1)

This initial stage is, in fact, more than a document review. It is a "desktop assessment" of the quality and safety assurance system. Ideally, all documentation relating to the scope of the audit should be reviewed by the auditor before an on-site assessment. This important activity may enable the design of an initial assessment checklist. Even a quick scan may allow the auditor to get an idea of the standards being applied, whether all the hazards have been identified by the supplier, and whether CCPs are well under control.

Typically, the documents reviewed might include:

- the site layout plan,
- the process flow diagram and specifications relating to it,
- a brief description of the GHP procedures, and
- the HACCP plan.

The preassessment documentation review may help the assessor identify the personnel required for detailed discussions, the specific questions to be asked, and which areas are to be focused on when carrying out the on-site assessment activities. If the auditor finds through the document review that there are obvious inadequacies, it may be decided to stop the assessment at this point instead of proceeding to the on-site audit. Based on these preliminary findings, the auditor may decide not to approve the supplier. Alternatively, the auditor may ask the potential supplier to review its quality and safety assurance system and make adjustments based on the preliminary findings.

4.7.1.2 Audit Protocol (Stage 2)

On-site Initial Meeting

It is helpful to have a brief opening meeting with the key people of the operation being assessed to confirm the scope, timetable, facilities and personnel requirements for the audit. The time and location of the next meeting could be confirmed and any additional documents required for the on-site document review could be requested.

Activities

It is important at an early stage of the audit to verify that the process flow diagram truly reflects what actually occurs. This is facilitated by an initial walk-through to observe the operation. The auditor will subsequently need to engage responsible staff in a range of questioning and investigative activities to assess the efficacy of the quality and safety assurance system, particularly the HACCP system and prerequisite control measures.

Often a form of memory aid or checklist may be used to focus these activities. The questions should cover, for example, the seven HACCP principles (see Chapter 3). Auditors must adapt other supplementary questions/activities to the scope and purpose of the audit.

The auditor should use a range of recognized questioning and auditing techniques to obtain the information required. For example, have previous audits been conducted, what were the findings and how has the operator responded? It is important that the auditor maintain sufficiently detailed records during the audit to support subsequent recommendations.

Such records could typically include:

- personnel interviewed
- records examined

- equipment examined
- product/process details
- areas not included within the assessment
- defects/inadequacies noted.

Auditors

Auditing large operations may require a team of auditors because the complexity of the operation may require a range of skills. In smaller operations of less complexity, the scope of the audit may be limited and an individual auditor is often used.

However, certain principles always apply:

- auditors should be independent even if employed by the supplier's organization; and
- auditors should be sufficiently skilled for the activity in question (e.g., proven auditing skills, knowledge of quality and safety assurance systems, experience with the technology being applied in the operation, HACCP as applied in commercial operations, and relevant GHP procedures).

Outcome and Follow-up

A closing meeting to discuss and agree on initial findings and identify corrective actions should be held with the responsible manager. It also is necessary to discuss and agree on the mechanism and timetable, accompanying documentation, and product liability concerns. If certain improvements must be made before products are accepted or resumed, a timetable should be set and, if necessary, plans made for a follow-up visit.

Frequency of Audits

An initial audit is conducted as part of a supplier approval system.
Further audits may be triggered by:

- changes in the buyer's or supplier's products/process/formulation, etc.;
- poor performance;
- a minimum baseline frequency for audits regardless of other factors.

4.7.2 Example of Auditing a Supplier When an FSO Has Been Established

Manufacturers that produce food for which an FSO has been established must assure that performance and process criteria are being met for their specific processes and conditions. The process criteria are typically used as critical limits for CCPs in the HACCP plan. Other criteria for less critical factors may appear as GHPs or prerequisite procedures.

Although most processes and practices for which process criteria have to be established will normally be the responsibility of a food operator, some may be controlled by suppliers of critical raw materials. For example, a producer of dried milk used as an ingredient in a dry mixed infant formula will have to assure that the number of *Salmonella* is below a certain level. If an FSO for infant formula was established as < 1 *Salmonella* in 10^8 g of product, then the performance criterion for the milk powder would be, for example, < 1 *Salmonella* in 10^9 g of product. Meeting such a criterion would require pasteurization of the milk at a temperature of 80 °C for 5 s as a minimum before spray drying (see Chapter 15). Recontamination must be prevented through the application of GHP. For some GHPs (e.g., air filtration), a process criterion can be set, but for others (e.g., dry cleaning practices), criteria cannot easily be established and monitored.

The selection of an acceptable supplier of a sensitive raw material is of great importance to assure the safety of a product such as a dry mixed infant formula. Audits are essential for selecting reliable suppliers with effective control measures. Once a supplier is approved, it is necessary that buyer and supplier agree on the practices to be followed. Audits at regular intervals would be a normal aspect of the partnership.

The checklist in Exhibit 4–3 demonstrates how a general audit can be targeted for a specific food operation. During an initial audit to determine the acceptability of a supplier, emphasis will be on whether the supplier can be trusted to deliver ingredients or foods needed to produce a safe product. Follow-up audits will clearly be targeted toward confirming that the supplier can indeed be trusted. The checklist outlined in Exhibit 4–3 shows how an audit can be adapted to these specific needs and provides examples of questions that should be raised. The production of dry milk (see Book 4, Chapter 11.7 and Chapter 15 in this book) is used as the example to illustrate some of these points.

While this example describes the approval of a potential supplier of an ingredient, the same procedures can be applied to other situations (e.g., auditing to approve a contract manufacturer or a manufacturer of a food for retail sale or institutional use). In addition, the concepts could serve as a basis for assessments of food operations by control authorities.

Exhibit 4–3 General and Specific Questions that Could be Considered during an Audit of a Potential Supplier of Dried Milk to be Used in the Manufacture of a Dry Mixed Infant Formula

Stage of audit	General questions to be considered when conducting an audit	Additional specific questions when auditing a dried milk operation
Stage 1: Off-site preparation for audit, study of documentation	What evidence is there of management commitment to comply with requirements? Does the company have a food safety policy? Have the food safety requirements been clearly defined? Are written documents on GHP and HACCP available and up to date? Who is on the HACCP team? Are all appropriate disciplines represented? What is the likely knowledge level of the individuals (evidence of training, qualifications, experience, etc.)? Was external expertise sought? Have performance and process criteria been properly established and validated? Are monitoring and verification data available?	Is the dried milk producer a supplier to other manufacturers of products similar to dry mixed infant formula or where dried milk is considered a sensitive ingredient? Have critical aspects of the production been worked out together? Who validated the process and performance criteria? Are monitoring data obtained with reliable methods?
Stage 2: On-site audit	What is the general standard of GHP? Are the personnel properly trained? Do they apply the rules?	Are the wet and dry zones properly separated, with overpressure of air in critical areas?

continued

Exhibit 4–3 continued

Stage of audit	General questions to be considered when conducting an audit	Additional specific questions when auditing a dried milk operation
A. Good Hygienic Practices	Are the zones with different requirements sufficiently separated? Is the pest control program functioning effectively? Are chemicals being safely controlled? How is rework, if any, being handled? How is waste treated material being handled?	Is any wet cleaning being performed in dry areas? Is the air handling system adequate? Is there evidence of infestation with insects? Does it appear that the conditions will lead to recontamination with *Salmonella*?
B. HACCP plan a. Hazard Analysis	Have the hazards of significance been correctly identified? Have any hazards been overlooked that may have an adverse affect when used in the buyer's product? Will the product meet an established FSO? Has the product been properly described? Is the PFD complete and accurate? How and when was the PFD last verified for accuracy and by whom? Are all raw materials, process/storage activities, and the use of rework included in the flow diagram? Have changes in the operation been properly analyzed to identify new hazards?	Were hazards other than *Salmonella* identified? Does the process flow diagram (PFD) indicate the various zones and personnel movements that are important to control of *Salmonella*? Does it appear that product from this operation will meet the FSO for dried milk?
b. Control measures	Are the CCPs adequately identified and do they control the hazard(s) of importance? How were the critical limits established? Are they adequate to meet any performance criteria that have been established? Have the critical limits been validated and how was this done? Are the monitoring procedures adequate to assure that deviations are detected and product is not released without appropriate review? What are the statistical methods used in monitoring? Have the monitoring methods been validated, and are the tests included in a proficiency testing program? Are the GHPs adequate to support the HACCP plan and control the hazard(s) of importance?	Is the pasteurizer identified as a CCP? What critical limits are being used for pasteurization? Are the times and temperatures of operation continuously recorded? Will the operating conditions achieve an 8 log reduction of *Salmonella*? What is the specification of the air filters for the cooling air, and is the effectiveness of the filters maintained?

continues

Exhibit 4–3 continued

Stage of audit	General questions to be considered when conducting an audit	Additional specific questions when auditing a dried milk operation
	What is the nature of the deviations that occur? Are corrective actions properly described and are the personnel adequately trained to carry them out? Does adequate documentation exist to show that when deviations occur they are detected and corrective actions are properly implemented to prevent their reoccurrence?	
c. Verification	Have verification procedures been clearly and appropriately established? Do they include tests for indicators and/or pathogens? Are the methods validated and is the laboratory included in a proficiency testing program? What are the microbiological criteria and actions taken when lots do not meet them? How are consumer complaint data being used within the verification system? Who is responsible for reviewing the HACCP verification data? How often are the HACCP records verified? What is the frequency of review of the HACCP plan?	Is the product analyzed for Enterobacteriaceae (as indicators) and for *Salmonella*; how often and how many samples? What is the sampling plan? Is the line-environment examined for the same bacteria; do tests indicate absence of *Salmonella*? What do the consumer or customer complaints indicate?
d. Packaging and shipping	Will the packaging material protect the product under expected conditions of shipping, storage, and use? Are the packaging materials as specified? Is the product correctly labeled for ingredient content and are production codes legible? Are storage conditions adequate? Are distribution procedures in-house or third party? Are good distribution practices being maintained? Is a recall procedure in place and has it been recently tested?	Are the bags water tight? Are pallets in good order, preventing damage to the bags of dried milk? Can the supplier trace the source of raw milk, production codes containing different sources/lots of milk, and where dried milk has been shipped? Have any incidents occurred where unacceptable product was shipped and, if so, were customers timely informed?

4.8 REFERENCES

CAC (Codex Alimentarius Commission) (1997a). *Joint FAO/WHO Food Standards Programme, Codex Committee on Food Hygiene. Food Hygiene, Supplement to Volume 1B-1997. Principles for the Establishment and Application of Microbiological Criteria for Foods.* CAC/GL 21–1997. Secretariat of the Joint FAO/WHO Food Standards Programme. Rome: Food and Agriculture Organization of the United Nations.

CAC (Codex Alimentarius Commission) (1997b). *Joint FAO/WHO Food Standards Programme, Codex Committee on Food Hygiene. Criteria for a Generic Certificate for the Export of Food and Food Products. Alinorm 97/30A, Appendix III, Annex 1.* Rome: Food and Agriculture Organization of the United Nations.

CAC (Codex Alimentarius Commission) (2000). *Proposed Draft Framework for Determining the Equivalency of Sanitary Measures Associated with Food Inspection and Certification Systems.* CCFICS/CX/FICS 00/6, Attachment 1. Joint FAO/WHO Food Standards Programme, Codex Committee on Food Import and Export Inspection and Certification Systems. Rome: Food and Agriculture Organization of the United Nations.

Schultz, W. B. (1997). Draft guidance on equivalence criteria for food. *Fed Register* 62, 30593–30600.

WHO (World Health Organization) (1998). *Guidance on Regulatory Assessment of HACCP, Report of a Joint FAO/WHO Consultation on the Role of Government Agencies in Assessing HACCP.* WFO/FSF/FOS/98.5. Geneva, Switzerland: WHO.

Chapter 5

Establishment of Microbiological Criteria for Lot Acceptance

5.1 Introduction
5.2 Purposes and Application of Microbiological Criteria for Foods
5.3 Definition of Microbiological Criterion
5.4 Types of Microbiological Criteria
5.5 Application of Microbiological Criteria
5.6 Principles for the Establishment of Microbiological Criteria
5.7 Components of Microbiological Criteria for Foods
5.8 Examples of Microbiological Criteria
5.9 References

5.1 INTRODUCTION

As described in the previous chapter, microbiological criteria represent one form of acceptance criteria. This chapter will focus on the use of microbiological criteria as a basis for determining acceptability of individual lots or consignments of food. Other uses of microbiological testing will not be addressed in this chapter.

Ideally, criteria that address food safety concerns should be based upon a food safety objective (FSO). The FSO is a statement of the maximum frequency and/or concentration of a microbiological hazard in a food at the time of consumption that provides the appropriate level of protection. FSOs are very suitable for the design and control of a food operation, but they are not intended for lot acceptance determinations. FSOs, however, can be used as the basis for establishing microbiological criteria. Indeed, it is important that microbiological criteria are compatible with the FSO for a food. Microbiological criteria should not be so lenient that the intended public health goals cannot be achieved. Conversely, a microbiological criterion that is excessively stringent relative to an FSO may result in rejection of food even though it has been produced under conditions that provide an acceptable level of protection. While very similar at first impression, FSOs and microbiological criteria differ considerably in function and content as can be seen in Exhibit 5–1.

An FSO can be used as a basis for the establishment of microbiological criteria. Certain information will be known from the FSO—namely, the food, the microbial hazard, and the maximum frequency or concentration that is considered tolerable. To fulfill the requirements of a microbiological criterion, additional information must be provided. The information provided in this chapter and in Chapters 6, 7, and 8 must be considered before a meaningful microbiological criterion can be established from an FSO.

Exhibit 5–1 Characteristics of Microbiological FSOs and Microbiological Criteria

FSO	Microbiological Criterion
A goal upon which food processes can be designed so the resulting food will be acceptable	A statement that defines acceptability of a food product or lot of food
Applied to food processing operations	Applied to individual lots or consignments of food
Components: —Maximum frequency or concentration of a microbiological hazard	Components: —Microorganism of concern and/or their toxins/metabolites —Sampling plan —Analytical unit —Analytical method —Microbiological limits —Number of analytical units that must conform to the limits
Can be used to establish microbiological criteria	Cannot be used to establish an FSO
Used only for food safety	Used for food safety or quality characteristics
Based on what risk managers believe will ensure the safety of a food	Based on an FSO or what risk managers believe will ensure a food to be safe or acceptable for the intended use
Can be used to cause change in processing conditions for a food commodity and improve its safety	Can be used to cause change in processing conditions so individual lots or consignments will meet established criteria
Very suitable for assessing a food safety system of a food operation	Not very suitable for assessing a food safety system of a food operation; sampling lots of food from an operation provides a snapshot of the level of control at the time of collection, but provides little confidence for past or future lots
A high level of confidence is possible when processes are designed and validated to meet an FSO	Confidence may be less certain if an FSO is not used when processes are designed and validated to meet a microbiological criterion

Since they are a relatively new concept, FSOs are not yet in broad use. In the interim, traditional approaches must be used by governments and industry to ensure food safety. The traditional approach is similar to that used in connection with an FSO. However, traditional methods often permit varying interpretations that may directly lead to non-tariff trade barriers and add to the burden of irrelevant specifications between trade partners. According to the World Trade Organization Agreement on the Application of Sanitary and PhytoSanitary Measures (WTO/SPS agreement) (WTO, 1995), every criterion that has not been established through the application of Codex principles can be challenged when its application leads to a trade barrier.

The information in this book should make it evident that microbiological testing of product is of limited value for determining the safety of a lot of food, particularly when the prevalence and concentration of a pathogen are low. This limitation was stressed by the Codex Alimentarius Commission through the following statement: "Mandatory microbiological criteria shall apply to those products and/or points of the food chain where no other more effective tools are available, and where they are expected to improve the degree of protection offered to the consumer" (CAC, 1997a, p. 28).

This chapter will discuss the principles for establishing microbiological criteria for determining lot acceptance and leave for later chapters the selection of sampling plans. A microbiological criterion is not complete without a sampling plan that is appropriate for the risk and severity of the hazard.

5.2 PURPOSES AND APPLICATION OF MICROBIOLOGICAL CRITERIA FOR FOODS

Developing meaningful microbiological criteria for a food or ingredient is a complex process that requires considerable effort and resources. Therefore, a microbiological criterion should be established only when there is a need and when it can be shown to be effective and practical. The criterion must be capable of accomplishing one or more clearly defined objectives, such as to assess:

- the safety of a food;
- adherence to Good Hygienic Practices (GHP);
- the utility (suitability) of a food or ingredient for a particular purpose;
- the shelf-life of certain perishable foods; and/or
- the acceptability of a food or ingredient from another country or region where the conditions of production are unknown or uncertain.

In the absence of an FSO, microbiological criteria may be used to formulate design requirements and to indicate the required microbiological status of raw materials, ingredients, and end products at any stage of the food chain as appropriate. Microbiological criteria may be relevant to the examination of foods, including raw materials and ingredients, of unknown or uncertain origin or when other means of verifying the efficacy of GHP and Hazard Analysis Critical Control Point (HACCP)-based systems are not available. Generally, microbiological criteria may be applied to define the distinction between acceptable and unacceptable raw materials, ingredients, products, or lots of food by regulatory authorities and/or food business operators. Microbiological criteria may also be used to determine whether processes are consistent with the General Principles of Food Hygiene (CAC, 1997a).

Three basic principles should be applied when establishing microbiological criteria. Microbiological criteria should:

- accomplish what they purport to do (e.g., prevent or reduce foodborne disease);
- be technically attainable through the application of GHP and HACCP; and
- be administratively feasible (NRC, 1964; NRC, 1985).

5.3 DEFINITION OF MICROBIOLOGICAL CRITERION

A microbiological criterion defines the acceptability of a product or a food lot, based on the absence or presence or number of microorganisms, including parasites, and/or quantity of their toxins/metabolites, per unit(s) of mass, volume, area, or lot.

5.4 TYPES OF MICROBIOLOGICAL CRITERIA

Microbiological criteria for use in lot acceptance determinations fall into three categories (although Codex recognizes only one category; see also Chapter 4).

- Microbiological Standard—a mandatory criterion that is incorporated into a law or ordinance.
- Microbiological Guideline—an advisory criterion used to inform food operators and others of the microbial content that can be expected in a food when best practices are applied.
- Microbiological Specification—part of a purchasing agreement between a buyer and a supplier of a food; such criteria may be mandatory or advisory according to use.

5.4.1 Microbiological Standards

Microbiological standards are used to determine the acceptability of a food with regard to a regulation or policy. They are established by control authorities and define the microbiological content that foods must meet to be in compliance with a regulation or policy. Foods not meeting a standard are in noncompliance and would be subject to removal from the market. Standards may be established for a wide variety of reasons, but are best applied when risk is sufficiently high and compliance is essential for public health protection. Microbiological standards should be established following the principles outlined by Codex (CAC, 1997b) and as further elaborated in this book. Ideally, standards should be developed following establishment of an FSO for a hazard in a food.

5.4.2 Microbiological Guidelines

Microbiological guidelines may be established by a regulatory authority, industry trade association, or a company to indicate what is expected for the microbial content of a food when best practices are applied. Food operators use microbiological guidelines as a basis to design their control systems. Guidelines are advisory in nature and may not lead to rejection of a food. Microbiological guidelines may be a preferred means to inform and direct food operators within a segment of the industry to improve its practices, particularly when insufficient data prevent the establishment of a microbiological standard. A microbiological guideline along with guidance material on what constitutes best practices may be adequate to bring about the changes necessary to improve food safety and quality. This may negate the need to establish a standard.

5.4.3 Microbiological Specifications

Buyers of food may be in industry or government (e.g., schools, institutions, military). Buyers establish purchase specifications to reduce the likelihood of accepting an ingredient or food that may be unacceptable in terms of safety or quality. Microbiological specifications define the microbiological limits for an ingredient, for example, so that when it is used, the final product will meet all requirements for safety and quality. It is common practice for buyers along the food chain to establish microbiological specifications for the materials they purchase. In most cases, the specifications are advisory and the materials are sampled only on an as needed basis. In other cases (e.g., sensitive ingredients), each incoming lot may be tested.

5.5 APPLICATION OF MICROBIOLOGICAL CRITERIA

5.5.1 Application by Regulatory Authorities

Mandatory microbiological criteria shall apply to those products and/or points of the food chain where no other more effective tools are available and where they are expected to improve the degree of consumer protection. Where these are appropriate, they should be product-specific and only applied at the point of the food chain as specified in the regulation.

In situations of noncompliance with microbiological criteria, the regulatory control actions may lead to sorting, reprocessing, rejection, or destruction of product, and/or further investigation to determine the appropriate actions to be taken. The appropriate action(s) will depend on the assessment of risk to the consumer, the point in the food chain, the product, and its intended use.

Governments establish microbiological standards only when they are deemed appropriate to ensure the safety of the foods for which they have regulatory responsibility. Thus, food control authorities serve as risk managers and, through the risk analysis process, may ultimately decide that a microbiological criterion is necessary for a food at one or more points in the food chain. Ideally, the standards will be based upon a tolerable level of risk and FSO for the hazard of concern. Microbiological standards may be used by both industry and government as the basis for assessing the acceptability of a lot or consignment of food. Microbiological standards should be applied equally to domestic and imported foods. Lots that do not meet the microbiological standard would be considered unacceptable and subject to removal from the market or rejection at port-of-entry.

5.5.2 Application by a Food Business Operator

In addition to checking compliance with regulatory provisions (see sections 5.4.1 and 5.5.1), microbiological criteria may be applied by food business operators to formulate design requirements and to examine products as one of the measures to verify and/or validate the efficacy of elements of their GHP and HACCP plans (see Chapters 3 and 4). Such criteria will be specific for the product and the stage in a process or in the food chain at which they apply. They may be stricter than the criteria used for regulatory purposes and should, as such, not be used for legal action.

Any buyer of food can enter into an agreement with a supplier to ensure that food being purchased will meet mutually agreed upon criteria. Food operators frequently establish purchase specifications for food ingredients and other materials that will ensure compliance with microbiological standards and as a means to ensure product quality. The factors outlined in Exhibit 4–2 influence whether a buyer will sample ingredients or foods upon receipt. For example, an audit of an operation may suggest that a supplier can not consistently meet purchase specifications. This may lead the buyer to sample incoming lots of purchased material until a follow-up audit indicates improvement. Lots that fail to meet the established specifications may be rejected.

5.5.2.1 Application in GHP

Microbiological criteria can be used to check certain aspects of GHP. An example would be verifying the acceptability of water if not supplied by a source tested by others. Also, food operators can establish microbial limits that are to be met when the cleaning and disinfecting routines are properly performed. These criteria often consist of aerobic counts or indicator organisms and reflect experience of what is attainable with the equipment, materials, and conditions that exist for the operation. Another option to assess GHP is to sample product at selected times and steps in an operation and analyze for aerobic count or other indicators. Evidence of an increase in the microbial concentration on/in a product may be due to microorganisms acquired from contact with equipment during processing. In-process criteria should be based on knowledge of the conditions that influence microbial content during processing.

5.5.2.2 Application in HACCP

Microbiological criteria are not normally suitable for monitoring critical limits, as defined in the publication *Hazard Analysis and Critical Control Point System and Guidelines for its Application* (CAC, 1997c). Monitoring procedures must be able to detect loss of control at a critical control point (CCP). Monitoring should provide this information in time for corrective actions to be taken to regain control before there is a need to reject the product. Consequently, on-line measurements of physical and chemical parameters are often preferred to microbiological testing because results are often available more rapidly and at the production site. Moreover, the establishment of critical limits may require other considerations than those described in this document.

5.6 PRINCIPLES FOR THE ESTABLISHMENT OF MICROBIOLOGICAL CRITERIA

Principles for the establishment of microbiological criteria have been developed by Codex (CAC, 1997b). The principles, initially developed through World Health Organization (WHO)/Food and Agriculture Organization (FAO) consultations (Christian, 1983), have continued to evolve through a number of revisions, each with input from the International Commission on Microbiological Specifications for Foods (ICMSF). Thus, it is appropriate that this chapter be based on the principles outlined in the Codex document. This chapter is intended to give guidance for the establishment and application of microbiological criteria for foods at any point in the food chain from primary production to final consumption.

The safety of foods is principally assured by control at the source, product design and process control, and the application of Good Hygienic Practices during production, processing (including labeling), handling, distribution, storage, sale, preparation, and use, in conjunction with the application of the HACCP system. This preventive approach offers more control than microbiological testing because the effectiveness of microbiological examination to assess the safety of foods is limited. This view was emphasized by the U.S. Food and Drug Administration (FDA) in its discussion of the control measures considered necessary to achieve a required level of protection. This statement is in reference to all forms of product testing (e.g., chemical, physical, microbiological).

> End product testing, which measures outcome, cannot generally be relied upon exclusively to provide an adequate level of protection because it only tests for a specific risk or group of risks on a particular day. The results of end product sampling may or may not be representative of the actual, continuing risk, depending upon product uniformity, the amount of sampling, and other factors. Processing controls coupled with adequate verification by a regulatory authority provide an essential assurance that food will not present unacceptable risks.

Processing controls can assure that the level of protection is met in many circumstances where end-product testing alone realistically cannot (Schultz, 1997, p. 30597).

Microbiological criteria should be established according to the following principles and be based on scientific analysis and advice and, where sufficient data are available, a risk assessment appropriate to the foodstuff and its use. Microbiological criteria should be developed in a transparent fashion and meet the requirements of fair trade. They should be reviewed periodically for relevance with respect to emerging pathogens, changing technologies, and new understandings of science.

A microbiological criterion should be established and applied only where there is a definite need and where its application is practical. Such need is demonstrated, for example, by epidemiologic evidence that the food under consideration may represent a public health risk and that a criterion is meaningful for consumer protection, or as the result of a risk assessment. A criterion should be established only when it is the best, or only, means of ensuring a food will be safe or will not spoil. Application of the criterion should be practical and technically attainable by applying GHP and HACCP. The criterion should be capable of accomplishing the intended purpose (i.e., reduce foodborne disease or spoilage).

To fulfill the purpose of a microbiological criterion, consideration should be given to:

- evidence of actual or potential hazards to health;
- microbiological status of the raw materials;
- effect of processing on the microbiological status of the food;
- likelihood and consequences of microbial contamination and/or growth during subsequent handling, storage, and use;
- intended use of the food;
- category(s) of consumers concerned;
- cost-benefit ratio associated with the application of the criterion; and
- need to inform personnel along entire food chain.

The number and size of analytical units per lot tested should be as stated in the sampling plan and should not be modified. Unacceptable lots should not be subjected to repeated testing to bring the lot into compliance.

The intended use is an important consideration. Food safety is: "assurance that food will not cause harm to the consumer when it is prepared and/or eaten according to its intended use" (FAO, 1997). Thus, a raw agricultural commodity that is intended to be thoroughly heated before consumption could contain *Salmonella* and with adequate processing, not cause harm. A stringent sampling plan for *Salmonella* in such a food would normally have no value. Proper labeling with adequate instructions for preparation and use would be more effective.

Consideration of the intended use should also include who is going to prepare the product (e.g., professional caterer or consumer). Even more important is for which group of consumers is the food intended. Babies, ill people, the elderly, and immunosuppressed individuals are more vulnerable, for example, than normal healthy adults; thus, greater care is essential for the production or preparation of food for these consumers. This should be reflected in the stringency of microbiological criteria and accompanying sampling plans.

Consideration of the cost-benefit ratio should assess whether establishing and enforcing a criterion would be an effective means of using the available resources. In addition, the criteria should be administratively feasible by the control authority. These considerations fall under the activity of choosing an adequate risk management option within the risk management concept, where other options are considered as well.

5.7 COMPONENTS OF MICROBIOLOGICAL CRITERIA FOR FOODS

A microbiological criterion consists of:

- a statement of the microorganisms of concern and/or their toxins/metabolites and the reason for that concern (see section 5.7.1);
- microbiological limits considered appropriate to the food at the specified point(s) of the food chain;
- the number of analytical units that should conform to these limits (see section 5.7.2);
- a sampling plan defining the number of field samples to be taken, the method of sampling and handling, and the size of the analytical unit (see section 5.7.3); and
- the analytical methods of detection and/or quantification (see section 5.7.4).

A microbiological criterion should also state:

- the food to which the criterion applies;
- the point(s) in the food chain where the criterion applies; and
- any actions to be taken when the criterion is not met.

When applying a microbiological criterion to assess products, it is essential, in order to make the best use of money and personnel, that only appropriate tests be applied to those foods and at those points in the food chain that offer maximum benefit in providing consumers with foods that are safe and suitable for consumption.

5.7.1 Microorganisms, Parasites, and Their Toxins/Metabolites of Importance in a Particular Food

Microorganisms of concern and/or their toxins/metabolites include:

- bacteria, viruses, yeasts, molds, and algae;
- parasitic protozoa and helminths; and
- microbial toxins/metabolites.

The microorganisms included in a criterion should be widely accepted as relevant—as pathogens, as indicator organisms, or as spoilage organisms—to the particular food and technology. Organisms whose significance in the specified food is doubtful should not be included in a criterion.

The mere finding, with a presence-absence test, of certain microorganisms known to cause foodborne illness (e.g., *Clostridium perfringens*, *Staphylococcus aureus*, and *Vibrio parahaemolyticus*) does not necessarily indicate a threat to public health.

Where pathogens can be detected directly and reliably, consideration should be given to testing for them in preference to testing for indicator organisms. If a test for an indicator organism is applied, there should be a clear statement whether the test is used to indicate unsatisfactory hygienic practices or a health hazard.

5.7.2 Microbiological Limits

A microbiological limit is the frequency or maximum concentration of a microorganism, microbial toxin, or metabolite that can be used to differentiate acceptable from unacceptable lots of food. Limits are established as a basis to assess the safety or quality of a food. Microbiological limits should be appropriate for the food and be applicable at one or more points in the food chain. Limits for specific hazards in food should be compatible with any FSOs that may have been established.

Limits used in criteria should be based on microbiological data appropriate to the food and should be applicable to a variety of similar products. The process of establishing limits for use as standards should include collecting and analyzing data from a variety of operations to determine what can be expected for foods produced under acceptable conditions of GHP and HACCP. The data can be used to establish limits that can be met by all who operate under acceptable conditions. Alternatively, the limit may be made more stringent if improvement is deemed necessary in a certain segment of industry to reduce the likelihood of a hazard. This assumes operators can adapt by making certain practical modifications. If, however, the technology does not exist or is not affordable, then the more stringent limit will fail and the desired improvement will not be achieved. The process of establishing microbiological criteria, and other acceptance criteria, should be transparent and allow input from all interested parties.

Microbiological limits can be related only to the time and place of sampling and not to the presumed number of microorganisms at an earlier or a later stage. Because GHP aims at producing foods with microbiological characteristics significantly better than those required by public health considerations, a numerical limit in a guideline may be more stringent than in a standard or an end-product specification.

In the establishment of microbiological limits, any changes in the microflora likely to occur during storage and distribution (e.g., decrease or increase in the numbers) should be taken into account. Microbiological limits should take into consideration the risk associated with the microorganisms and the conditions under which the food is expected to be handled and consumed. These considerations are discussed in Chapter 8. Microbiological limits should also take account of the likelihood of uneven distribution of microorganisms in the food (see Chapters 6 and 7) and the inherent variability of the analytical procedure (see Chapter 12).

Whether foods are acceptable or not is defined in a criterion by the:

- microbiological limit(s);
- number of samples examined;
- size of the analytical unit; and
- number of units that should conform to the limits.

If a criterion requires the absence of a particular microorganism, the size and number of analytical units (as well as the number of analytical sample units) should be indicated. It should be borne in mind that no feasible sampling plan can ensure the absence of a particular organism in the entire contents of the lot.

The microbial populations in many foods processed under GHP and HACCP are generally not explicitly expressed because they are as low as is reasonably achievable. In some cases, the levels cannot be measured due to the technical problems involved. For instance, a lot of canned food that has received a "bot cook" will most probably not contain surviving spores of *C. botulinum* in 10^{10} or 10^{11} g of product. Likewise, enteric pathogens are not likely to be found in many kilograms of pasteurized products. Setting limits for pathogens in processed foods in which a validated kill step is included during processing should not be an arbitrary activity, but should be done only if there is a need to detect contaminated product.

Limits for use in purchase specifications are best established from data collected during normal production when the operation is under control. It is not uncommon for a company to establish for its own use criteria more stringent than required to ensure compliance with customer and regulatory requirements.

5.7.2.1 Indicators

Indicators are frequently used to examine foods or ingredients. Indicators can be very useful, but their selection and application must be done with care and thorough understanding of how accurately to interpret the analytical results. An important use of indicators is to verify process control and identify opportunities for process improvements. Most considerations used to set a microbiological criterion are applicable for indicators; however, indicators are not used solely for pathogenic concerns. Microorganisms, their cellular components, or their metabolic products used as indicators may indicate:

- the possible presence of a pathogen or toxin (*Staphylococcus aureus* for potential enterotoxin and/or excessive human handling in cooked crab meat).
- the possibility that faulty practices occurred during production, processing, storage, and/or distribution (coliforms in pasteurized milk).
- the suitability of a food or ingredient for a desired purpose (*E. coli* in nuts for ice cream).
- an estimate of the shelf-life of perishable foods during the expected conditions of handling and storage (yeast in yogurt).
- the possibility of changes in the food through microbial activity that would result in the food becoming more hazardous (acetophiles in spices for acidic sauces).
- the effectiveness of cleaning and disinfection (ATP (adenosine triphosphate) for residual soil).

Indicator organisms and agents can be divided into indicators of potential (a) human contamination, (b) fecal contamination, (c) survival of a pathogen or spoilage organism, and (d) post-processing contamination.

Examples of microbial indicators that can be analyzed quantitatively or qualitatively include aerobic bacteria, coliforms, *Enterobacteriaceae*, *Escherichia coli*, yeasts, molds, proteolytic bacteria, and thermophilic bacteria. Examples of cellular components that can be used as indicators include ATP, ribonucleic acid (RNA), endotoxins (e.g., limulus lysate test for cellular polysaccharides), and various enzymes (e.g., thermonuclease). Examples of metabolic products used as indicators include hydrogen sulfide (early putrefaction), carbon dioxide (spoilage by *Zygosaccharomyces bailii*), lactic acid (in certain meat products, including ham), ethanol (in fruit juice), diacetyl (in fruit juice or beer), and ergosterol (indicates mold in grain).

Exhibit 5–2 lists some of the characteristics of an ideal indicator organism. It must be recognized that indicators often represent a compromise that is less than the ideal of being able to test for the microorganism(s) or toxin(s) of concern. They offer considerable advantages, however, that will ensure their continued use. Indicators should be selected that provide the best information with the least amount of compromise.

In practice, indicators seldom, if ever, prove the presence or the absence of a target microorganism; they merely indicate the possibility. For purposes of process control, the absence of, or low numbers of, an indicator microorganism can be useful as a sign that a process is under control, and therefore, there will be a lower likelihood of occurrence of an unacceptable target microorganism. However, a pathogen (e.g., salmonellae) may be

Exhibit 5–2 Factors To Consider When Selecting an Indicator for a Specific Purpose

- Presence indicates potential for spoilage, faulty practice, or faulty process
- Easily detected and/or quantified
- Survival or stability (including inactivation kinetics) similar to or greater than the hazard or spoilage microorganism
- Growth capabilities (requirements) similar or faster than the hazard or spoilage microorganism
- Identifiable characteristics of indicator are stable
- Methods are rapid, inexpensive, reliable, sensitive, validated, and verified with positive control
- Quantitative results have a correlation between indicator concentration and level of hazard or spoilage microorganism
- Results are applicable to process control
- Analyst health is not at risk
- Analytical method does not jeopardize the environment and is suitable for in-plant use

present independent of an indicator. This is more likely to occur if the pathogen can establish itself in a processing environment and multiply. *E. coli* and coliforms are not reliable indicators for the presence of salmonellae where this can occur.

Indicators can be very useful, but their selection and application must be done with care and thorough understanding of how to interpret the analytical results correctly.

5.7.3 Sampling Plans, Sampling Method, and Handling of Samples Prior to Analysis

Sampling plans include the sampling procedure and the decision criteria to be applied to a lot, based on examination of a prescribed number of sample units and subsequent analytical units of a stated size by defined methods. A well-designed sampling plan defines the probability of detecting microorganisms in a lot, but it should be borne in mind that no sampling plan can ensure the absence of a particular organism. Sampling plans should be administratively and economically feasible.

In particular, the choice of sampling plans should take into account:

- risks to public health associated with the hazard (severity and likelihood of occurrence of the hazard);
- the susceptibility of the target group of consumers (very young or old, immunocompromised, etc.);
- the heterogeneity of distribution of microorganisms where variable sampling plans are employed;
- the randomness of sampling;
- the acceptable quality level (i.e., percentage of nonconforming or defective sample units tolerated); and
- the desired statistical probability of accepting or rejecting a nonconforming lot.

The information needed for the first two points could be obtained through a risk assessment; however, good epidemiologic data often suffice. For many applications, two or three class attribute sampling plans may prove useful. For a detailed discussion of establishing sampling plans, see Chapters 6, 7, and 8. In general, the greater the risk, the more stringent (e.g., greater number of samples) should be the sampling plan.

The statistical performance characteristics or operating characteristic curve should be provided in the sampling plan. Performance characteristics provide specific information to estimate the probability of accepting a nonconforming lot. The sampling method should be defined in the sampling plan. The time between taking the field samples and analysis should be as short as reasonably possible and, during transport to the laboratory, the conditions (e.g., temperature) should not allow increase or decrease of the numbers of the target organism. By controlling these conditions, the results should reflect, within the limitations given by the sampling plan, the microbiological conditions of the lot.

5.7.4 Microbiological Methods

Whenever possible, only methods for which the reliability (accuracy, reproducibility, inter- and intra-laboratory variation) has been statistically established by comparative or collaborative studies in several laboratories should be used. Moreover, preference should be given to methods that have been validated for the commodity concerned, preferably in relation to reference methods elaborated by international organizations. While methods should be the most sensitive and reproducible for the purpose, methods to be used for in-plant testing might often sacrifice to some degree sensitivity and reproducibility in the interest of speed and simplicity. They should, however, have been proven to give a sufficiently reliable estimate of the information needed.

Methods used to determine the suitability for consumption of highly perishable foods (i.e., foods with a short shelf-life) should be chosen so that results are available before the foods are consumed or exceed their shelf-life.

The microbiological methods specified should be reasonable with regard to complexity, availability of media, equipment, etc., ease of interpretation, time required, and costs. Additional information on methods and their reliability is provided in Chapter 12.

5.7.5 Reporting

The test report shall give the information needed for complete identification of the sample, the sampling plan, the test method, and, if appropriate, the interpretation of the results.

5.7.6 Disposal of Unacceptable Lots

When a lot fails to meet a microbiological criterion, the lot must be disposed of in an acceptable manner. If the lot has failed a microbiological standard, the food is not in compliance with an existing regulation and would be subject to current regulatory policy. A lot failing a guideline may be acceptable depending on the circumstances. If the lot has failed a specification, depending on risk to consumers, its intended use, potential product failure due to spoilage, or other effects, the food may be rejected by the buyer. In each case, however, the cause for failure to meet the criterion should be determined and corrective actions implemented.

The option selected for disposal of a rejected lot should be influenced by the risk to consumers, the type of food, applicable regulatory policy, and the food's original intended use. Options for disposal may include sorting, reprocessing, and destruction. In certain circumstances, the food might be used as an ingredient when a subsequent process will eliminate a hazard. The option selected must be in compliance with existing regulations.

5.8 EXAMPLES OF MICROBIOLOGICAL CRITERIA

Table 5-1 provides an example of microbiological criteria that might result from application of the principles described in this chapter. The example is derived from an earlier recommendation from ICSMF (ICMSF, 1986). The criteria include indicators [i.e., APC (aerobic plate count), coliforms], methods of analysis, number of samples to be collected, number of samples that must conform to the criteria, and the microbiological limits per g. This brief summary must be supplemented with additional information such as a statement of why these criteria were selected and are considered necessary; the step(s) in the food chain where the criteria are to be applied; method of sample collection, handling, and preparation for analysis; the analytical unit (in this example, the unit for the *Salmonella* analysis is 25 g); whether the analytical units can be composited for analysis (in this example, the five 25 g analytical units could be combined into a single 125 g composite for analysis); and disposition of lots that do not meet the criteria. The criteria can be made more stringent if the eggs are intended for a sensitive population (e.g., hospitals, institutions for the elderly).

A second example of a microbiological criterion addresses the concern for *S. aureus* enterotoxin in cooked crustaceans. Cooked crustaceans are often hand-peeled, and *S. aureus* may be transferred from workers' hands to the cooked product during peeling. Subsequent temperature abuse could result in growth of *S. aureus*, particularly since a

Table 5-1 Example of Microbiological Criteria for Pasteurized Liquid, Frozen, and Dried Egg Products

Microorganism or group	Analytical Method	No. Samples (n)	c	Limit per g	
				m	M
APC	ISO 4833	5	2	5×10^4	10^6
Coliforms	ISO 4831	5	2	10^1	10^3
Salmonella spp.	ISO 6579	5*	0	0 [negative/125]	—
		10†	0	0 [negative/250]	—
		20††	0	0 [negative/500]	—

*,†,††The number of samples is to be chosen in accordance with whether the expected conditions of storage and use of the product will cause *)a reduction, †)no change, or ††)an increase in the number of *Salmonella* spp. between when the eggs are sampled and consumed.

c: the maximum allowable number of defective sample units (i.e., two-class plans for *Salmonella*) or marginally acceptable sample units (i.e., three-class plans for APC and coliforms). When more than this number are found in the sample, the lot is rejected.

m: a microbiological limit which, in a two-class plan, separates good quality from defective quality or, in a three-class plan, separates good quality from marginally acceptable quality. Values equal to m, or below, represent an acceptable product and values above it are either marginally acceptable or unacceptable. This value also could be seen as the value considered by those establishing the criteria to be acceptable and attainable through application of GHP and/or HACCP.

M: a microbiological limit which, in a three-class plan, separates marginally acceptable quality from defective quality. Values above M are unacceptable.

competitive flora may be lacking in the cooked, peeled product. For these reasons, cooked, peeled shrimp and other crustaceans have been associated with staphylococcal foodborne illness. Microbiological testing can be used to assess the safety of this food category (e.g., at port-of-entry). Considering the high levels of *S. aureus* associated with risk and the relatively small number of samples required for assessment, application of a microbiological criterion is a practical option to evaluate the acceptability of these products with respect to the likely presence of staphylococcal enterotoxin. The European Union, for example, has adopted a three-class sampling plan with $n = 5$, $c = 2$, $m = 100$ cfu g^{-1}, and $M = 1,000$ cfu g^{-1} for *S. aureus* in cooked crustaceans.

Chapters 6, 7, and 8 discuss the factors that must be considered when establishing microbiological criteria and provide further discussion of n, c, m, and M, and choice of sampling plan. In addition, a decision tree is provided in Figure 8–1 to help decide when microbiological criteria can be effective in relation to other control options.

5.9 REFERENCES

CAC (Codex Alimentarius Commission) (1997a). *Joint FAO/WHO Food Standards Programme, Codex Committee on Food Hygiene. Food Hygiene, Supplement to Volume 1B-1997. Principles for the Establishment and Application of Microbiological Criteria for Foods.* CAC/GL 21–1997. Secretariat of the Joint FAO/WHO Food Standards Programme. Rome: Food and Agriculture Organization of the United Nations.

CAC (Codex Alimentarius Commission) (1997b). *Joint FAO/WHO Food Standards Programme, Codex Committee on Food Hygiene. Food Hygiene, Supplement to Volume 1B-1997. Recommended International Code of Practice, General Principles of Food Hygiene.* CAC/RCP 1–1969, Rev. 3.

CAC (Codex Alimentarius Commission) (1997c). *Joint FAO/WHO Food Standards Programme, Codex Committee on Food Hygiene. Food Hygiene, Supplement to Volume 1B-1997. Hazard Analysis and Critical Control Point (HACCP) System and Guidelines for Its Application.* CAC/RCP 1–1969, Rev. 3.

Christian, J. H. B. (1983). *Microbiological Criteria for Foods. Summary of Recommendations of FAO/WHO Expert Consultations and Working Groups 1975–1981.* VPH/83.54, Geneva, Switzerland: World Health Organization.

FAO (Food and Agriculture Organization of the United Nations) (1997). *Codex Alimentarius, Food Hygiene Basic Texts.* Joint FAO/WHO Food Standards Programme. Rome: Food and Agriculture Organization of the United Nations.

ICMSF (International Commission on Microbiological Specifications for Foods) (1986). *Microorganisms in Foods 2. Sampling for Microbiological Analysis: Principles and Specific Applications,* 2nd edn. Toronto: University of Toronto Press.

NRC (National Research Council) (1964). *An Evaluation of Public Health Hazards from Microbiological Contamination of Foods.* Food Protection Committee. Washington, DC: National Academy of Sciences.

NRC (National Research Council) (1985). *An Evaluation of the Role of Microbiological Criteria for Foods and Ingredients.* Subcommittee on Microbiological Criteria. Washington, DC: National Academy Press.

Schultz, W. B. (1997). Draft guidance on equivalence criteria for food. *Fed Register* 62, 30593–30600.

WTO (World Trade Organization) (1995). The WTO Agreement on the Application of Sanitary and Phytosanitary Measures (SPS Agreement). http://www.wto.org/english/tratop_e/sps_e/spsagr_e.htm.

Chapter 6

Concepts of Probability and Principles of Sampling

6.1	Introduction	6.8	Stringency and Discrimination
6.2	Probability	6.9	Acceptance and Rejection
6.3	Population and Sample of the Population	6.10	What Is a Lot?
		6.11	What Is a Representative Sample?
6.4	Choosing the Sample Units	6.12	Confidence in Interpretation of Results
6.5	The Sampling Plan		
6.6	The Operating Characteristic Function	6.13	Practical Considerations
		6.14	References
6.7	Consumer Risk and Producer Risk		

6.1 INTRODUCTION

It is important to recognize that the management of food safety using the approaches outlined in the first five chapters, based on controlling hazards through Good Hygienic Practices (GHP) and Hazard Analysis Critical Control Point (HACCP), is much more effective than trying to ensure safety by lot acceptance through end-product testing. This chapter discusses the concepts of probability and sampling that determine the limitations of end-product testing. Those concepts form the basis of the design of statistically based sampling plans (Chapter 7) and the establishment of cases (Chapter 8).

In Chapters 6 and 7, it is assumed that the microbiological test will give an absolutely accurate result, with no false positives or false negatives. This is not a realistic assumption, and problems arising from sampling, sample transport and preparation for analysis, and the imprecision of microbiological methods are addressed in Chapters 10 and 12 and elsewhere in the text.

6.2 PROBABILITY

Consider a trial or a test, the outcome of which is doubtful, such as a test for the presence of an organism in a food. After the standard procedure has been followed, the test either shows the presence of the organism or it does not; thus we have a positive or a negative result. If many of the organisms were present in the food sampled, we would expect relatively many such tests to yield a positive result. But if there were few organisms present, we would expect that fewer tests would yield a positive result. In the two cases, the probability of a positive result would be high and low, respectively.

The probability of a positive result is, in fact, the long-run proportion of times a positive result occurs out of all the times we test the food. Thus, if a positive result occurs 112 times in 1000 tests, we estimate the probability to be 112/1000 = 0.112, while if a positive result occurs 914 times in 1000 trials, we estimate the probability to be 914/1000 = 0.914. The word "estimate" is used because if we were to run 1000 trials again on the same material and using the same procedure, we could not be sure of observing 112 and 914 positive results, respectively. The results should, however, be close to this proportion because 1000 is a large number of trials. Thus, the estimated probability of a positive result, or any other outcome in some other type of test, is the proportion of times a positive outcome occurred among the trials or tests actually made. A probability can be anywhere from 0 to 1. It will be zero if the organism is absent from the food (assuming a test procedure that will detect any organism present) and 1 if every one of a large number of tests provides a positive result.

What does an observed proportion of trials such as 0.112 signify? Suppose we divided the entire lot of food into small sample units, say perhaps 10,000,000 one-gram units, and then went through the test procedure on each of these sample units. Suppose 1,051,200 gave positive results. Then the ratio 1,051,200/10,000,000 (the actual proportion of positives) = 0.10512 is the measure of the probability. But what kind of measure? This is no longer an estimate of the probability of a positive result. Instead, it is the *true probability* or *population probability*. Of course, this approach is not practical, because of the test time needed, and because there would be no food left to eat! But it is useful to have this concept in mind. The population probability determines the kind of sample probabilities or estimated probabilities we may expect from a given number of sample units examined. If that number is low, the estimated probability is not likely to be precise. In general, we never know the true or population probability. But we do know that the larger the number of units included in the sample of the population, the closer the estimated probability is likely to be to the population probability.

6.3 POPULATION AND SAMPLE OF THE POPULATION

The preceding section introduced the two basic concepts: (i) population as a whole and (ii) sample of the population. In relation to results of an experimental test, these are, respectively, the whole set of results of the test made on each and every unit in the lot of food, and the partial collection of such results, derived only from the group of sample units actually examined. In terms of standard plate counts, for example, these concepts would be represented by (i) all the counts that could be made by examining every unit in the lot and (ii) those counts actually made on the few sample units examined. Statisticians use the word *sample* for the group of units which is withdrawn to estimate the character of a population, while analysts or bacteriologists would call any one of these units a sample. To try to minimize the confusion, we will speak of the sample of the population, being the whole group of units withdrawn, and the *sample units* of which the sample is composed. The assumption is also made that a sample unit is an identifiable unit that can be repeatedly recognized (e.g., a batch of uniform sized beef burgers or a unit of packaging, or the contents of a defined sampling tool). The *analytical unit* may then be a portion of the sample unit. In this chapter, the sample unit and the analytical unit are treated as though they are the same.

6.4 CHOOSING THE SAMPLE UNITS

Sections 6.10 and 6.11 describe how to choose the material to be tested from the total amount in the lot or shipment, and Chapter 12 describes the practical aspects of collecting and handling samples. The important point is to avoid bias, so that the sample of the population may represent the lot as well as possible. Random choice is one way of achieving this. Thus, if we can think of the lot as made up of a population of 10 g blocks called sample units, and we decide upon a sample from this population of 10 such units, then we should choose these units in such a way that each sample unit in the lot has the same chance of being included among the sample units chosen. In practice, it is often difficult to ensure such random sampling, and this can be particularly significant for populations with incomplete mixing or of unknown origin. At the very least, however, we should try to draw test material from all parts of the lot.

In this book, all statistical calculations assume that the sample has been drawn randomly, unless otherwise stated.

6.5 THE SAMPLING PLAN

The sample units so drawn will, after examination, yield results that will be compared with certain criteria to reach a decision as to whether the entire lot should be accepted or rejected (see section 6.9 for explanation of rejection). The particular choice of sampling procedure and the decision criteria make up the *sampling plan*. Proper drawing of samples is essential if the decision making criteria of the sampling plan are to give unbiased results. By sampling randomly, we are able to reduce the risk of making biased decisions.

A simple example of a sampling plan follows. Take 10 sample units of a food from a lot and subject them to appropriate laboratory procedure to test for the presence or absence of a microorganism. If two or fewer of the 10 sample units show presence of the organism, that is, give a positive result, then the whole lot of food is acceptable (relative to this organism). But if three or more give a positive result, the whole lot is to be rejected. This plan is described by the use of two terms $n = 10$ and $c = 2$, where n is the number of sample units chosen separately and independently, and where c is the maximum allowable number of sample units yielding a positive result.

6.6 THE OPERATING CHARACTERISTIC FUNCTION

In section 6.5, we described what we mean by a sampling plan and gave an example of a simple sampling plan based upon positive and negative indications of a microorganism. This plan was described by the two numbers $n = 10$, $c = 2$. If we are going to use this plan, we want to know what assurance it will give us. How discriminating is it? It is possible when using such a plan that it will sometimes accept a poor lot; it is also possible for the plan to reject a good lot. There is no way to avoid some degree of risk in each acceptance and in each rejection (see also section 6.7), unless we test the entire lot, in which case no edible food will be left. We can make these risks smaller by testing more sample units; that is, by using a larger number of samples (larger n). In fact, we can reduce the risks to any desired level by making n sufficiently large. But in practice, we usually have to seek a compromise between (a) large n (many sample units) and small risks, and (b) small n (few sample units) and large risks.

We normally use an *operating characteristic* function to describe the performance of a sampling plan. This is often depicted as an operating characteristic curve (OC curve). Such a curve has two scales. The horizontal scale shows a measure of lot quality. The most common measure of lot quality is the true probability or percentage of times that a test on a sample unit from a lot of this quality would show an unsatisfactory result (e.g., a positive, or a count above some number m). This is called p, and it can be anywhere between 0 and 100%. This probability is commonly expressed as percentage of *defectives*. The vertical scale gives the probability of acceptance, P_a; that is, P_a is the expected proportion of occasions that the results will indicate that the lot is acceptable out of the number of times a lot of this given quality is sampled to find a decision. Let us see how the probability runs $n = 10$, $c = 2$ (the calculations are carried out by using the so-called binomial distribution):

$p(\%)$	0	10	20	30	40	50	60
P_a	1.00	0.93	0.68	0.38	0.17	0.05	0.01

Figure 6–1 shows the full curve. We see that the greater the probability (p) that a test unit from a given lot yields a positive result to indicate the presence of the target microorganism, the lower is the probability of acceptance (P_a) of that lot. If, for example, we set a limit of 20% defectives (i.e., $p = 20\%$), then $P_a = 0.68$. This means that on 68 of every 100 occasions when we sample a lot containing 20% defectives, we may expect to have 2 or fewer of the 10 tests showing the presence of the organism, and thus calling for acceptance, while on 32 of every 100 there will be 3 or more positives, calling for nonacceptance. But if $p = 10\%$ (i.e., 10% positive units in the lot), such lots will be accepted 93 out of 100 times; if $p = 40\%$, such lots will be accepted 17 out of 100 times. Thus

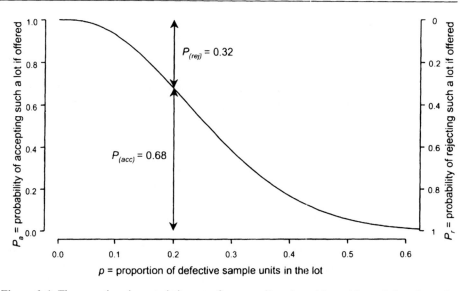

Figure 6–1 The operating characteristic curve for a sampling plan with $n = 10$, $c = 2$, i.e., the probability of accepting lots, in relation to the proportion defective among the sample units in the lot being examined. If, for example, the lot comprises 20% ($p = 0.2$) defective sample units, the lot will be accepted with a probability of 0.68 and rejected with a probability of 0.32.

lots with 10% of the sample units defective will be accepted most of the time, but with 40% defective, rather seldom.

6.7 CONSUMER RISK AND PRODUCER RISK

Since decisions to accept or reject lots are made on samples drawn from the lots, occasions will arise when the sample results do not reflect the true condition of the lot. The *producer's risk* describes the probability that an acceptable lot if offered will be falsely rejected. The *consumer's risk* describes the probability that an unacceptable lot when offered will be falsely accepted. Consumer's risk, for the purpose of this text, is considered to be the probability of accepting a lot whose actual microbial content is substandard as specified in the plan, even though the determined values indicate acceptable quality. This is expressed by the probability of acceptance (P_a). The producer's risk is expressed by the probability of rejection $1 - P_a$ (Figure 6–1).

6.8 STRINGENCY AND DISCRIMINATION

When sampling plans are compared and their stringency in making decisions is considered, different aspects of their performance can be addressed.

Figure 6–2 illustrates an OC curve for an idealized sampling plan. This allows for total discrimination between acceptable and unacceptable lots as acceptance probabilities fall from 100% to 0 just at the chosen limit of acceptability. In practice, no sampling plan can achieve this ideal, but the steeper the curve, the closer the plan comes to approaching the ideal. Generally steeper curves can only be achieved by increasing the number of sample units (n) to be drawn from a lot (Figure 6–3).

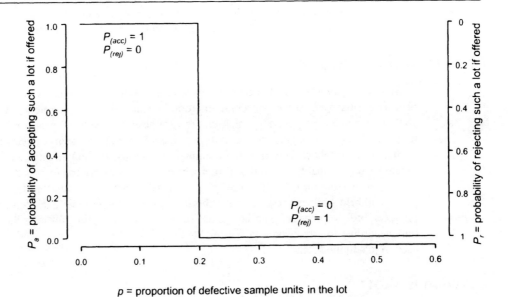

Figure 6.2 The operating characteristic curve (OC) for the idealized situation of complete discrimination between lots with a proportion of defective sample units below 20% and lots with such a proportion above 20%.

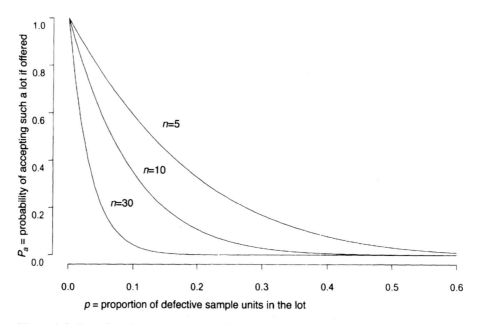

Figure 6–3 Operating characteristic curves for sampling plans with $n = 5$, $n = 10$, $n = 30$ and $c = 0$. With increasing numbers of sample units examined (n), the slope of the OC curve becomes steeper.

The ability to discriminate between acceptable and unacceptable lots should be clearly distinguished from a shift of the OC curve that is achieved by decreasing the acceptance number c. A lower acceptance number will result in a general reduction of consumer's risk by stating a different limit of acceptability, although the producer's risk will be increased.

6.9 ACCEPTANCE AND REJECTION

We have been speaking of acceptance or rejection of a lot on the basis of a sampling plan associated with some particular microbiological test. This judgment, of course, applies only to the purpose for which that test (or several such tests) was performed. A food unsuitable for one purpose may still be suitable for another; for example, if rejected for humans, it might still be suitable for animals. Or a rejected food might even, if sorted to remove objectionable material or if reprocessed, be so improved as to pass the test and become acceptable for the original purpose. Normally, therefore, a rejected lot will simply be withheld while the responsible authority decides what to do with it: return it to the producer, order reprocessing, forbid its use for human consumption, or order its destruction, according to circumstances. Throughout this text, the terms *accept* and *reject* are used in this limited sense.

6.10 WHAT IS A LOT?

Ideally, a lot is a quantity of food or food units produced and handled under uniform conditions. Under certain circumstances, it has been shown that the logarithms of the counts from a batch of food are likely to be normally distributed (Baird-Parker, 1980; Kilsby & Baird-Parker, 1983 [their Table 2]; Ingram & Roberts, 1976). The implication is that there is homogeneity within a lot, which, when considering microbial levels and distributions, rarely occurs in practice. In most instances, the distribution of microorganisms within lots of food is heterogeneous. Even in food factories, this heterogeneity makes the interpretation of sample results difficult, and it becomes even more so in circumstances where a lot is less well defined, such as in a consignment of a very large quantity of food.

It is therefore helpful if suppliers can be persuaded to give identifiable code numbers to batches (lots) of food produced over short time periods (e.g., a day or part of a day) for particular processes. The choice of coding system will vary from process to process depending on the type of process and degree of homogeneity within the batch. For example, a continuous process may produce a relatively homogeneous product, in which case a lot could include units produced continuously over a relatively long time, whereas a batch process, such as batch retorting, may require the coding of relatively few product units as one lot (e.g., one retort load).

If a consignment consisting of a mixture of production batches is treated as one lot, a rejection of an acceptable lot (i.e., the situation the producer's risk refers to) can have severe consequences, as the whole lot will be affected by this decision, even though only a few of the production batches within the lot may be of poor quality. Treating the individual production batches as lots, and coding appropriately, permits more precise identification of poor quality food and, at the expense of more analyses, can result in the rejection of less units from the whole consignment.

A lot should be composed of food produced with as little variation as possible for a given process or commodity. Because of the uncertainties in identifying a lot commercially, however, the use of the word *lot* in this book is usually in its statistical sense, as a collection of units of a product, the acceptability of which is determined by examining a sample drawn from it.

6.11 WHAT IS A REPRESENTATIVE SAMPLE?

A *representative sample* reflects, as far as is possible, the composition of the lot from which it is drawn.

How then should a representative sample be drawn? It is important to avoid bias and to draw a sufficient number of sample units to confidently make a judgment about a lot. Sampling at random is the universally recognized way of avoiding bias and is more reliable than consciously trying to draw the sample units from the various parts of a lot. The units (cartons or containers, particular weights of solid, or volumes of liquid) are drawn by using random numbers. There is, of course, no guarantee that a random sample has characteristics identical with those of the lot, but the randomness of sampling is the basis for calculation of the probability that a sample will give a biased result.

It is also possible, and may be desirable, to use a *stratified random sampling* approach, drawing a random sample of a given number of sample units from each *stratum* (e.g., each sublot or carton). The distribution of sample units on strata should correspond to the distribution of lot units on strata. This means that if each stratum contains the same quantity as every other one, the number of sample units per stratum should be the same. Otherwise, they should differ just as the proportions of the lot contained in the strata differ. Stratification is a device for handling known sources of variation and may be used where one has prior knowledge that the consignment is potentially not of uniform quality. A consignment might not be of uniform quality if portions of it are shipped in different vehicles or in different holds of a vessel, or if it is known that it is really composed of several lots representing, for example, different days of production from the same plant or different plants of the same company. The results for different strata should be assessed separately and then pooled if they appear to be homogeneous.

The calculation of acceptance probabilities is based on the assumption that random samples have been drawn and analyzed. Where physical or practical constraints prevent random sampling, such calculated probabilities may not be valid. Their reliability depends on how close reality approaches the ideal situation of random sampling. This fact should be borne in mind when subsequently interpreting results.

6.12 CONFIDENCE IN INTERPRETATION OF RESULTS

Decisions based on sampling plans are not suited to give total assurance that the decision will be correct. There are many factors that can affect our confidence in the sampling results obtained. Many of these are related to the methodologies we use. In this chapter, it is assumed that the analytical procedure is certain to detect a contaminant if present, and that it will give no false positives. In practice, these assumptions are never correct and, even for standard microbiological methods, precision values for sensitivity and specificity or repeatability and reproducibility of the analytical procedure may still be missing. Knowledge of these values sometimes would lead to a modified calculation of the OC function or to different critical limits.

The way in which the sample is handled and manipulated can also have an effect on our ability to identify accurately any contaminants present in it. The choice of handling procedure will vary according to the material under examination, the preservation system used, and the method of analysis. The choice of such handling procedures is a microbiological judgment to be made only by experts.

The statistical considerations are related to our sampling plan assumptions, as outlined previously. Particular attention should be given to the sampling plan OC curve (stringency and discrimination) and to the method of sampling, since the calculation of acceptance probabilities is only reliable if sample units are taken at random. With regard to quantitative analytical results, the design of sampling plans relies on knowledge about underlying distributions and the typical variability between sample units. This is reflected either by the calculation of acceptance probabilities or by the choice of critical limits. For further discussion of these issues, see Chapter 7. In addition, it is important to consider whether the whole sampling protocol and assumptions implied are suitable for the intended or desired decision to be made about the lot.

Finally, it is important to remember that many nonstatistical constraints will be brought into play. Issues, such as the economic cost of sampling and the need for rapid results with perishable or commercially sensitive foods, can result in selection of methods that are not statistically ideal or sampling plans with little discrimination. The usual numbers of sample units that can be found in many sampling plans in food microbiology, e.g., $n = 5$ or even $n = 1$, are mainly motivated by such issues. These low numbers represent self-imposed, nonstatistical constraints that limit the confidence we can have in the correctness of decisions based on the according sampling plans.

To be fit for the intended purpose, a sampling plan should be designed with regard to these constraints and the validity of assumptions made. The choice of the sampling plan type, critical limits, and the number of sample units to be analyzed should be chosen according to the intended purpose. Especially the last aspect, choice of n, should be reflected and guided in the light of the desired stringency of the decision-making process.

The recommended procedure would be to fix desired acceptance and rejection probabilities for lots of defined acceptable and unacceptable qualities first, and then to derive the number of sample units required for this purpose.

6.13 PRACTICAL CONSIDERATIONS

To use available resources to the best effect, all food cannot receive the same attention. The most important foods and lots should receive the most time and effort, which means that they should be subjected to the most intensive sampling. What factors govern this decision?

Hazard. The most important factor is the hazard involved. How hazardous is the type (or types) of microorganisms present? And how hazardous are the numbers likely to be present? These questions are discussed more fully in Chapter 8. The probability of accepting a lot that should be rejected should be reduced with increasing hazard.

Uniformity. If the food has relatively few loci of contamination, even if the number of contaminants is very high, then the chance of detection is low. On the other hand, if the same food is thoroughly mixed, and the contaminants are more uniformly spread, they will appear in more sample units. The chance of detection is then higher with any given sampling plan.

Stratification. If it is known that there is stratification within lots of the food (see section 6.11), a corresponding stratification can be used in selecting sampling units.

Record of consistency. A consistently good record for a food from a specific source indicates that its control is good and reliable, and hence may justify reduced sampling or even omission of sampling on occasional lots. Discretion must be used in making this decision, with increasing confidence as the record accumulates. Confidence will be maintained if the producer provides detailed records of control procedures. Skip-lot sampling is a statistical procedure that can be used under such circumstances.

Practical limitations. Since regulatory agencies never have the resources to test all imported lots, they must reduce the sampling plans to a feasible level. Most microbiological tests are laborious and slow; yet regulatory agencies cannot hold highly perishable foods pending results of analysis. Political or administrative pressures to reduce sampling will increase the probability of error.

In considering the above factors, it must be remembered that the ability to distinguish between acceptable and unacceptable lots often improves relatively little compared with the increase in the number of sample units withdrawn from the population. Indeed, in many instances, reliability only increases roughly as the square root of the number of sample units, so that multiplying the sample number by four will only halve the likelihood of making wrong decisions.

6.14 REFERENCES

Baird-Parker, A. C. (1980). The role of industry in the microbiological safety of foods. *Food Tech Aust* 32, 254–260.

Ingram, M. & Roberts, T. A. (1976). The microbiology of the red meat carcass and the slaughterhouse. *Royal Soc Health J* 96, 270–276.

Kilsby, D. C. & Baird-Parker, A. C. (1983). Sampling programmes for the microbiological analysis of food. In *Food Microbiology, Advances and Prospects*, pp. 307–315. Edited by T. A. Roberts & F. A. Skinner. Society for Applied Bacteriology Symposium Series No. 11. London: Academic Press.

Chapter 7

Sampling Plans

7.1 Introduction
7.2 Attributes Plans
7.3 Variables Plans
7.4 Comparison of Sampling Plans
7.5 References

7.1 INTRODUCTION

This text is concerned primarily with plans that may be applied to lots of food presented for acceptance at ports or other points of entry. Often, little or no information is available to the receiving agency about the method by which the food was processed or the record of previous performance by the same processor. Under these circumstances, attributes plans are appropriate. Variables sampling plans (see section 7.3), which depend upon the nature of the frequency distribution of microorganisms within lots of foods, are suitable only if this distribution is known. This limits severely their usefulness in port-of-entry sampling, but they may be particularly helpful to food producers monitoring their own production.

7.2 ATTRIBUTES PLANS

7.2.1 Two-Class Attributes Plans

A simple way to decide whether to accept or reject a food lot may be based on some microbiological test performed on several (n) sample units. This will usually be a test for the presence (positive result) or absence (negative result) of an organism. Concentrations of microorganisms can be assigned to a particular attribute class by determining whether they are above (positive) or below (negative) some preset concentration.

As explained in Chapter 6, the decision-making process is defined by two numbers. The first is represented by the letter n, and defines the number of sample units required for testing. The second number, denoted c, is the maximum allowable number of sample units yielding unsatisfactory test results, for example, the presence of the organism, or a count above the defined concentration, denoted by m, which in a two-class plan separates good quality from defective quality (Figure 7–1a). Thus, in a presence/absence decision on a lot, the sampling plan $n = 10$, $c = 2$ means that 10 sample units are taken and a test made on each; if 2 or fewer show the presence of the organism, the lot is accepted (with respect to this characteristic); but if 3 or more of the 10 show the presence of the organism, the lot is rejected, although not necessarily destroyed (see Chapter 6, section 6.9, for an explanation of rejection).

Only under circumstances where the lot is intimately mixed can data obtained from the application of a two-class plan based on presence/absence of a microorganism be used to give an approximate guide to the likely concentration of that microbe within the lot. Such

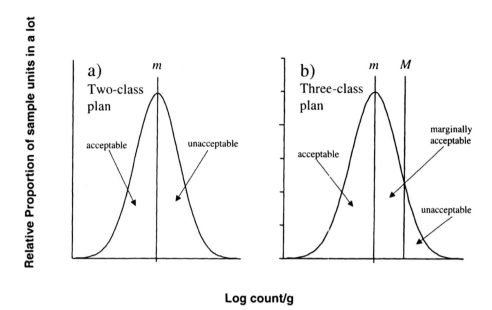

Figure 7–1 Two- and three-class attributes plans.

techniques should only be applied when the analyst is confident that there is a uniform distribution of the microorganisms of concern.

The performance of the sampling plan depends upon n and c (see Chapter 6, section 6.8). The larger the value of n at a given value of c, the better is the discrimination between lots of acceptable and unacceptable quality. Thus, compared with $n = 10$, $c = 2$, the plan $n = 15$, $c = 2$ is more discriminating, while the plan $n = 5$, $c = 2$ is more lenient. On the other hand, for given sample size n, if c is decreased, the better the food quality must be to have the same chance of being passed. Conversely, if c is increased, the plan becomes more lenient and will more often pass food lots of a given quality whenever offered for test (i.e., there is a higher probability of acceptance, P_a). Probabilities of acceptance for a set of plans are given in Tables 7–1 and 7–2. Figure 7–2 displays the operating characteristic (OC) curves for a few of these plans to illustrate the characteristics of various two-class attributes plans.

The importance of the sample size, n, is also stressed when we take a different approach to assess lot quality. Suppose a sampling plan was applied not only to decide between acceptance and rejection of the lot, but also to give an estimate of the proportion of defectives in the lot. Such estimates should be reported in terms of confidence intervals, giving a range of values and supported by the actual sampling results. The width of these ranges, i.e., the precision of the estimates, is greatly influenced by n. For illustration purposes, lower and upper limits of 95% confidence intervals are given in Table 7–3 for some combinations of sample sizes, n, and actual numbers of positive sample units, k.

Another aspect that might be considered is the influence of the size of the lot. A sample size of, say, $n = 30$ sample units might be taken at random from a lot of any size (or consignment if appropriate—see Chapter 6, section 6.10 and Chapter 17, section 17.5.3.2) for use in a two-class plan with $c = 0$. If the lot contains a very large number

Table 7-1 Two-Class Plans (c = 0): Probabilities of Acceptance (P_a) of Lots Containing Indicated Proportions of Acceptable and Defective Sample Units

Composition of lot		Number of sample units tested from the population (n)							
% acceptable (100 − p)	% defective (p)	3	5	10	15	20	30	60	100
98	2	0.94	0.90	0.82	0.74	0.67	0.55	0.30	0.13
95	5	0.86	0.77	0.60	0.46	0.36	0.21	0.05	0.01
90	10	0.73	0.59	0.35	0.21	0.12	0.04	<	<
80	20	0.51	0.33	0.11	0.04	0.01	<		
70	30	0.34	0.17	0.03	<	<			
60	40	0.22	0.08	0.01					
50	50	0.13	0.03	<					
40	60	0.06	0.01						
30	70	0.03	<						
20	80	0.01							
10	90	<*							

*< means P_a < 0.005.

of sample units, one obtains an OC curve such as that shown in Figure 7–2c. According to that OC curve, a lot with one defective unit out of 40 (proportion defective, $p = 0.025$) will be accepted about half the time ($P_a = 0.5$) when using the plan $n = 30$, $c = 0$. Calculation of this probability is based on the binomial distribution model. If the lot contains a smaller number of sample units, the probabilities of acceptance P_a for various values of p should, in principle, be calculated using a different distribution model, because the 30 sample units might now constitute an appreciable fraction of the whole lot. However, acceptance probabilities would be only slightly smaller than those shown in Figure 7–2c. This effect only becomes important when one-quarter to one-half of the lot is taken as a sample, a circumstance that rarely occurs in bacteriological analysis of lots of food.

Suppose that instead of $n = 30$, $c = 0$, we use $n = 5$, $c = 0$. The latter plan is undiscriminating, but the associated probabilities of acceptance are, again, nearly independent of the consignment size.

In summary, a sampling plan of the form described gives nearly the same degree of protection against acceptance of unacceptable lots as against rejection of acceptable ones, regardless of the size of the lot. The attributes schemes given in this book do not increase stringency as lot size increases. The sampling plan is, therefore, independent of the size of the lot, provided the lot is large in comparison to sample size. If the important criterion is the actual number of defective units, rather than the proportion of the lot that is defective, then the sampling plans given are not appropriate.

7.2.2 Three-Class Attributes Plans

Three-class attributes plans (see Bray, Lyon, & Burr, 1973) were devised for situations where the quality of the product can be divided into three attribute classes, depending upon the concentration of microorganisms within the sample units. As in two-class plans based on quantitative analytical results, counts above a concentration m, which in a three-

Table 7-2 Two-Class Plans (Selected c Values): Probabilities of Acceptance (P_a) of Lots Containing Indicated Proportions of Acceptable and Defective Sample Units

Composition of lot		n = 5			n = 10				n = 15			n = 20		
% acceptable (100 − p)	% defective (p)	c = 3	c = 2	c = 1	c = 3	c = 2	c = 1	c = 4	c = 2	c = 1	c = 9	c = 4	c = 1
98	2	1.00	1.00	1.00	1.00	1.00	0.98	1.00	1.00	0.96	1.00	1.00	0.94
95	5	1.00	1.00	0.98	1.00	0.99	0.91	1.00	0.96	0.83	1.00	1.00	0.74
90	10	1.00	0.99	0.92	0.99	0.93	0.74	0.99	0.82	0.55	1.00	0.96	0.39
80	20	0.99	0.94	0.74	0.88	0.68	0.38	0.84	0.40	0.17	1.00	0.63	0.07
70	30	0.97	0.84	0.53	0.65	0.38	0.15	0.52	0.13	0.04	0.95	0.24	0.01
60	40	0.91	0.68	0.34	0.38	0.17	0.05	0.22	0.03	0.01	0.76	0.05	<
50	50	0.81	0.50	0.19	0.17	0.05	0.01	0.06	<		0.41	0.01	<
40	60	0.66	0.32	0.09	0.05	0.01	<	0.01			0.13	<	
30	70	0.47	0.16	0.03	0.01	<		<			0.02		
20	80	0.26	0.06	0.01	<						<		
10	90	0.08	0.01	<									
5	95	0.02	<*										

n = number of sample units tested from the population
*< means $P_a < 0.005$.

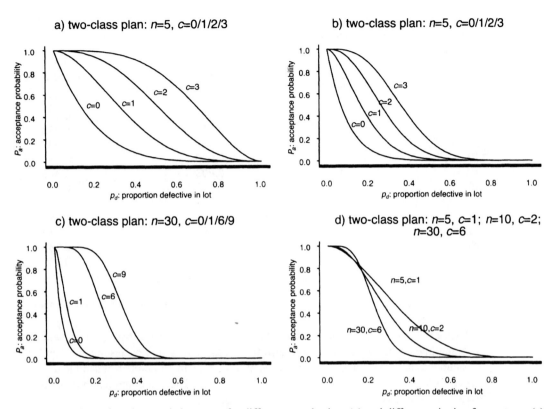

Figure 7–2 Operating characteristic curves for different sample sizes (n) and different criteria of acceptance (c) for two-class plans.

class plan separates good quality from marginally acceptable quality, are undesirable, but some can be accepted. However, a count above a second concentration M for any sample unit is unacceptable, and if any count for the n sample units from a lot exceeds M, that lot is rejected (Figure 7–1b). This concept is based on the idea that analytical results for sample units drawn from a lot are of a quantitative nature. In this case, quantities of microorganisms in sample units can be described in terms of frequency distributions that can be characterized by some measures of location and spread.

Figure 7–3 illustrates the effect of various frequency distributions of microbial content within a lot on the location of m and M for three-class plans. Curve 1 represents an entirely satisfactory lot, with low numbers of bacteria generally, and thus a low average count with little variation and no counts exceeding m. Curve 2 represents a lot with a similar average count, but with a much wider variation, so that a small proportion of sample units would have counts exceeding m, though none exceed M. If the proportion in the range m to M were small, the situation would be acceptable; if this proportion were larger, it might still be acceptable, but it would serve as a warning call to the producer, as tending toward the situation shown in curve 3. Curve 3 represents a lot with a higher average count and larger variation, such that a small proportion of sample units exceeds M and would result in immediate rejection, while a substantial proportion falls in the range m to M, which itself could also suffice to justify rejection (see Chapter 6, section 6.9, for

Table 7–3 Lower and Upper Limits of 95% Confidence Intervals for the Estimated Proportion Defective Based on k Positive Results When n Sample Units Are Analyzed

n	k	Lower limit	Upper limit
5	0	0.000	0.451
5	1	0.005	0.716
5	2	0.053	0.853
5	3	0.147	0.947
10	0	0.000	0.259
10	1	0.003	0.445
10	2	0.025	0.556
10	3	0.067	0.652
15	0	0.000	0.181
15	1	0.002	0.319
20	0	0.000	0.139
20	1	0.001	0.249

explanation of the term *rejection*). Curve 4 represents a lot of even greater unacceptability, requiring rejection.

Hence the definition of a three-class sampling plan incorporates two limits, m and M, M being higher than m, which distinguish three classes of sampling results, the number of sample units to be drawn from the lot n, and the maximum number of sample units c that are allowed to fall into the region between m and M. The maximum number that may exceed M is always set to 0 in the plans in this book.

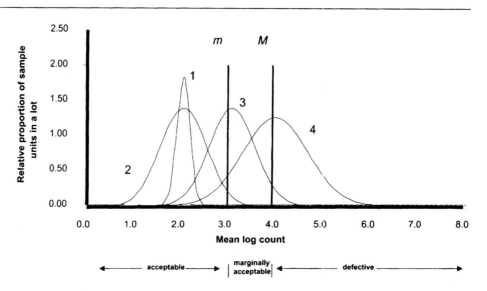

Figure 7–3 Three-class attributes plans.
m = count above a defined concentration, separating good quality from marginally acceptable quality.
M = count above a second defined concentration, separating marginally acceptable quality from unacceptable quality.

Accordingly, in the three-class plans there are again only two numbers, n and c, from which it is possible to find the probability of acceptance, P_a, for a food lot of given microbiological quality. To describe the lot quality, we consider all sample units that could be drawn from the lot, which must yield counts in three classes: 0 to m, m to M, and above M. Since the proportions in the lot for the three classes must total 1, one need only specify two of them in describing lot quality. We might call these proportions the proportion *defective*, i.e., above M (p_d), and the proportion *marginally acceptable*, i.e., m to M (p_m). The proportion *acceptable*, equal to or less than m, must be 100% minus the sum of p_d and p_m. By appropriate calculations, we can find the probability of acceptance, P_a, for a given lot quality for any specified sampling plan. For example, for the plan $n = 10$, $c = 2$, P_a will be 0.21 for a lot distribution for which 20% of the sample counts are marginally acceptable ($p_m = 20\%$) and 10% defective ($p_d = 10\%$). That is, on the basis of the particular values decided upon for m and M, only about 21 lots out of 100 of that quality will be accepted, because they have no defective counts and two or fewer marginally acceptable counts out of the 10 samples chosen from the lot. The other lots will all be rejected.

Probabilities associated with a collection of three-class plans are shown in Table 7–4 for various lot qualities. The scheme depends for acceptance or rejection not only on the proportion of defective material (p_d), but also on the proportion of marginally acceptable product (p_m). For example, a lot with 20% marginal quality and 10% defective quality with a sampling plan based on $n = 10$, $c = 2$ will, on average, be accepted 21% of the times it is put forward for testing. Using the identical plan $n = 10$, $c = 2$, a lot containing 0% defective material, but 40% marginal quality has a lower chance of acceptance (17%).

Table 7–4 gives only some examples of acceptance probabilities for selected combinations of proportions p_m and p_d. To gain an impression of the overall behavior of a three-class attributes plan, the complete operating characteristic function should be referred to.

Compared with two-class plans, the OC functions of three-class plans are more complex and more difficult to visualize since their values depend on combinations of two proportions, p_m and p_d, and not only on one. Because of these dependencies and the variety of lot qualities that can occur, for a three-class sampling plan, the OC function should be plotted as an OC surface, either in a three-dimensional graph, or as a contour map in the two-dimensional (p_m, p_d) area with contour lines for selected acceptance probabilities. Such OC function maps are shown in Figure 7–4 for three-class sampling plans with $n = 5$ and different acceptance numbers $c = 0$, $c = 1$, $c = 2$, and $c = 3$.

All lots with combinations of p_m and p_d lying on the same contour line in such a graph have the same probability of acceptance that is indicated at the end of the line. If, for instance, the three-class plan $n = 5$, $c = 1$ is applied, all kinds of lots with (p_m, p_d) combinations on the outermost line are accepted with a probability of only 0.1, or 10% of the times such a lot is examined. Thus, the three-class attributes scheme is affected to some extent by the frequency distribution of microorganisms within the batch, but the advantages of the scheme are its simplicity and general applicability, which make it appropriate to port-of-entry sampling.

However, there is a need to elaborate sound methods to set the values of m and M. These should be related to actual concentrations of microorganisms and the frequency distributions of analytical results. There are statistically based techniques for achieving this, although assumptions must be made. However, one example, based upon assumptions that can readily be checked and have been found (historically) to be reasonable,

Table 7-4 Three-Class Plans: Probabilities of Acceptance (P_a) of Lots Containing Indicated Proportions for Selected Numbers of Sample Units and c Values (p_d = percent defective, p_m = percent marginal)*

p_d	p_m								
	10	20	30	40	50	60	70	80	90
n = 5, c = 3									
50	0.03*	0.03	0.02	0.01	<†				
40	0.08	0.07	0.06	0.04	0.02	<			
30	0.17	0.16	0.15	0.12	0.07	0.03	<		
20	0.33	0.32	0.31	0.27	0.20	0.12	0.04	<	
10	0.59	0.58	0.56	0.52	0.43	0.32	0.18	0.06	<
5	0.77	0.77	0.75	0.69	0.60	0.47	0.31	0.14	0.02
0	1.00	0.99	0.97	0.91	0.81	0.66	0.47	0.26	0.08
n = 5, c = 2									
50	0.03	0.02	0.01	<					
40	0.08	0.06	0.04	0.02	<				
30	0.16	0.14	0.11	0.06	0.02	<			
20	0.32	0.29	0.24	0.16	0.09	0.03	0.01	<	
10	0.58	0.55	0.47	0.36	0.23	0.12	0.05	0.01	<
5	0.77	0.72	0.63	0.50	0.35	0.20	0.09	0.02	<
0	0.99	0.94	0.84	0.68	0.50	0.32	0.16	0.06	0.01
n = 5, c = 1									
50	0.02	0.01							
40	0.06	0.04	0.01	<					
30	0.14	0.09	0.05	0.02	<				
20	0.29	0.21	0.13	0.06	0.02	0.01	<		
10	0.53	0.41	0.27	0.16	0.07	0.03	0.01	<	
5	0.70	0.55	0.38	0.23	0.12	0.05	0.01	<	
0	0.92	0.74	0.53	0.34	0.19	0.09	0.03	0.01	
n = 10, c = 3									
40	0.01								
30	0.03	0.02	0.01	<					
20	0.10	0.08	0.05	0.02	<				
10	0.34	0.29	0.20	0.10	0.03	0.01	<		
5	0.59	0.51	0.36	0.20	0.08	0.02	<		
0	0.99	0.88	0.65	0.38	0.17	0.05	0.01	<	
n = 10, c = 2									
30	0.02	0.01	<						
20	0.09	0.06	0.02	0.01	<				
10	0.32	0.21	0.10	0.04	0.01	<			
5	0.55	0.39	0.20	0.08	0.02	<			
0	0.93	0.68	0.38	0.17	0.05	0.01	<		
n = 10, c = 1									
30	0.02	<							
20	0.07	0.03	0.01	<					
10	0.24	0.11	0.04	0.01	<				
5	0.43	0.21	0.08	0.02	<				
0	0.74	0.38	0.15	0.05	0.01	<			

*Each of these blocks of numbers, relating P_a to p_m and p_d represents a three-dimensional relation called an OC surface, corresponding to the two-dimensional OC curve.
†< means $P_a < 0.005$.

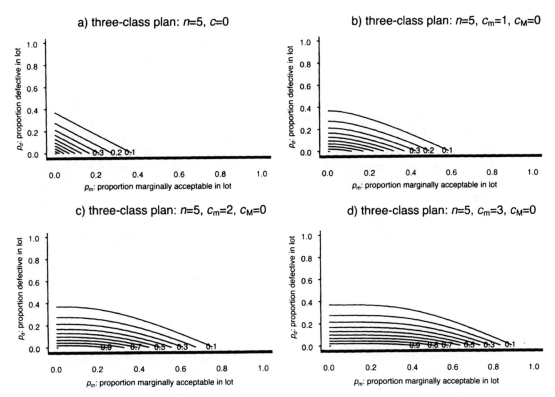

Figure 7-4 Contour maps of operating characteristic functions of three-class attributes plans for sample size $n = 5$ and different criteria of acceptance (c).
*Numbers within graphs (e.g., 0.1, 0.2, 0.3 in graph (a)) represent probability of acceptance.

is found in Dahms & Hildebrandt (1998), which is explained in more detail in section 7.4.4.

7.3 VARIABLES PLANS

Where the underlying distribution of microorganisms within the lot is known, or can be assumed, there is the option to use variables sampling plans. When properly applied, such plans can prove more useful under some conditions than the attributes plans for a particular purpose. Variables plans make full use of the microbial counts, rather than ascribing the counts to categories or ranges.

7.3.1 Identification

To define a variables plan, the underlying frequency distribution for the microorganisms of concern within the sample units in a lot must be known. There are many types of frequency distributions, and they differ in complexity. Some of the simpler distributions are symmetrical in shape and can be described by their mean and some measure of the distribution about that mean. The normal, or Gaussian, distribution is such an example. The normal distribution is defined by its mean value (which is also the median) for the range of concentrations found, and a measure σ, the standard deviation, which defines the possibility of finding any other concentration in a unit. More complex distributions may not be symmetrical, so that the mean is unlikely to equal the median.

It is not the purpose of this book to describe the options available to define frequency distributions. What is essential to apply a variables plan is that the frequency distribution is known. To achieve this, data may be required to support the application of a particular distribution. The more complex the frequency distribution (i.e., more parameters are needed to define it), the more data are needed to gain confidence that it is an appropriate distribution to apply. An example of a distribution often observed for microorganisms in foods, the Log-Normal distribution, is used below (section 7.3.3). It is worth noting that the scale has been adjusted to obtain a normal distribution. This adjustment results in a symmetrical, two parameter (mean and standard deviation) distribution, achieved by using log(concentration) instead of concentration.

It is important to remember that any measurement of the parameters for a distribution is based upon a sample and is therefore an estimate of those parameters. These measured parameters must accommodate the uncertainty implicit in the measurement. The smaller the sample, the larger the likely error could be. The example given in section 7.3.3 illustrates this principle by allowing for sample size in its decision matrix.

7.3.2 Prescribing Confidence in Decisions

Making critical microbiological decisions about the safety or quality of a lot of food involves three steps. The first is to define the acceptable limits for the lot, the second is to specify the confidence with which we wish to identify acceptable and unacceptable lots, and the third is to choose the appropriate sampling plan. The following is an example of the way in which a variables plan may be designed. In this case, the decision rule is based upon an assumption that the underlying distribution of contaminants in the lot is log-normal (i.e., the log of the concentrations is normally distributed). While this assumption is often correct, in practice, its justification needs to be clear and recorded. Assuming a log-normal distribution, sampling plans based on the characteristics of this distribution can be used to develop acceptance sampling plans.

7.3.3 Operation

It is necessary to obtain and handle samples and sample units in the same way as for attributes plans. The log-transformation of the concentration measurements is used to compute the sample mean ($\bar{x}$) and standard deviation (s). These two values are then used to make the decision whether to accept or reject the lot. The lot is rejected if $\bar{x} + k_1 s > V$, where V is a log-concentration related to safety/quality limits. The value k_1 is obtained by reference to appropriate tables and is chosen to define the stringency of the plan for a given number of sample units, n.

Selection of k_1

Table 7–5 contains a range of k_1 values for sample unit numbers between 3–10. To choose k_1, it is necessary to decide on the maximum proportion p_d of the units in the lot that can be accepted with a concentration above the limit value, V. Having selected p_d, the desired probability P can be chosen, where P is the probability of rejecting a lot that contains at least a proportion p_d above V.

Table 7–5 k_1 Values Calculated Using the Non-Central t-Distribution–Safety/Quality Specification (Reject if $\bar{x} + k_1 s > V^*$)

Probability (p) of rejection	Proportion (p_d) exceeding V	Number of sample units n							
		3	4	5	6	7	8	9	10
0.95	0.05	7.7	5.1	4.2	3.7	3.4	3.2	3.0	2.9
	0.1	6.2	4.2	3.4	3.0	2.8	2.6	2.4	2.4
	0.3	3.3	2.3	1.9	1.6	1.5	1.4	1.3	1.3
0.90	0.1	4.3	3.2	2.7	2.5	2.3	2.2	2.1	2.1
	0.25	2.6	2.0	1.7	1.5	1.4	1.4	1.3	1.3

*$\bar{x}$ = sample mean; V = log-concentration related to safety/quality limits.

For example, if five sample units are analyzed per lot, then the k_1 value can be chosen from Table 7–5. If a lot in which 10% of sample units exceeded V is to be rejected with a probability of 0.95, then the k_1 value 3.4 would be used.

In practice, the two values p_d and P will be selected along with the value V. The scheme then allows n to be selected over the range 3–10. The larger n becomes, the lower the chance of rejection of an acceptable lot.

Selection of the limit value, V

The limit value V is selected by the microbiologist as the safety or quality limit, expressed as log-concentration. This value is likely to be numerically very similar to the value M used in the three-class attributes plans (section 7.2.2).

Table 7–6 gives the results for the aerobic plate count (APC) analyses of five sample units obtained from a lot of poultry. An appropriate variables sampling plan might be $P = 0.90$, $p_d = 0.25$, with a limit value of $V = 7$. The k_1 value, obtained from Table 7–5, is 1.7. Applying the formula $\bar{x} + k_1 s$ gives $5.039 + (1.7 \times 0.378)$, which equals 5.682. This value is less than the limit value of 7, and the lot is therefore accepted.

The use of variables plan for good manufacturing practice

Food producers often find it advantageous to specify a good manufacturing practice (GMP) standard. It may be possible to apply the variables plan under these circumstances, applying the formula outlined previously. The criterion is to accept the lot if $\bar{x} + k_2 s < v$. The k_2 value for the GMP plan is obtained from Table 7–7. The values P and p_d are selected as before and the appropriate k_2 value is obtained. The limit value, v, will be very similar numerically to the limit value, m, used in the three-class attribute plan.

For a more extensive treatment of this topic, see Kilsby (1982), Kilsby et al. (1979), and Malcolm (1984).

Table 7–6 An Example of Aerobic Plate Counts for a Sample of Poultry ($n = 5$)

APC	$\log_{10}$ (APC)	mean log ($\bar{x}$)	standard deviation (s)
40,000	4.602		
69,000	4.839		
81,000	4.909	5.039	0.378
200,000	5.301		
350,000	5.544		

Table 7–7 k_2 Values Calculated Using the Non-Central t-Distribution–GMP Limit (Accept if $\bar{x} + k_2 s < v^\dagger$)

Probability (P) of acceptance	Proportion (p_d) exceeding v	Number of sample units n							
		3	4	5	6	7	8	9	10
0.90	0.05	0.84	0.92	0.98	1.03	1.07	1.10	1.12	1.15
	0.10	0.53	0.62	0.68	0.72	0.75	0.78	0.81	0.83
	0.20	0.11	0.21	0.27	0.32	0.35	0.38	0.41	0.43
	0.30	0.26*	0.13*	0.05*	0.01	0.04	0.07	0.10	0.12
	0.40	0.65*	0.46*	0.36*	0.30*	0.25*	0.21*	0.17*	0.16*
	0.50	1.09*	0.82*	0.69*	0.60*	0.54*	0.50*	0.47*	0.44*
0.75	0.01	1.87	1.90	1.92	1.94	1.96	1.98	2.00	2.01
	0.05	1.25	1.28	1.31	1.33	1.34	1.36	1.37	1.38
	0.10	0.91	0.94	0.97	0.99	1.01	1.02	1.03	1.04
	0.25	0.31	0.35	0.38	0.41	0.42	0.44	0.45	0.46
	0.50	0.47*	0.38*	0.33*	0.30*	0.27*	0.25*	0.24*	0.22*

*Negative values
$^\dagger \bar{x}$ = sample mean; v = log-concentration limit value

7.4 COMPARISON OF SAMPLING PLANS

7.4.1 General Remarks

The decision for, or the design of, a suitable sampling plan depends on the given purpose, i.e., on the sampling material, the type of microbiological result being assessed, and on available prior information on production processes and frequency distributions of sampling results in lots. In the following paragraphs, some statistical aspects of the choice of a sampling plan will be discussed comparing two-class with three-class attributes plans and three-class attributes plans with variables plans.

Only when the result of a microbiological analysis is given as a count, or in another quantitative manner, is there a choice between types of sampling plans. For mere qualitative results (presence-absence tests), only two-class plans are applicable.

When dealing with quantitative analytical results for sample units in a lot, questions arise concerning the frequency distributions of sample results and whether there is any previous information on shape, location, and spread of these distributions. Is a typical distribution/shape expected to occur? Is the production process known and well documented? Especially for the design of variables sampling plans, some knowledge and data concerning the production processes and variations in distributions that may occur between lots is required. Because variables plans are based on the assumption that log-transformed sampling results follow a normal distribution, they should only be used when this assumption can be justified. For such situations, the performance of attributes and variables plans can be compared.

The following considerations are restricted to such a situation. Sampling plans will be compared by means of their OC function, which is calculated and plotted for various scenarios. Because lot distributions are assumed to be of the normal type (after log-transformation), lot quality is described by the mean concentration of microbes for all units in the lot, μ, and the standard deviation, σ, as a measure of variation. Therefore, acceptance probabilities are calculated for lots with varying μ and σ. With σ held fixed, the OC curve of a sampling plan can be plotted as a function of varying mean concentrations μ for all three types of sampling plans.

7.4.2 Determining the Concentration of Microorganisms Controlled by Attributes Sampling Plans

As outlined in Chapter 3, to compare the equivalence of different control measures, it is necessary to be able to relate their performance in terms of achieving an FSO. In other words, the performance of each control measure needs to be expressed as the resultant frequency or concentration of a microbiological hazard in a food. This is essential if the use of microbiological criteria is to be validated as a control measure to achieve a given FSO through controlling the initial level of a hazard (see Chapter 3, Section 3.5.1, Example 5).

The method for relating the performance of attributes plans to concentration is to use the frequency distribution of analytical results in sample units to establish the proportion of defective samples, as proposed by Hildebrandt et al. (1995). A normal distribution for the log concentration of microbes is assumed, and the area under the normal density function below m is used to define the value for the proportion acceptable. The area between m and M defines the value for proportion marginally acceptable (p_m), and the area above M (or m for a two-class plan) defines the value for proportion defective (p_d). The mathematics involved in calculating the three proportions for a given mean log concentration are detailed in Legan et al. (2001).

OC curves expressed in terms of mean concentration are developed by fixing the standard deviation σ, and then increasing the mean of a normal distribution through a range of values. Figure 7–5 illustrates this for a two-class plan with $n = 5$, $c = 0$, $m = 1.0$ log cfu g^{-1}. A distribution with σ = 0.8 and three different means is shown in Figure 7–5a. The σ value of 0.8 is chosen based on published concentrations of mesophilic *Clostridium* spores in raw pork, beef, and chicken (Greenberg et al., 1966) and similar observations in other food materials. All parts of the distribution above m in each position are defective. The proportion defective in the distribution in each position (or in a position defined by any other mean) is plotted against mean log count to show how the proportion defective increases with mean log count (Figure 7–5b). Finally, the operating characteristic curve for the specified plan is used to determine the probability of acceptance from the proportion defective at each mean log count. This probability of acceptance is plotted against mean concentration (Figure 7–5c).

The protection each sampling plan gives can then be expressed in terms of the mean concentration of microorganisms associated with a defined probability of accepting the sampled material. In Chapter 8, this approach will be used to compare the concentration of microorganisms controlled under different cases.

7.4.3 Comparison of Two-Class and Three-Class Attributes Plans

Table 7–8 compares the operating characteristics of the two types of attributes plans recommended in this text, on the basis of equal sample sizes n, acceptance numbers c, and lot qualities. To facilitate comparison, lot quality is measured as the proportion of the lot worse than level m and, correspondingly, the same value of c is taken for the two-class plan as the c of marginally acceptable quality for the three-class plan.

The two-class plans do not distinguish values between m and M from those above M. If not more than c sample units give results above m, the lot is acceptable, regardless of

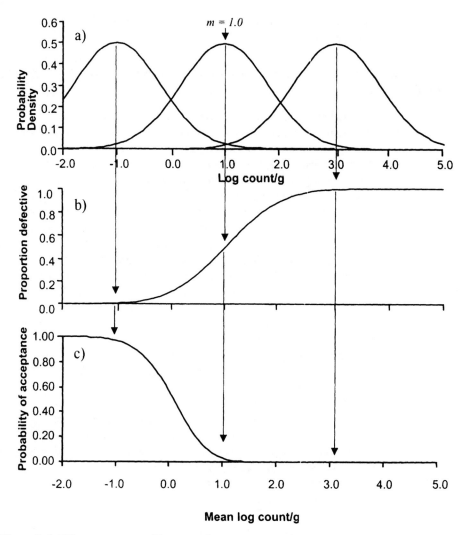

Figure 7–5 OC curves expressed in terms of mean concentration.

how far individual results exceed m. But the corresponding three-class plan does make a distinction, by including an extra subdivision of lot quality since the limit M separates marginally acceptable from defective quality.

By comparing OC surfaces for three-class plans with a fixed number of sample units n, but with varying values for c (Figure 7–4), it becomes obvious that the surface heights change mainly in the p_m-direction, i.e., for varying proportions of marginally acceptable quality in the lot. The reason is that the number of sample units that is allowed to exceed M remains constant at 0. In fact a three-class plan might be interpreted as a mixture of two two-class plans: a two-class plan (n, c) referring to the limit m, and a two-class plan $(n, 0)$ referring to the limit M. In extreme situations, one of these two-class plans can be dominating the decision process. Generally, however, the actual performance of a three-class plan depends on the variety of combinations of p_m and p_d that are likely to occur in practice.

Table 7-8 Probabilities of Acceptance (P_a) for Two- and Three-Class Attributes Sampling Plans When the Lot Examined Contains 5%, 20%, or 50% of Sample Units > m

Sampling plan		5% > m			20% > m				50% > m			
		Two-class plan	Three-class plan		Two-class plan	Three-class plan			Two-class plan	Three-class plan		
n	c	plan	0% > M*	5% > M	plan	0% > M	10% > M	20% > M	plan	0% > M	25% > M	50% > M
5	0	0.77	0.77	0.77	0.33	0.33	0.33	0.33	0.03	0.03	0.03	0.03
5	1	0.98	0.98	0.77	0.74	0.74	0.53	0.33	0.19	0.19	0.11	0.03
5	2	1.00	1.00	0.77	0.94	0.90	0.58	0.33	0.50	0.50	0.19	0.03
5	3	1.00	1.00	0.77	0.99	0.99	0.59	0.33	0.81	0.81	0.23	0.03
10	0	0.60	0.60	0.60	0.11	0.11	0.11	0.11	<	<	<	<
10	1	0.91	0.91	0.60	0.38	0.38	0.24	0.11	0.01	0.01	0.01	<
10	2	0.99	0.99	0.60	0.68	0.68	0.32	0.11	0.05	0.05	0.02	<
10	3	1.00	1.00	0.60	0.88	0.88	0.34	0.11	0.17	0.17	0.03	<
15	0	0.46	0.46	0.46	0.04	0.04	0.04	0.04	<	<	<	<
20	0	0.36	0.36	0.36	0.01	0.01	0.01	0.01	<	<	<	<
60	0	0.05	0.05	0.05	<	<	<	<	<	<	<	<

*Note that any % > M is included in the % > m; e.g., for the case where 20% > m and 10% > M, it follows that 10% is between m and M. < means P_a < 0.005.

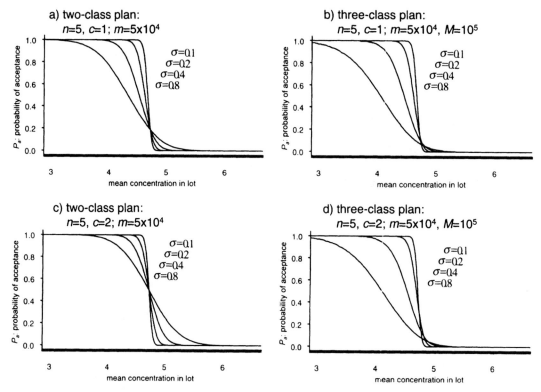

Figure 7–6 Comparison of two- and three-class attributes plan with different criteria of acceptance (c) when distribution of log-transformed analytical results is assumed to be normal.

Hildebrandt et al. (1995) studied the performance aspects of two-class and three-class sampling plans in the case that log-normal lot distributions can be assumed by comparing the two-class plan $n = 5$, $c = 1$; $m = 5 \times 10^4$ cfu ml$^-$ with the three-class plan $n = 5$, $c = 1$; $m = 5 \times 10^4$, $M = 10^5$ cfu ml^{-1}.

Four different types of lots characterized by lot standard deviations of $\sigma = 0.1$, $\sigma = 0.2$, $\sigma = 0.4$, and $\sigma = 0.8$ were considered to study the impact of the standard deviation of the lot in relation to the distance between m and M on the operation of the three-class sampling plan. In Figure 7–6(a), the OC curves for the two-class plan are shown; Figure 7–6(b) contains the OC curves for the three-class plan. The OC function values are calculated by deriving the proportions of marginal and defective quality, p_m and p_d, resulting from the various possible combinations of μ and σ with given microbiological limits $m = 5 \times 10^4$ and $M = 10^5$.

As can be seen for lot distributions with a low standard deviation ($\sigma = 0.1$) in relation to the distance between m and M, there is hardly any difference in performance between the two-class and the three-class sampling plans. As the within-lot standard deviation is increased in relation to the distance between m and M, the OC difference becomes larger, showing reduced acceptance probabilities when the three-class plan is applied.

If the number of acceptable units in the sample, c, is changed from $c = 1$ to $c = 2$, the effect on the operation of two-class plans is much more obvious than the effect on three-class plans (Figures 7–6(c) and 7–6(d)), especially for lots with higher standard devia-

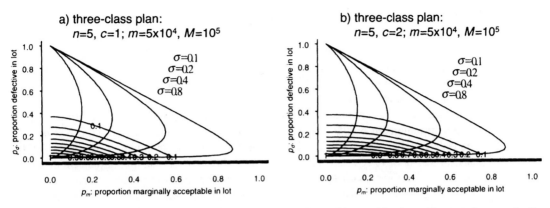

Figure 7-7 Contour maps for three-class attributes plan and possible combinations of proportion marginally acceptable (p_m) and defective (p_d) in lot, for various standard deviations (σ) when distribution of analytical results is assumed to be normal.

tion. This example demonstrates the dominance of the $c_M = 0$ rule (the null-tolerance concerning M, incorporated in three-class plans when lots are examined that are characterized by high standard deviation in relation to the distance between m and M).

A plot of the possible combinations of p_m and p_d in a given sampling situation together with the contour lines of the three-class plan serves to summarize these points (Figure 7-7). When lots are homogenous (standard deviation is low in relation to the distance between m and M, or m and M are different in relation to lot standard deviations likely to occur), the three-class plan (n, c) will operate like a simple two-class plan with (n, c) accordingly. When lots are heterogeneous (standard deviation is high in relation to the distance between m and M, or the distance from m to M is narrow in relation to lot standard deviations likely to occur), the three-class plan will operate like a two-class plan with (n, $c_M = 0$). Hence the performance of three-class plans depends not only on the combination of n and c, but on m and M and their distance in relation to lot heterogeneity.

7.4.4 Construction of Three-Class Plans Using Previous Information

The considerations in the preceding section started with given microbiological limits m and M without questioning the reasons for their selection. Often, however, these limits are described as the maximum level of target organisms m acceptable under conditions of good manufacturing practice (GMP) and the limit of target organisms M, that, if exceeded, is considered unacceptable, i.e., defective.

This reference to GMP conditions and to empirical studies describing frequency distributions for quantitative analytical results that are achievable under these conditions implies that the design of sampling plans could be based on knowledge about production technologies leading to values for m and M that take into account the maximum mean concentration of contaminants, as well as the maximum extent of heterogeneity, under conditions of GMP.

As an example, guidance for defining the microbiological limits with regard to GMP conditions has been given by Dahms & Hildebrandt (1998). They addressed the problem that there might be an unduly high probability of rejecting a lot still meeting GMP conditions if the distance between m and M is too narrow in relation to the acceptable heterogeneity. Especially with regard to nonpathogenic microorganisms such as total bacterial count or indicators of hygiene that represent no health risk for the consumer, it is questionable whether to reject a lot meeting GMP requirements solely because of a single sampling result exceeding M. In this context, the difference $(M - m)$ should be chosen such that lots characterized by marginal concentrations of microorganisms and by acceptable or even unavoidable heterogeneity run only a minor, and known, risk of rejection due to a single sample lying above M.

Based on the assumption that sampling results (after log-transformation) follow a normal distribution, an *indifferent lot* for the two-class sampling plan $n = 5$, $c = 2$ is first considered. The term indifferent lot indicates that there is an equal probability of 0.5 to accept the lot or to reject it. The definition of this two-class sampling plan implies that an indifferent lot is characterized by a mean concentration of microorganisms just at the limit m: $\mu = m$, i.e., just at the maximum acceptable mean concentration under conditions of GMP. As soon as lot mean concentrations exceed this limit, the probability of rejecting the lot will be greater than the probability of accepting it. Therefore, the hypothesis tested with this two-class plan is one concerning the mean concentration of contaminants of the lot being examined.

Application of a three-class plan to the same lot with $n = 5$, $c = 2$, i.e., the introduction of a second limit M and the requirement $c_M = 0$, can lead to a reduction of acceptance probabilities. However, whether the resulting difference between using the two-class and the three-class plans is relevant or not depends on the distance between m and M in relation to lot heterogeneity (section 7.4.3). With regard to these relationships, it is proposed to define the additional risk of rejecting an indifferent lot with a given, acceptable, heterogeneity (standard deviation) as a first step, i.e., to define the required reduction of acceptance probability for a lot with marginal mean concentration of contaminants $\mu = m$ and marginal spread σ. As a second step, a value for the upper limit M should be chosen that meets this requirement.

For the situation in which previous information indicates that log-transformed sampling results follow a normal distribution with a known standard deviation σ that is achievable under GMP conditions, Dahms & Hildebrandt (1998) derived a formula to calculate the required distance between m and M:

$$M - m = u_{1-p_d^*} \sigma$$

Here $u_{1-p_d^*}$ is the $(1-p_d^*)$-quantile of the standard normal distribution, with p_d^* being the marginal acceptable proportion of defectives exceeding M. This value can be calculated as:

$$p_d^* = \frac{5}{8} - \sqrt{\frac{25}{64} - \frac{4}{5} \cdot a}$$

for $n = 5$, $c = 2$ and the additional risk to reject an indifferent lot given as a. Table 7–9 lists some values for the distance between m and M for different combinations of a and σ.

These considerations illustrate how to select M, once m has been set. This procedure is mainly oriented toward the design of sampling plans for nonpathogenic microorganisms such as indicators of hygiene. However, the relationship between M and m could be used in a similar way for pathogens. One would simply start by first setting M with regard to safety, and choosing m accordingly.

Following a procedure like this, a three-class sampling plan can be constructed to meet defined requirements concerning its stringency in comparison with the equivalent two-

Table 7–9 Distances between *m* and *M* for Given Additional Risk *a* To Reject Indifferent Lots with Acceptable Heterogeneity (Standard Deviation) σ for a Three-Class Sampling Plan with *n* = 5, *c* = 2

a	p_d^*	$u_{1-p_d^*}$	M − m $\sigma = 0.2$	$\sigma = 0.4$	$\sigma = 0.8$
0.01 (1%)	0.006	2.51	0.502	1.004	2.008
0.05 (5%)	0.033	1.84	0.368	0.736	1.472
0.10 (10%)	0.068	1.49	0.298	0.596	1.192

class plan. A characteristic of the three-class plan is the statement of two hypotheses that are implicitly tested when this sampling plan is applied—one concerning the marginal mean concentration of contaminants of an acceptable lot as *m*, and the other concerning the marginal spread that is acceptable by fixing the distance between *m* and *M*.

7.4.5 Comparison of Three-Class Attributes and Variables Plans

The three-class plan is a simple plan to apply. It relies upon ascribing a concentration measurement to one of three broad concentration bands. To achieve this simplicity, the plan sacrifices discrimination. For example, if *m* is 1,000 and *M* is 10,000, then the three-class attributes plan assigns the same level of concern to 1,001 and 9,999 contaminants in a sample unit. Also, it assigns a totally different level of concern to 999 and 1,000 contaminants. The variables plan described above has the advantage that it has high discrimination between individual concentration measurements. On the other hand, the variables plan is mathematically more complex to operate and to understand, and its performance depends upon the validity of the assumptions made about the frequency distribution.

Taking the three-class plan $n = 5$, $c = 2$; $m = 5 \times 10^4$, $M = 10^5$ as an example, the operation characteristic of this attributes sampling plan can be compared with that of a variables sampling plan with $n = 5$, $V = 10^5 = M$, $p_d = 0.05$, and $P = 0.95$. Figure 7–8 shows the OC curves for the variables sampling plans when the lot standard deviation σ is assumed to be known in comparison with the OC curves for the three-class sampling plan already discussed under section 7.4.3. It should be noted that the decision rule of the variables plan is unable to respond if the actual standard deviation is larger than the assumed σ, whereas the decision rule of the three-class plan remains the same, responding to higher measured concentrations. In this comparison, we can see that for homogenous lots, both plans have a high stringency in discriminating between lots of acceptable and unacceptable quality. However, whereas the three-class plan discriminates at the marginal lot mean concentration *m*, the variables plan discriminates at mean concentrations nearer to $V = M$, as this is the starting point to define the marginal mean contamination $V − k \times \sigma$ that is relevant for this type of plan.

With increasing lot standard deviation, the slopes of both OC curves become less steep, this effect being stronger for the three-class plan. The variables plan remains more stringent, i.e., acceptance probabilities for lots of acceptable quality remain quite high, whereas they fall more rapidly than those of the three-class plan as soon as lot quality changes from acceptable to unacceptable.

Closer operation comparisons of these types of sampling plans can be achieved if previous information is used in the design of the three-class plan, especially in choosing the appropriate distance between *m* and *M*. However, it is important to realize that these two

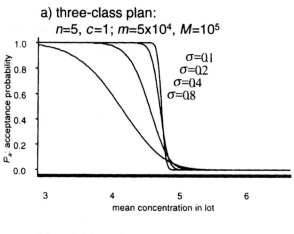

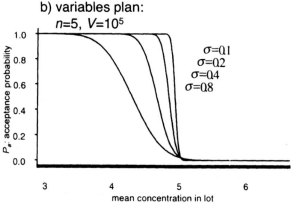

Figure 7–8 Comparison of three-class attributes plan and variables plan with standard deviation of concentrations in lot assumed to be known.

approaches are fundamentally different. Because the bases for decision making are fundamentally different, the two approaches should not be expected to be generally equivalent. The attributes schemes were designed for situations in which no assumptions about underlying distributions could be made (e.g., port-of-entry), and they will perform well under those circumstances against their set parameters. When it is possible to have high confidence in underlying frequency distributions (e.g., in food manufacturing units), alternative approaches may allow for different sets of assumptions and more cost-effective sampling plans related to these assumptions. Therefore, choice of sampling plan should depend upon knowledge of the lot and its intended use.

7.5 REFERENCES

Bray, D. F., Lyon, D. A. & Burr, I. W. (1973). Three-class attributes plans in acceptance sampling. *Technometrics* 15, 575–585.

Dahms, S. & Hildebrandt, G. (1998). Some remarks on the design of three-class sampling plans. *J Food Prot* 61, 757–761.

Greenberg, R. A., Tompkin, R. B., Bladel, B. O. *et al.* (1966). Incidence of mesophilic *Clostridium* spores in raw pork, beef and chicken in processing plants in the United States and Canada. *Appl Microbiol* 14, 789–793.

Hildebrandt, G., Böhmer, L. & Dahms, S. (1995). Three-class attributes plans in microbiological quality control: Contribution to the discussion. *J Food Prot* 58, 784–790.

Kilsby, D. (1982). Sampling schemes and limits. In *Meat Microbiology*, pp. 387–421. Edited by M. H. Brown. London: Applied Science Publishers.

Kilsby, D., Aspinall, L. J. & Baird-Parker, A. C. (1979). A system for setting numerical microbiological specifications for foods. *J Appl Bacteriol* 46, 591–599.

Legan, J. D., Vandeven, M. H., Dahms, S. & Cole, M. B. (2001). Determining the concentration of microorganisms controlled by attributes sampling plans. *Food Control* 12, 137–147.

Malcolm, S. (1984). A note on the use of the non-central *t*-distribution in setting numerical microbiological specifications for foods. *J Appl Bacteriol* 57, 175–177.

Chapter 8

Selection of Cases and Attributes Plans

8.1	Introduction	8.7	Determining Values for m and M
8.2	Microbial Criteria: Utility, Indicator, and Pathogens	8.8	Specific Knowledge about the Lot
		8.9	What Is a Satisfactory Probability of Acceptance?
8.3	Factors Affecting the Risk Associated with Pathogens	8.10	Selecting n and c
8.4	Categorizing Microbial Hazards According to Risk	8.11	Sampling Plan Performance of the Cases
8.5	Definition of Cases	8.12	References
8.6	Deciding between Two-Class and Three-Class Attributes Sampling Plans		

8.1 INTRODUCTION

While the overall philosophy of this book is to control microbial hazards through raw material selection, Good Hygienic Practices (GHP) and Hazard Analysis Critical Control Point (HACCP), and not to rely on microbiological testing, there are occasions when testing might be considered. If it is concluded that testing is appropriate, this chapter provides guidance on the choice of sampling plan and discusses their limitations. The recommended sampling plans are based on statistical considerations in Chapters 6 and 7, the severity of the hazard, and the potential for change in risk (decrease, no change, or increase) before a food is consumed. The International Commission on Microbiological Specifications for Foods (ICMSF) recommended 15 cases that reflect different levels of risk (ICMSF, 1974, 1986). The greater the risk, the higher the case number, the more stringent the sampling plan (see section 8.5 and Table 8–1).

The principles for establishing microbiological criteria were described in Chapter 5. Microbiological criteria can be used to assess:

- the safety of food;
- adherence to GHP;
- the utility (suitability) of a food or ingredient for a particular purpose;
- the keeping quality (shelf-life) of certain perishable foods; and/or
- the acceptability of a food or ingredient from another country or region where the conditions of production are unknown or uncertain.

Chapter 5, however, did not describe how to choose a sampling plan, an essential component of all microbiological criteria. Due to the importance and complexity of choosing a sampling plan, Chapter 6 discussed the concept of probability and factors to consider when collecting representative samples from a lot or consignment of food. In Chapter 7, two basic sampling plans (two-class and three-class attributes plans) were described. The choice of sampling plan should take into account a number of factors, including the risk to public health associated with each hazard and susceptibility of the target group of consumers. This chapter incorporates information from the previous chapters and provides a scheme that can be used in deciding on a sampling plan that is based on risk.

The stringency of sampling plans for foods should be based on the hazard to the consumer from pathogenic microorganisms and their toxins or toxic metabolites, or on the potential for quality deterioration to an unacceptable state. Plans should also take account of the types of microorganisms present and their numbers. Some microorganisms merely spoil a food, others can cause illness, and still others are taken to indicate the likelihood of contamination by pathogens. Some pathogens cause mild illnesses that seldom spread; others cause mild illnesses that spread rapidly; and yet others cause severe illness. The degree of foodborne hazard is often increased if the organism has grown to high numbers in the food and, conversely, is usually reduced if the number is reduced. In some cases, the food merely acts as a vehicle for transmission of the infectious microorganism. Treatment in the normal course of distribution, storage, and preparation for consumption may decrease, leave unchanged, or increase numbers of microorganisms, while labile toxins would decompose and stable toxins remain.

The choice of sampling plan for microbiological criteria should first reflect the purpose. Is the microbiological criterion intended to assess the general quality and acceptability of a food (i.e., utility) or is it intended to assess microbiological safety either indirectly (i.e., indicator) or directly (i.e., pathogen, toxin, toxic metabolite)?

8.2 MICROBIAL CRITERIA: UTILITY, INDICATOR, AND PATHOGENS

8.2.1 Utility Tests

Some microbiological tests provide information regarding general contamination, incipient spoilage, or reduced shelf-life. Evidence should support the use of a utility test for the intended purpose. For example, evidence should support the use of a total aerobic count as a measure of incipient spoilage. Such tests may be useful indicators of product quality. However, utility tests are not related to health hazards, but rather to economic and aesthetic considerations; therefore, the level of concern is low. Utility tests are included in cases 1–3 (see Table 8–3) and satisfied by relatively lenient sampling plans. They may involve direct microscopic counts, yeast and mold counts, aerobic plate counts, or specialized tests, such as for cold tolerant organisms or for species causing a particular type of spoilage (e.g., lactobacilli in mayonnaise or thermophilic spore-formers in sugar).

8.2.2 Indicator Tests

Microorganisms that are not normally harmful but may indicate the presence of pathogenic microorganisms may be used as indirect indicators of a health hazard. For example, for dried egg products, Enterobacteriaceae can be used as indicators of the presence of salmonellae. In these products, any practically applicable sampling plan cannot detect the low level of salmonellae. It is important to recognize that relationships between pathogen and indicators are not universal and are influenced by the product and process.

Care must be taken when selecting indicator organisms. For instance, "coliform counts" have been widely used as universal indicators of hygiene, but in many products, psychrotrophic Enterobacteriaceae will inevitably be present, and the apparently high coliform counts do not necessarily indicate hygienic failure or consumer risk. Microorganisms naturally present in the product may also interfere with the analysis and result in meaningless counts. For instance, aeromonads are often detected by coliform count procedures, rendering counts of coliforms meaningless for many seafood products where aeromonads are common members of the microflora.

Indicator organisms may be useful in other situations, e.g., when assessing efficiency of cleaning and disinfection or in investigational sampling. A food plant laboratory may prefer not to test for a specific pathogen (salmonellae or *Listeria monocytogenes*) because culturing those organisms in the laboratory could increase the risk of the organism being introduced in the food processing environment. Therefore, a generic test, e.g. for *Listeria*-like organisms, may be considered safer because the moist, cold conditions known to favor the dominance of *L. monocytogenes* will also favor generic, nonpathogenic strains of *Listeria*.

An inspector at a port-of-entry may know very little about the history of a consignment of food, e.g., whether a heat process adequate to kill relevant microorganisms was used in processing, whether the food was contaminated after processing, or whether the consignment was temperature-abused during shipment. Tests for relevant microorganisms can indicate whether certain foods have been underprocessed. For instance, high numbers of mesophilic spore-forming bacteria in low-acid, shelf-stable canned foods indicate probable underprocessing when it is certain the container is not leaking. The presence of Enterobacteriaceae or coliforms in some properly pasteurized foods indicates recontamination after heat processing. *Escherichia coli* in water indicates recent fecal contamination, and *Staphylococcus aureus* in cooked foods can indicate contamination from the human skin or nose. Because of the uncertain relationship between indicators and specific pathogens, the level of concern is moderate and it is inappropriate to apply sampling plans with high stringency for indicator microorganisms.

8.2.3 Pathogen Tests

There are occasions when testing for a pathogen may help to ensure food safety. Using the decision tree in Figure 8–1 will assist in deciding whether testing will be useful. Tests for specific pathogens can be applied:

- in routine sampling when experience indicates that testing is an effective means of consumer protection;
- for verification of GHP/HACCP systems when a suitable indicator microorganism is not available; and
- in investigational sampling when the epidemiology of a foodborne disease outbreak points to a particular lot of food as the cause of illness, or when there are other circumstances creating suspicion of the presence of a pathogen or toxic metabolite.

8.3 FACTORS AFFECTING THE RISK ASSOCIATED WITH PATHOGENS

Microbiological criteria and sampling plans should reflect the severity of the disease and be appropriate for the food. Certain well-known food-pathogen combinations have

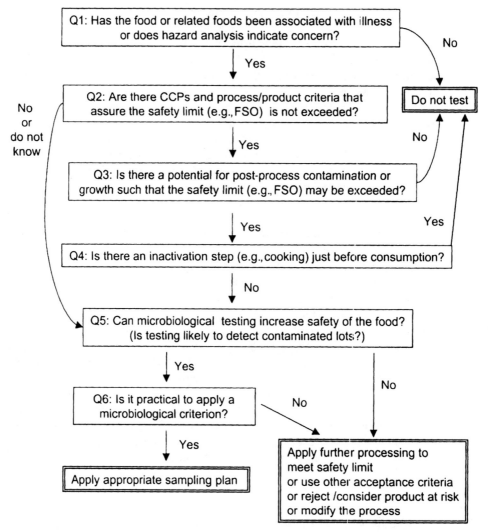

Figure 8–1 Decision tree to help decide when sampling and testing for foodborne pathogens can be used as lot acceptance criteria.

become recognized. Some understanding of the conditions that determine whether a food is likely to contain pathogens or their toxic metabolites is necessary. Frequently, there are strong regional and cultural influences on these associations.

8.3.1 Epidemiologic Considerations

Water and some types of seafood have been shown to be common vehicles in outbreaks of typhoid, cholera, and hepatitis A infection. Meat and poultry are commonly identified as vehicles in outbreaks of salmonellosis. Ham and cream-filled pastries are frequently implicated in outbreaks of staphylococcal foodborne illness. Outbreaks of gastroenteritis caused by *Vibrio parahaemolyticus* are usually associated with seafood. Cooked meat and cooked poultry or stews and gravy that have been subjected to time-temperature abuse after cooking are the usual vehicles in outbreaks of enteritis caused by *Clostridium perfringens*. Botulism is a rare disease usually associated with the ingestion of inadequately processed home-preserved foods, particularly cured pork products, fermented fish, fish eggs or marine mammals, and low-acid foods, including vegetables. Histamine poisoning, which is rarely a serious disease, is typically associated with scombroid fish species. Raw milk is commonly identified as the vehicle of campylobacteriosis, brucellosis, salmonellosis, and more recently enterohemorrhagic *E. coli* infection. In addition, cheese made from raw milk has been the cause of brucellosis, staphylococcal intoxication, bloody diarrhea, and hemolytic uremic syndrome caused by enterohemorrhagic *E. coli*.

The association between gastroenteritis caused by *Bacillus cereus* and temperature-abused cooked rice is well established. Undercooked ground beef is a vehicle of enterohemorrhagic *E. coli* O157:H7 infections, although recent outbreaks have been associated with fresh produce, fermented meat products, contaminated water, and unpasteurized dairy products. Produce such as raspberries has been associated with outbreaks of cyclosporiasis, whereas contaminated water has been the principal vehicle of cryptosporidiosis.

8.3.2 Ecological Features

The primary sources of foodborne microbial pathogens include a variety of animal, human, and environmental reservoirs. After contamination of food, behavior of the pathogens is influenced by the food composition, the presence of other microflora, and the environmental conditions of the food.

Many of the pathogens that affect humans are widely distributed in the agricultural environment, for example, *Salmonella* spp., *Campylobacter* spp., *L. monocytogenes*, *Yersinia enterocolitica*, pathogenic *E. coli*, *C. perfringens*, and *S. aureus*. Although foodborne disease has long been primarily associated with animal products (e.g., meat, seafood, and dairy products), in recent years many large outbreaks have been traced to produce, including lettuce, sprouts, and raspberries (Beuchat, 1996). Humans are also a reservoir of certain foodborne pathogens, some of which may persist for weeks or months in the carrier state, for example, *S. typhi*, *Shigella* spp., and hepatitis A and Small Round Structured Viruses (SRSV) such as Norwalk-like viruses (ACMSF, 1995).

Particular products present greater risk than others due to possible contamination during production and harvest, their intrinsic properties that affect microbial growth and survival, traditional preparation and handling practices specific to that food, and, often, the absence of a critical control point that will eliminate the hazard. For example, foods consumed raw, such as oysters, present high risks to susceptible consumers since they may be contaminated with SRSVs or *V. vulnificus* at harvest. Ready-to-eat foods may be recontaminated with *L. monocytogenes* that may grow during subsequent refrigeration unless there is a competitive anti-listerial microflora in the product.

Local customs and standards of community hygiene, especially those related to food, water supply, and sanitation, are important determinants of the extent and variety of foodborne illnesses. Effectiveness of prevailing standards for safeguarding water supplies, milk supplies, and shellfish harvesting areas warrants consideration. Control of food processing; detecting, recalling, or condemning contaminated foods; vermin control; public health supervision of food service establishments; and appropriate use of refrigeration in processing plants, food service establishments, and homes all play a role in reducing the incidence of foodborne illness and influence the selection of sampling plans for particular commodities from particular sources.

Dietary customs specific to a region also influence the foodborne hazards. For example, the Japanese custom of eating raw fish has contributed to the relatively high incidence of *V. parahaemolyticus* infection in that country. Similarly, the various fermented traditional marine foods consumed among the Inuits in Alaska and Canada are likely to contribute to incidences of type E botulism in these regions.

Members of the *Aeromonas hydrophila* group occur in raw fish, raw meats, and other foods. Although high counts of *A. hydrophila* can occur in patients with various types of diarrhea, its role as a cause of diarrhea remains in dispute. *Plesiomonas shigelloides* can be isolated from water and raw aquatic products. High numbers of *P. shigelloides* have occasionally been demonstrated in patients with diarrhea, but its role in foodborne or waterborne illness remains in dispute.

8.3.3 Clinical Features

Certain foodborne microorganisms are inherently associated with severe illnesses in humans. *C. botulinum* types A, B, E, and F, for instance, can produce toxins that cause neurological illness in healthy people, even when very small amounts are ingested. If not effectively treated with antitoxins and provided with respiratory support, the case fatality rate may exceed 50%. Virulence properties of *S. typhi*, *S. dysenteriae* I, *V. cholerae*, certain strains of *S. typhimurium*, *E. coli* O157:H7, and *C. perfringens* type C enable these pathogens to cause severe disease, even death. Cholera often presents a medical emergency in the developing world because 50–70% of dehydrated cholera patients who are malnourished die unless they are appropriately treated by oral or intravenous fluid and electrolyte replacement. *L. monocytogenes*, causing listeriosis, mainly affects susceptible, typically immunocompromised, populations; however, among these patients, the mortality can be as high as 25%. People with underlying chronic disease, in particular males with a history of high alcohol consumption, are prone to infection by *V. vulnificus* that is associated with iron overload in the patient.

Initially, pathogenic *E. coli* were considered to be strains of specific O serogroups causing diarrhea, mainly in infants, and referred to as the "classical" enteropathogenic *E. coli*. However, several other types of *E. coli* have been recognized and are concerns of today's food industry. Enterotoxigenic *E. coli* is a major cause of infantile diarrhea in developing countries and a leading cause of traveler's diarrhea. Enteroinvasive *E. coli* closely resemble *Shigella* in pathogenicity and antigenicity. Enterohemorrhagic *E. coli* (EHEC), such as *E. coli* O157:H7, were first identified as pathogens in 1982 and produce Shiga-like toxins. Food-associated outbreaks attributed to *E. coli* O157:H7 have been well documented and are of great concern because the infective dose is low (<100 cells) and the illness is severe, sometimes leading to kidney failure and death, especially in young children and the elderly.

Low doses (10–100 cells) of *S. dysenteriae* can cause shigellosis. The infective dose of other *Shigella* spp., *V. cholerae*, and some salmonellae may also be low in highly susceptible individuals, such as infants, and malnourished and immunocompromised persons. Also, the severity of enteritis caused by salmonellae, *Shigella* spp., and *E. coli* is greater (and probably the infective dose lower) for the very young, the aged, immunocompromised and persons with concomitant diseases than for healthy young adults. In these groups of people, even the usually moderate gastroenteritis caused by *V. parahaemolyticus*, staphylococcal enterotoxin, *B. cereus*, or *C. perfringens* sometimes become severe.

The relatively rare instances where beta-hemolytic streptococci are foodborne may lead to tonsillitis complicated by severe sequelae of glomerulonephritis and arthritis, while cardiovascular disabilities (such as rheumatic fever) may follow. Convalescence after some other foodborne illnesses (particularly typhoid and paratyphoid fevers, brucellosis, and viral hepatitis) may be lengthy.

Many foodborne illnesses are associated with chronic secondary sequelae that can linger long after the acute effects of enteric infections. Examples include Guillain-Barré syndrome, a rapidly ascending paralysis that can lead to death, which is in part associated with antecedent *Campylobacter jejuni/coli* infections; reactive arthritis, which follows enteric infections caused by salmonellae, *Shigella* spp., *Y. enterocolitica*, and thermophilic campylobacters; hemolytic uremic syndrome, which is associated with *E. coli* O157:H7 infection; depression from chronic diarrhea caused by *Toxoplasma*; and septic arthritis following salmonellosis.

8.3.4 Diagnostic Considerations

Physicians' experience and laboratory procedures play a crucial role in diagnosing foodborne illness. For example, few physicians are likely to have previously encountered botulism; hence misdiagnoses may occur even when symptoms are typical or occasionally when the symptoms are very mild or appear similar to other illnesses. Laboratory isolation of specific pathogens is the only way certain enteric foodborne diseases can be diagnosed because the clinical syndromes of many of these diseases may be similar, e.g., bloody diarrhea may be symptomatic of bacillary or amoebic dysenteries and Shiga-like toxin-producing *E. coli*.

When a previously unrecognized foodborne disease is reported by a laboratory, awareness often increases and other incidents are revealed. The recognition of foodborne campylobacteriosis and EHEC O157:H7 infection illustrates this point.

In public health and food laboratories, completely satisfactory procedures are not yet available for the routine isolation or detection of several foodborne pathogens from foods (such as *Shigella* spp., *C. jejuni*, *Y. enterocolitica*, EHEC non-O157, *Cyclospora*, *Cryptosporidium*, and foodborne viruses). Laboratory methodology, therefore, limits the precision with which their presence can be measured.

8.4 CATEGORIZING MICROBIAL HAZARDS ACCORDING TO RISK

In this book, the term *hazard* is limited to microbiological concerns associated with foodborne illness. These are bacterial pathogens and their associated toxins or toxic metabolites, viruses, parasites, and toxigenic fungi. The risks associated with microbial hazards vary greatly, ranging from quite mild symptoms of short duration to very severe, life-threatening illnesses. When deciding on the level of concern, health hazards generally fall into three categories.

8.4.1 Moderate Hazards

Moderate hazards are rarely life threatening, do not result in sequelae, are normally of short duration, and cause symptoms that are usually self-limiting, but can result in severe discomfort. Some microorganisms can be both severe hazards for specific populations and mild hazards for the general population. For example, *L. monocytogenes* can cause abortion in pregnant women and life-threatening disease among immunocompromised people, whereas flu-like symptoms of diarrhea of short duration occur in the general population.

8.4.2 Serious Hazards, Incapacitating But Not Life Threatening

These hazards result in disease of moderate duration and do not normally cause sequelae. Some pathogens, such as *C. jejuni* and other thermophilic campylobacters, occur most commonly in the lower, moderate category of hazard, but some strains of *C. jejuni* cause severe illness, e.g., Guillain-Barré Syndrome in susceptible persons. The majority of strains cause only mild diarrhea of moderate duration.

8.4.3 Severe Hazards, Life Threatening

These microbial hazards can result in substantial chronic sequelae or the effects can be of long duration. They can affect either the general population or be specific to populations at high risk. Factors influencing the development of illness in high-risk populations include specific host susceptibility to infection such as listeriosis in pregnant women, cultural practices such as consumption of potentially hazardous foods unique to specific subpopulations (e.g., *Burkholderia cocovenenans* associated with coconut tempeh), and geographic influences such as fumonisin intoxication associated with regions in which moldy maize is consumed.

The major microbial pathogens and toxins associated with foods in relation to their impact on public health, their frequency of involvement in disease, the types of foods that have served as vehicles, and significant factors contributing to disease are listed in Appendix 8-A. This table is not intended to be all-inclusive and no attempt has been made to arrange these pathogens and toxins according to frequency with which they cause outbreaks or cases of foodborne illness; this varies with locality. Table 1–1 (Chapter 1) indicates whether microbiological testing of foods (e.g., at port-of-entry) or other control measures has been instrumental in controlling the hazard and ensuring food safety.

8.5 DEFINITION OF CASES

The foregoing information can be used to establish sampling plans that consider the risk associated with a hazard. Thus, the choice of sampling plan must consider:

- the significance of the test result in relation to the type and severity of disease, indicator of a microbial hazard, or its commercial utility, and
- the conditions under which the food is expected to be handled and consumed after sampling.

Table 8–1 classifies 15 different cases of sampling plans on a two-dimensional grid taking into account these factors. In the table, the stringency of the sampling plan increases with the type and degree of hazard: from a situation of no health hazard but of utility only, through a low indirect health hazard (as implied by the presence of indicator organisms), to direct health risks related to diseases of moderate or severe implication. The stringency of the sampling plan also changes according to the conditions under which the food is expected to be handled. Hazards may remain unchanged, be reduced by

Table 8-1 Sampling Plan Stringency (Case) in Relation to Degree of Risk and Conditions of Use

Type of hazard	Conditions in which food is expected to be handled and consumed after sampling, in the usual course of events		
	Reduce risk	Cause no change in risk	May increase risk
Utility (e.g., general contamination, reduced shelf-life, spoilage)	Case 1	Case 2	Case 3
Indicator; low, indirect hazard	Case 4	Case 5	Case 6
Moderate hazard, not usually life threatening, usually no sequelae, normally of short duration, symptoms are self-limiting, can be severe discomfort	Case 7	Case 8	Case 9
Serious hazard, incapacitating but not usually life threatening, sequelae rare, moderate duration	Case 10	Case 11	Case 12
Severe hazard for (a) the general population or (b) restricted populations, causing life-threatening or substantial chronic sequelae or illness of long duration	Case 13	Case 14	Case 15

cooking, or increase because of subsequent growth of microorganisms. The most lenient plan is case 1. Stringency increases from left to right and from top to bottom of the table, so that the sampling plan for case 15 is the most stringent.

8.5.1 Choosing the Case

The choice of case to apply depends on whether the hazard may increase, not change, or decrease between when a food is sampled (e.g., at port-of-entry) and when the food is consumed. Thus, the value of microbiological testing as a method of consumer protection depends on knowledge of the type of food to be sampled. For example, it is helpful generally to understand a food's normal method of production/harvesting, processing, composition, packaging, and the conditions to which it would normally be exposed during storage and preparation. In addition, some understanding of pathogen-food interactions and the intended consumer is needed. Information of this nature is needed before an examiner can choose an appropriate case. The following illustrates such considerations.

In general, foods that have received an adequate heat treatment during a manufacturing process are generally safer than those that have not. Risk increases when heat-treated foods (i) become contaminated after processing, (ii) are exposed to conditions that permit multiplication of pathogens, and (iii) are not recooked shortly before consumption.

If a food is expected to be fully cooked before consumption and because cooking reduces the hazard, one would choose case 4, 7, 10, or 13, depending on the degree of the hazard. Raw poultry, fresh dry pasta, cake mix, and dried soup mix with beans are examples of foods in this category.

If conditions of anticipated use would not permit a change in the number of relevant bacteria (e.g., frozen storage), the appropriate case would be 5, 8, 11, or 14, depending on the type of hazard. Ice cream is an example of a food that would be classified in one of these cases, because it is ordinarily maintained and consumed frozen.

If the food is ordinarily subjected to conditions that permit growth or an increase in the hazard, thereby increasing risk, the case would be 6, 9, 12, or 15, depending on the type of the hazard (Table 8–1). For example, one of these cases would apply to dried milk since contaminating pathogens may multiply after reconstitution.

Preservation Conditions

The preservation conditions (e.g., salt concentration, a_w, pH, temperature) of the food should be considered in relation to the growth requirements of the relevant microorganism(s). Foods with a brine concentration of approximately 10% may support the growth of staphylococci, but not salmonellae. Salmonellae, however, may survive for an extended period of time on dried meats. Hence, such products (if not refrigerated) might be classified in case 6 for staphylococci and case 11 for *Salmonella* spp.

Fresh meat supports the growth of various pathogens, whereas dried meat with a brine concentration at ≥16% in the water phase does not. Hence, if fresh meat is stored at temperatures allowing multiplication, the risk would increase corresponding to case 6, 9, 12, or 15, whereas for dried beef there would be no change in risk, so case 5, 8, 11, or 14 would apply.

Storage Temperature

Temperature is especially important. Microbial numbers, and the associated risks, generally increase at 10–20 °C and even more rapidly at warmer temperatures. In contrast, refrigeration to below 10 °C tends to control most hazards, because many pathogens do not multiply or do so more slowly at low temperatures. For example, for ham kept below 6 °C (at which temperature staphylococci do not produce toxin), case 8 rather than case 9 would apply. For foods in which psychrotrophic pathogens, such as *L. monocytogenes*, *Y. enterocolitica*, and nonproteolytic *C. botulinum* can multiply, the potential for growth will depend on decreasing the storage temperature to closer to 0 °C.

Competitive Flora

Growth of pathogens can sometimes be prevented by competition from other microorganisms. While salmonellae grow in most foods of appropriate pH, a_w, and temperature, growth of staphylococci is often restricted by the associated spoilage flora. Fresh raw meats and bacon are not normally associated with staphylococcal food poisoning because they also carry large numbers of competing microorganisms that suppress the growth of *S. aureus*. The hazard of enterotoxin formation usually arises in foods that have been processed in some way to reduce the microbial population, and then the food is contaminated with staphylococci (e.g., cooked ham contaminated after cooking).

Eating Customs

Custom also affects hazard and the choice of case. For example, *V. parahaemolyticus* grows readily on raw fish unless it is refrigerated. It is a major cause of foodborne illness in Japan, where raw fish is commonly consumed, but in other countries, though widely distributed, *V. parahaemolyticus* is a much less common cause of illness because fish is cooked before consumption. Hence, for Japan, case 8 or 9 would be appropriate for this pathogen, whereas in another country with different dietary customs, case 7 would be suitable.

Reconstituted Dried Foods

Foods that are pasteurized before distribution (e.g., powdered eggs, dried milk) are sometimes eaten without cooking when distributed in relief areas. If a food is intended for consumers with unusually high susceptibility to foodborne illnesses, the hazard will be increased (Table 8–2).

Type of Hazard

Certain microbiological hazards can increase (e.g., salmonellae, *L. monocytogenes*) if the above conditions of temperature and food composition are favorable. Many hazards such as toxins and toxic metabolites tend to be quite resistant to environmental conditions, including normal cooking, and remain stable. Other hazards, such as viruses and parasites cannot increase, but may decline in concentration, depending on the conditions to which they are exposed.

Susceptibility of the Intended Consumer

If a food is intended for consumers with unusually high susceptibility to foodborne disease, risk will be increased. Examples of special foods intended for high-risk consumer groups are described in Table 8–2.

Storage and Preparation for Serving

Only the usual conditions to which the food is expected to be exposed between when the lot is sampled and when the food is consumed should be considered. For example, a frozen food will ordinarily be kept frozen until it is cooked or reheated for serving. If a

Table 8–2 Special Foods for Consumer Groups with Increased Susceptibility

Food class	Reason for a more stringent sampling plan
Baby food	High susceptibility of the consumer population to enteric pathogens; severe response to infections and toxins; increased risk of fatality
Dietetic food	Infection is a severe risk for diabetics
Foods for hospitals	Patients may be prone to infection and to serious sequelae after enteric disease because of stresses from other disabilities and from immunosuppressive treatment, and intensive care; interference with convalescence from other disease; staff and patients need to be protected because of their potential for spreading disease within the hospital
Foods for AIDS patients	Immunocompromised populations are highly susceptible to enteric pathogens
Relief foods, especially dehydrated high-protein foods	Populations needing relief foods are usually highly susceptible and prone to serious complications because of malnutrition and other stressful conditions; also increased risk for person-to-person spread of disease because of confinement of the population in crowded areas often having poor sanitary conditions; particular hazards are reconstitution with contaminated water, unhygienic handling, and poor storage conditions leading to rapid bacterial growth

food is unexpectedly abused after having been sampled and approved, the sampling plan may not provide the level of protection expected.

Method of Food Preparation

An important consideration is the method of food preparation (i.e., normally eaten raw, warmed, or cooked).

8.5.2 Examples of Choosing a Case

The following examples illustrate how knowledge about microbial ecology, and food storage and use are integrated in choosing the case.

Salmonellae are serious hazards and often occur in raw protein foods (e.g., liquid eggs), but they are destroyed by pasteurization. However, recontamination of pasteurized products with salmonellae has occurred, and subsequent drying or freezing cannot be relied upon to destroy these bacteria. If such a dried food is consumed in the dry state, there is no change in hazard (case 11); if use after reconstitution is delayed, and heating does not take place before consumption (a practice that is highly undesirable with many such products), the case would be 12. Cooking promptly after reconstitution will reduce the hazard, hence case 10 would then be appropriate.

When applying the decision tree (Figure 8–1) to a raw food (e.g., raw meat or poultry) that is to be cooked, testing for *S. aureus* would not be appropriate. If, however, the food has been cooked (e.g., cooked crustaceans or whole chickens) and the food is then handled (e.g., peeling shrimp, removing skin and bones from the chickens), then contamination with *S. aureus* is a concern and case 9 would be appropriate. In certain salted foods in which salt-tolerant *S. aureus* can grow, the competing flora are inhibited by the reduced a_w. Hence, the case for staphylococci would be similar to a pasteurized food (case 9).

Foods to which a significant amount of salt has been added also may be of increased concern. For example, the addition of salt to cheese curd or to sausage batter during the manufacture of cheddar cheese or fermented sausage will retard the growth of the indigenous flora, but not the more salt tolerant *S. aureus*. Under normal circumstances, the addition of a starter culture will control the growth of *S. aureus*; but, if the starter culture is rendered inactive by the addition of too much salt or the presence of an antibiotic, extensive *S. aureus* can occur. Considering this information, case 9 could be applied to these foods, but with caution, because the number of viable *S. aureus* will decline over time and may not reflect the likelihood of enterotoxin in the food.

B. cereus and *C. perfringens* are also moderate hazards, differing from *S. aureus* in that they produce spores that survive mild heating. Few processes will reduce the hazard provided by these bacteria, so case 8 or 9 is usually appropriate. Consideration has to be given to subsequent use of the food. For example, if a dehydrated food is eaten immediately after reconstitution, then testing would not be appropriate, compared with a food for which a delay would be expected between when the food is rehydrated and when it is consumed.

The foregoing examples illustrate and emphasize the need for some knowledge of the microbial ecology and history of a food before an examiner can choose an appropriate case or even test for a particular purpose. When choosing cases on the basis of hazard as described above, one must consider the many possible uses for the consignment of food. Some uses may be of higher risk than others, and the selection of case should reflect this possibility.

Microbial hazards are related to the presence of numbers of undesirable microorganisms or the occurrence and concentration of a toxic metabolite in a food. After choosing the category of hazard (category I, II, or III in Appendix 8–A) and the effect of subsequent conditions of handling and preparing the food on the hazard, the appropriate case is selected (Table 8–1).

8.6 DECIDING BETWEEN TWO-CLASS AND THREE-CLASS ATTRIBUTES SAMPLING PLANS

To decide whether the plan should be two-class or three-class, one must consider whether any positives (e.g., *Salmonella* or aerobic plate count levels above those reflected by GHP) can be permitted in any of the sample units. If the answer is no, a two-class plan with $c = 0$ should be used. If the answer is yes, a two- or three-class plan can apply, but if the number of microbes in a unit-volume or mass can be obtained, a three-class plan is recommended (Figure 8–2).

Three-class sampling plans may, for the following reasons, be more appropriate than two-class plans.

1. To accept a proportion of sample units yielding test values in the marginally acceptable interval (between acceptable and defective), as these plans do, is in keeping

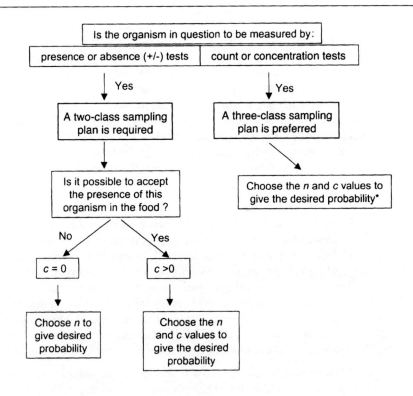

*A variables plan may be applicable; see Chapter 7.

Figure 8–2 Deciding between two- and three-class sampling plans.

with practical experience where, even under good manufacturing conditions, a few sample units may well reveal test values beyond those normally encountered without causing any consequent problem. This situation applies especially to counts of indicator organisms.
2. Sufficient experience also defines a different level, beyond which counts indicate substantial likelihood of health or utility hazard, a level that will not be attained if control has been adequate. This, designated M, should remain stable unless new experience reveals an error in its initial placement. Obviously, greater stability of criteria of acceptance may be expected to promote wider adoption of sampling plans.
3. A three-class plan allows a regulatory authority to carry out a limited form of trend analysis. For example, an increasing proportion of values within the marginally acceptable (m to M) range may indicate a lack of control.

8.7 DETERMINING VALUES FOR m and M

Definitions of m and M from a statistical point of view have been given in Chapter 7. From a microbiological point of view, the level of the test organism that is acceptable and attainable in the food is traditionally defined as m. The value of m reflects implementation of GHP by an importing country or, more often, for domestic production of a food, or is the hazardous level in two-class plans. If the test organism is a pathogen for which there is no level of tolerance (i.e., for two-class plans), m may be zero or, more correctly, absence in all analytical units when tested by a specified method. This will reflect the detection limits of the test, e.g., < 0.04 cells g^{-1}. Hence, the value of m for two-class plans is usually 0. For three-class plans, m will usually be assigned some nonzero value.

M, used only for three-class plans, defines an unacceptable level. All lots with values exceeding M should be rejected (i.e., they should be withheld pending further investigation to determine whether the product can be used after further treatment or is unacceptable as food). Investigation of the processor's facilities should be undertaken when such lots are detected. In international commerce, this presupposes a system whereby the analytical results can be conveyed to appropriate officials of the country of origin.

There are several approaches to choosing the value of M:

1. As a utility (spoilage or shelf-life) index. Relate microbial levels to detectable spoilage (odor, flavor) or to a decrease in shelf-life to an unacceptable short period.
2. As a general hygiene indicator. Relate levels of the indicator bacteria to a clearly unacceptable condition of hygiene—whether contamination or growth, or both.
3. As a health hazard. Relate levels of pathogens to illness. In choosing the value of M for a pathogen, use epidemiologic and laboratory data in combination, experimental animal feeding or inoculation data, human feeding data, laboratory analyses for toxin related to levels of organisms, or other guides that indicate the level at which there is a serious concern for the hazard. For this purpose, consider the maximum amount of food likely to be eaten at one time and the susceptibility of the group of persons likely to eat the food (see also Table 8–2).

It is important to realize that M is defined by the hazard. The value of m is available and defined by GHP in three-class plans and may change with time. In Chapter 7, a procedure for establishing a relationship between m and M was outlined. However, the values for m and M do not necessarily need to have a constant relationship.

In an unusual situation, one might find a consignment composed of two widely dissimilar lots. One lot may be entirely acceptable and the other entirely unacceptable (see for example curve 1 and curve 4 in Figure 7–3); this situation would have a high proportion of units falling below m and above M, but few between m and M.

These examples illustrate how the choice of m and M, in relation to the microbial quality of the lot, influences the types of lots likely to be rejected. Moreover, they illustrate the importance to producers of limiting variation in numbers of microbes present in food, i.e., of keeping production microbiologically under control, to avoid lots of otherwise satisfactory average quality being rejected. With good or bad control, the average microbial levels are likely to be respectively low or high, and the distribution of the data around the average (expressed logarithmically) respectively narrow or broad. The spread of the distribution may be nearly as important as the average level in determining compliance or noncompliance with a particular criterion, as indicated in Figure 7–3.

The decision to accept or reject any particular lot should be based upon the laboratory results obtained and is subject to the error associated with the laboratory procedure used (see Chapter 12). It is especially important when using two-class plans with $m = 0$ and $c = 0$ that the methods employed provide accurate results, for whatever false-negative and false-positive results occur will lead to corresponding wrong decisions about lots because of the inaccurate methods.

Unfortunately, documentation of the inherent variation within a given method is rare, though experienced analytical microbiologists develop estimates that become included in their choice of criteria. Some methods, for example, most probable number methods for coliforms, have great inherent variability (see Silliker et al., 1979). Similarly, it is common that familiarity of an analyst with a particular method (e.g., for *Salmonella* spp. or *S. aureus*) enables that method to be most productive for them. This is a cogent argument for widely accepted use of a standard method so that all analysts become experienced in its use: another method with different sensitivity might require change of criteria. The testing method is critical in establishing the acceptance criteria of sampling plans, hence it is not desirable to deviate from a standard method unless there is a major advantage.

All probability computations in this book (e.g., operating characteristic curves and surfaces) for sampling plans assume that the laboratory results are obtained without error.

8.8 SPECIFIC KNOWLEDGE ABOUT THE LOT

The relatively long delay between collection of samples and issue of the corresponding laboratory result may create a need for costly storage. For products with an extensive history of good quality, the need for storage can be avoided by releasing consignments as soon as samples have been taken (provided, of course, that records are adequate to trace for recall if a lot should prove to be unacceptable). If the analysis should reveal an unsatisfactory condition, future consignments could be held at the port until the test results proved satisfactory. Upon reestablishing a satisfactory record (perhaps three consecutive consignments), the food could again be released as soon as samples have been taken. These systems have been in use by many countries.

Application of a sampling plan with a comparatively small routine sample (e.g., $n = 5$), and thus with low stringency, is only appropriate to detect lots with a high proportion of defectives. When a food product is processed and shipped under uniform, adequate controls and there is a favorable history of compliance and low risk associated with a defective lot that is not sampled, it could be argued whether acceptance sampling serves any purpose. Until such assurance has been obtained through a good history of test results and/or auditing (see Chapter 4), the only way to obtain satisfactory protection (i.e., substantial discrimination between acceptable lots and unacceptable lots) is by increasing the sample size (larger n). However, as noted previously (Chapters 6 and 7), to double the reliability may require four times as many sample units. The cost of the additional testing should be balanced against the potential gain in discriminatory power; hence a practical decision should be made. However, it should be balanced as well against the impact on the actual level of risk a wrong decision would have.

It may not be feasible to sample at random over the entire consignment. It may only be possible to sample randomly from a portion of the consignment from which the sample units can be drawn. If so, results apply only to the portion of the consignment sampled, not to the whole consignment, and it is necessary to consider whether the results can be applied to the entire consignment. For example, accessible containers may be those nearest the door of a vehicle, nearest the hatch of a ship's hold, or on the periphery of a stack in a warehouse. The sample units chosen may sometimes represent the portion of a lot or consignment that has been exposed to greater hazard, through contamination or conditions that permitted microbial growth. Convenient as it may be, this practice may also provide greater consumer protection through the selection from the portion of the lot most likely to be hazardous. But this would not be so for a perishable food that is packed too warm and cools slowly in the boxes at the center of a pallet.

Another aspect that can only be addressed by referring to specific knowledge about the lot is the distribution of the microorganism(s) that are of concern (i.e., their frequency distributions). As discussed in Chapter 7, knowledge of these distributions, especially their standard deviations, is needed to evaluate the performance of sampling plans based on quantitative criteria. The calculation of sampling plan performance, to be discussed in section 8.11, assumes a standard deviation of 0.8. Frequency distributions having standard deviations above or below 0.8 will result in sampling plans that differ in their performance.

8.9 WHAT IS A SATISFACTORY PROBABILITY OF ACCEPTANCE?

Since decisions to accept or reject lots are made on samples drawn from the lots, occasions arise when the sample results do not reflect the true condition of the lot. The *producer's risk* describes the probability that an acceptable lot, if offered, will be falsely rejected. The *consumer's risk* describes the probability that an unacceptable lot, when offered, will be falsely accepted. Consumer's risk, for the purpose of this text, is considered to be the probability of accepting a lot whose actual microbial content is substandard as specified in the plan, even though the determined values indicate acceptable quality. This is expressed by the probability of acceptance (P_a) as given in Tables 7-1, 7-2, and 7-4. The producer's risk (the inverse of consumer's risk) is expressed by $1 - P_a$.

When acceptance sampling is chosen to measure whether a food safety objective (FSO)/performance criterion is met, besides the tolerable level of risk, from which the FSO is derived, an additional uncertainty is introduced, viz., the risk of incorrectly accepting a lot that does not meet the FSO/performance criterion. Such an incorrect decision may increase consumer risk. In principle, the minimally required probability to reject a lot not meeting the FSO/performance criterion (or the confidence in appropriate operation of the sampling plan) should be adapted to the level of concern. This would lead to increasing minimal rejection probabilities with higher case numbers. The stringency of a plan is measured by the probability of accepting lots in which a particular proportion of sample units is defective. A relatively lenient three-class plan ($n = 5, c = 3$) accepts a lot with a 5% proportion of defective units and 30% marginal units on about three occasions in four ($P_a = 0.75$). The most stringent two-class plan ($n = 60, c = 0$) would accept lots with the same 5% proportion of defective units on about one occasion in 20 ($P_a = 0.05$), and lots with 0.5% defectives on about 19 occasions in 20 ($P_a = 0.95$).

If it is important to determine correctly whether a 0.5% proportion of defective units is in compliance with an FSO/performance criterion, the question might be asked whether the plan can provide any worthwhile protection, even with a significant increase in the number of sample units. This demonstrates that microbiological testing will have very little value in process validation when the process is designed to produce a low proportion of defective units.

In practice, a probability of acceptance on three occasions in four ($P_a = 0.75$) means that one lot in every four will be rejected, a loss serious enough to compel a manufacturer to tighten microbial control to a level well below the limit(s) set in the test employed. Even the rejection of one lot in 20 ($P_a = 0.95$) may suffice to have this effect. Nevertheless, the protection conferred on the consumer for a particular lot is seriously limited when using small numbers of sample units such as $n = 5$; hence the recommendation is to use large values for n when a direct hazard is recognized. In these situations, following the decision tree on testing and use of microbiological criteria (Figure 8–1) will result in more reliable assurance of safety.

The sampling plan stringency must be considered in view of the associated criteria as well. Consider the example of a moderate health hazard, such as the presence of *S. aureus* in cooked, peeled shrimp. If the microbiological limit m were placed at a numerical level well below that likely to represent hazard (M), one could frequently accept lots containing a high proportion of marginally acceptable units, which would require only a lenient sampling plan. If, however, m were placed nearer the hazardous level, one would accept such lots infrequently, requiring the use of a relatively stringent sampling plan. If m were at a hazardous level, one would not accept lots containing any units exceeding that level, and two-class plans of high stringency would be required. Adjustment is possible by choosing limits known by experience to be associated with safety. When this is done, even though a high proportion of lots with substandard units will be accepted, the probability of consuming food that would cause illness is kept low.

8.10 SELECTING n AND c

The choice of n and c varies with the desired stringency (probability of rejection and power of discrimination) and hence with the cases in the grid of Table 8–1. For stringent cases, n is high and c is low; for lenient cases, n is low and c is high. As n decreases with the attributes plans proposed in this book, the chance of acceptance of unacceptable lots increases. This fact must be taken into account if the number of sample units (n) exceeds the analytical capability of a laboratory and n is reduced.

The procedure (see Chapters 6 and 7) for selecting the number of sample units, n, should be first to fix desired acceptance and rejection probabilities for lots of defined acceptable and unacceptable qualities. Then the number of sample units required for this purpose is derived. However, the choice of n is usually a compromise between what is an ideal probability of assurance of consumer safety and the workload the microbiology laboratory can handle, as well as the cost and time needed for the tests. Consider first the nature of the hazard; then decide the appropriate probabilities of acceptance and rejection for the hazard in question. Guidance on determining these probabilities requires establishing a relationship between the confidence desired and the impact a wrong decision (acceptance of a defective lot) would have on the actual risk. Determining the required number of sample units then depends on the desired confidence.

The stringency of any of the sampling plans can be increased by adjusting c and/or n. In this way, pressure can be brought to bear upon hygienic practices, the nature of purchasing specifications, the severity of processing, and the extent and nature of the quality control practiced within the food industry concerned. The desirable effect should be carefully weighed, and the decision made should be known and understood by producer, manufacturer, distributor, and control agency alike.

In summary, the final judgment on which a sampling plan is based should involve the relative weight to be placed on the above microbiological, epidemiologic, and ecologic factors, as well as the statistical probabilities of acceptance or rejection desired and economic considerations arising within the laboratory (e.g., the nature of its physical facilities, equipment, and work capacity of its personnel). Some degree of subjective judgment remains unavoidable because data adequate to allow fully objective decisions may not be available. Under these circumstances, individual judgments can vary widely and, if applied in international trade, could lead to widely differing sampling plans and inadvertent barriers to trade until the differences between trading partners are resolved.

8.11 SAMPLING PLAN PERFORMANCE OF THE CASES

The cases described in Table 8–1 have been widely adopted in sampling plans. The ICMSF sampling plans were developed based on past experience, available data, practical constraints and statistical considerations and have provided helpful guidance. It should be noted, however, that the sampling plans assigned for each case in Table 8–3 may not provide the desired stringency required to come to a reliable decision for lot acceptance or whether a performance standard or FSO has been met. The plans simply imply that the number of samples indicated for each case provides the desired level of protection; whether this assumption holds must be assessed by the user.

The techniques described in Chapter 7 can be used to calculate the stringency of the cases in terms of the probability of accepting a lot at various concentrations of organisms. The stringency of each case and recommended sampling plan has an associated level of performance. Table 8–4 shows the performance of the sampling plans for cases 4 to 15. The table shows the mean concentrations that would be associated with an acceptance probability of 0.05 (i.e., a rejection probability of 0.95) for the sampling plan in each case. For cases 4 to 9, the three-class attributes sampling plans involve quantitative analysis. For cases 10 to 15, the two-class attributes sampling plans involve qualitative analysis (presence/absence testing) using 25 g analytical units.

To demonstrate the relative sensitivities of cases 4 to 9, constant values were used for m (10^3 cfu g^{-1}) and M (10^4 cfu g^{-1}) and a standard deviation of 0.8 log cfu g^{-1} was

Table 8–3 Suggested Sampling Plans for Combinations of Degrees of Health Concern and Conditions of Use (i.e., the 15 "Cases")

Degree of concern relative to utility and health hazard	Conditions in which food is expected to be handled and consumed after sampling in the usual course of events*		
	Conditions reduce degree of concern	Conditions cause no change in concern	Conditions may increase concern
Utility; general contamination, reduced shelf-life, incipient spoilage	Increase shelf-life Case 1 Three-class $n=5$, $c=3$	No change Case 2 Three-class $n=5$, $c=2$	Reduce shelf-life Case 3 Three-class $n=5$, $c=1$
Indicator; Low, indirect hazard	Reduce hazard Case 4 Three-class $n=5$, $c=3$	No change Case 5 Three-class $n=5$, $c=2$	Increase hazard Case 6 Three-class $n=5$, $c=1$
Moderate hazard; direct, limited spread	Case 7 Three-class $n=5$, $c=2$	Case 8 Three-class $n=5$, $c=1$	Case 9 Three-class $n=10$, $c=1$
Serious hazard; incapacitating but not usually life threatening, sequelae are rare, moderate duration	Case 10 Two-class $n=5$, $c=0$	Case 11 Two-class $n=10$, $c=0$	Case 12 Two-class $n=20$, $c=0$
Severe hazard; for (a) the general population or (b) restricted populations, causing life threatening or substantial chronic sequelae or illness of long duration	Case 13 Two-class $n=15$, $c=0$	Case 14 Two-class $n=30$, $c=0$	Case 15 Two-class $n=60$, $c=0$

*More stringent sampling plans would generally be used for sensitive foods destined for susceptible populations. (See Table 8–2)

---

assumed. Similarly, cases 10 to 15 are compared for a standard deviation of 0.8. In cases 10 to 15, "$m=0/25$ g" indicates that each 25 g analytic unit should test negative for the hazard of concern. The number of samples tested is indicated by the value of n. The calculations indicate that lots having or exceeding the mean concentrations will be rejected with a probability of at least 0.95 when the indicated sampling plan is applied. Thus, the mean concentrations define the sensitivity of each sampling plan.

The information for sampling plan performance provided in Table 8–4 enables the examiner to better understand the expected performance of a given sampling plan. For example, the most stringent sampling plan chosen for case 15 ($n = 60$, $c = 0$ with no positive samples detected ($m = 0$)) will, with 95% probability, reject the lot if the mean concentration of the pathogen is at least 1.9 cfu 1000 g^{-1}. If lower mean concentrations of the pathogen are present (e.g., 1 cfu 5000 g^{-1}), the sampling plan will accept the lot with a probability higher than 5%. In the sampling plan for case 8 (three-class-plan, $n = 5$,

Table 8–4 Cases and Sampling Plan Performance, Assuming a Standard Deviation of 0.8. Lots Having the Calculated Mean Concentrations or Greater Will be Rejected with at Least 95% Probability

Type of hazard	Cases, sampling plans and calculation of their performance		
	Conditions reduce hazard	Conditions cause no change in hazard	Conditions may increase hazard
Indirect	Case 4 (three-class, n=5, c=3) e.g. m=1000/g, M=10000/g Mean conc. = 5128/g	Case 5 (three-class, n=5, c=2) e.g. m=1000/g, M=10000/g Mean conc. = 3311/g	Case 6 (three-class, n=5, c=1) e.g. m=1000/g, M=10000/g Mean conc. = 1819/g
III. Moderate	Case 7 (three-class, n=5, c=2) e.g. m=1000/g, M=10000/g Mean conc. = 3311/g	Case 8 (three-class, n=5, c=1) e.g. m=1000/g, M=10000/g Mean conc. = 1819/g	Case 9 (three-class, n=10, c=1) e.g. m=1000/g, M=10000/g Mean conc. = 575/g
II. Serious	Case 10 (two-class, n=5, c=0) e.g. m=0/25g Mean conc. = 32/1000g (1cfu/32g)	Case 11 (two-class, n=10, c=0) e.g. m=0/25g Mean conc. = 12/1000g (1 cfu/83g)	Case 12 (two-class, n=20, c=0) e.g. m=0/25g Mean conc. = 5.4/1000g (1 cfu/185g)
I. Severe	Case 13 (two-class, n=15, c=0) e.g., m=0/25g Mean conc. = 7.4/1000g (1 cfu/135g)	Case 14 (two-class, n=30, c=0) e.g. m=0/25g Mean conc. = 3.6/1000g (1 cfu/278g)	Case 15 (two-class, n=60, c=0) e.g. m=0/25g Mean conc. = 1.9/1000g (1 cfu/526g)

$c = 1$ with $m = 10^3$ and $M = 10^4$), a mean concentration of at least 1819 cfu g^{-1} is required for the plan to reject the lot with 95% probability. The basis for the calculations is described in Legan et al. (2001).

The poor performance of the sampling plans in detecting lots with low concentrations of pathogens demonstrates that lot acceptance testing is an unreliable approach to ensure consumer safety. It is for this and other reasons discussed throughout the text that greater emphasis should be placed on control systems, such as GHP and HACCP.

8.12 REFERENCES

ACMSF (Advisory Committee on Microbiological Safety of Food, UK) (1995). *Report on Foodborne Viral Infections.* London: HMSO.

Beuchat, L. R. (1996). Pathogenic microorganisms associated with fresh produce. *J Food Prot* 59, 204–216.

ICMSF (International Commission on Microbiological Specifications for Foods) (1974). *Microorganisms in Foods 2. Sampling for Microbiological Analysis: Principles and Specific Applications.* Toronto: University of Toronto Press.

ICMSF (International Commission on Microbiological Specifications for Foods) (1986). *Microorganisms in Foods 2. Sampling for Microbiological Analysis: Principles and Specific Applications,* 2nd edn. University of Toronto Press, Toronto, Canada.

Legan, J. D., Vandeven, M. H., Dahms, S. & Cole, M. B. (2001). Determining the concentration of microorganisms controlled by attributes sampling plans. *Food Control* 12, 137–147.

Silliker, J. H., Gabis, D. A. & May, A. (1979). ICMSF methods studies. XI. Collaborative/comparative studies on determination of coliforms using the Most Probable Number procedure. *J Food Prot* 42, 638–644.

Appendix 8–A

Ranking of Foodborne Pathogens or Toxins into Hazard Groups

Severity of threat to health	Frequency of involvement in foodborne disease	Examples of vehicles	Other factors contributing to significance
I.A. Severe hazard for general population, life threatening or substantial chronic sequelae or long duration			
Brucella melitensis, B. abortus, B. suis (brucellosis)	Common in endemic areas	Raw milk and cheese, especially from goats and sheep	Convalescence often prolonged
Botulinal neurotoxin (Clostridium botulinum, C. butyricum, C. barati) (botulism)	Rare	Improperly processed canned or preserved low-acid foods: home-cured meat products; smoked fish, other marine products; foil wrapped baked potato in salad, garlic in oil	Rapid recognition and treatment essential for patient survival; substantial mortality
Enterohemorrhagic E. coli (e.g., E. coli O157:H7, O111:NM) (hemorrhagic colitis and hemolytic uremic syndrome)	Sporadic; epidemic	Undercooked ground beef, unpasteurized apple juice, vegetable sprouts, lettuce, venison, yogurt, fermented sausage, untreated and recreational water, contact with farm animals	Very severe for children and elderly, severe complications including kidney failure and death, low infectious dose, acid tolerance
Salmonella typhi, S. paratyphi A, B (S. schotmulleri) and C (typhoid and paratyphoid fevers)	Endemic in many parts of the world; occasionally epidemic	Untreated water, raw milk, meat products, raw shellfish	Prolonged medical care required, asymptomatic chronic carrier state commonly occurs
Shigella dysenteriae I (shigellosis)	Sporadic; endemic	Salads, untreated water	High mortality rate, especially among children, low infectious dose
Burkholderia cocovenenans	Rare	Coconut tempeh	Restricted geographic populations

continues

Appendix 8-A continued

Severity of threat to health	Frequency of involvement in foodborne disease	Examples of vehicles	Other factors contributing to significance
Vibrio cholerae O1 and O139 (cholera)	Sporadic; endemic; sometimes epidemic	Raw seafood from polluted water; untreated water	Substantial mortality among dehydrated, untreated persons; moderate symptoms with available rehydration treatment
Mycobacterium bovis (tuberculosis)	Rare	Raw (unpasteurized) milk	
Aflatoxins, produced by *Aspergillus flavus* and *A. parasiticus* (aflatoxicosis)	Common in tropical regions	Nuts and oilseeds, especially peanuts and maize, figs, milk	Most potent liver carcinogens known; acutely toxic in high doses; carcinogenic, teratogenic, and probably immunosuppressive at low levels
v CJD/BSE; Protease Resistant protein (PrP); Bovine Spongiform encephalopathy (BSE); variant Creuzfeldt-Jakob disease	Sporadic	Specified bovine offal (brain, spinal cord, intestines, tonsils, thymus, spleen, spinal cord) from adult cattle	Severe central nervous system disorder resulting in death; no treatment or cure.

I.B. Severe hazard for restricted populations, life threatening or substantial chronic sequelae or long duration

Severity of threat to health	Frequency of involvement in foodborne disease	Examples of vehicles	Other factors contributing to significance
Campylobacter jejuni serovar O:19 and other serotypes associated with GBS (Guillain-Barré Syndrome)	Sporadic	Poultry, water	Associated with Guillain-Barré Syndrome
Enteropathogenic *E. coli* (EPEC) Enterotoxigenic *E. coli* (ETEC)	Frequent in certain regions	Untreated water; food contaminated by nonpotable water or infected food handler	Symptoms are mild but can be severe in infants, major cause of infant mortality in certain regions
Clostridium perfringens type C (enteritis necroticans)	Rare	Cooked pork	High mortality in protein-deficient persons, associated with malnutrition and a diet rich in trypsin inhibitors

Clostridium botulinum (types A and B)	Sporadic	Honey	Infant botulism
Enterobacter sakazakii	Rare	Dried milk powder in infant formula (temperature abuse of dehydrated infant formula)	Causes death in infant populations (up to 70% mortality rate among neonates)
Listeria monocytogenes	Sporadic; occasionally epidemic	Soft cheeses, paté, smoked fish, ready-to-eat food	High risk groups include immunocompromised persons and pregnant women; high mortality (ca. 25%) in high risk populations; infrequent illness in immunocompetent persons; low numbers of L. monocytogenes are frequently consumed in foods
Vibrio vulnificus	Sporadic	Raw oysters	High mortality (ca. 50%) among persons that have elevated levels of serum iron; liver disorder associated with high alcohol consumption
Hepatitis A virus	Common	Raw or underprocessed bivalve mollusks, salads, untreated water	Very severe for patients with liver disease, convalescence prolonged
Cryptosporidium parvum	Sporadic; endemic; occasionally epidemic	Untreated water; unpasteurized apple juice, contaminated produce, unpasteurized milk	Severe prolonged diarrhea that is life threatening in immunocompromised; prognosis is poor for AIDS patients; usually short-term diarrhea that resolves spontaneously in immunocompetent persons

continues

Appendix 8–A continued

II. Serious hazard; incapacitating but not life threatening; sequelae infrequent; moderate duration

Severity of threat to health	Frequency of involvement in foodborne disease	Examples of vehicles	Other factors contributing to significance
Salmonella enteritidis, *Salmonella typhimurium* and other *Salmonella* serovars (salmonellosis)	Very common; epidemic	Eggs; poultry; dairy products; wide range of other foods	Serious for young and elderly persons; cross contamination from raw meat and poultry; eggs and poultry meat can be internally contaminated during production; some serovars of *Salmonella* are highly virulent, reactive arthritis occurs in 1–2% of cases (Reiter's syndrome)
Yersinia enterocolitica (pathogenic), *Yersinia pseudotuberculosis* (yersiniosis)	Sporadic	Milk; chitterlings; untreated water	Most infections occur in children less than 5 years of age, with symptoms of mild gastroenteritis; in older children symptoms are severe, presenting a pseudo appendicular syndrome; only certain serovars of *Y. enterocolitica* are noteworthy pathogens; sequelae; arthritis can occur in genetically predisposed persons that carry the human leucocyte antigen (HLA-B27)
Shigella flexneri, *S. boydii*, *S. sonnei* (shigellosis) (nondysentery)	Sporadic in industrialized countries; sometimes endemic in developing countries	Foods subjected to contamination by infected persons or sewage-contaminated water, such as salads; untreated water	Serious for young and elderly persons; secondary infections among contacts; sometimes low infectious dose, HUS occasionally
Listeria monocytogenes	Sporadic, rare	Foods where multiplication has occurred during storage	Low numbers of *L. monocytogenes* are often consumed on a wide variety of foods

Hepatitis A	Common	Raw or undercooked bivalve mollusks, salads, untreated water, strawberries	Less severe if no liver disease, but convalescence prolonged
Arcobacter butzleri and A. cryaerophila	Sporadic	Untreated water; poultry	Appendicitis infrequently occurs
Cryptosporidium parvum	Sporadic, endemic, occasionally epidemic	Untreated water, unpasteurized apple juice, contaminated produce, unpasteurized milk	
Cyclospora cayetanensis	Sporadic; endemic; occasionally epidemic	Raspberries; lettuce; water	Severe, prolonged diarrhea
Trichothecene toxins, especially deoxynivalenol, nivalenol, and T-2 produced by *Fusarium graminearum* and related species	Not clear	Cereals, especially wheat and maize in temperate climates	Immunosuppressive, probably contributing to increased disease incidence in endemic areas
Zearalenone, produced by *Fusarium graminearum* and related species	Probably not common	Cereals, especially wheat and maize in temperate climates	Estrogenic effects, not commonly observed in humans
Fumonisins, produced by *Fusarium moniliforme* and related species	Not clear	Fungus endemic in maize, toxins present in staple diets in regions of high maize consumption	Immunosuppressive, carcinogenic to rats and probably humans, implicated in esophageal cancer
Ochratoxin A, produced by *Penicillium verrucosum*, *Aspergillus ochraceus*, and related species, *A. carbonarius* and perhaps related species	Not clear	Cereals and pig meats in cool temperate climates, dried fruit, wines, and coffee beans	Nephrotoxic, probably contributing to reduced life spans in parts of Europe

continues

Appendix 8-A continued

Severity of threat to health	Frequency of involvement in foodborne disease	Examples of vehicles	Other factors contributing to significance
III. Moderate, not usually life threatening; no sequelae; normally short duration; symptoms are self-limiting; can be severe discomfort			
Bacillus cereus (*B. cereus* gastroenteritis), including emetic toxin	Common	Fried and boiled rice; reconstituted cereal products; puddings, custards	Usually diarrhea and/or vomiting of short duration; death is rare
Clostridium perfringens type A (*C. perfringens*)	Common	Cooked, non-cured meats; poultry; gravy	Symptoms usually mild but are more serious in elderly or debilitated persons; death is uncommon
Escherichia coli (EPEC, ETEC)	Common in developing countries; infrequent in developed countries	Foods handled by persons carrying EPEC or ETEC; foods contaminated with nonpotable water	Usually diarrhea of short duration in general population
Staphylococcus enterotoxins (*Staphylococcus aureus*) (enterotoxicosis or food poisoning)	Common	Cooked foods handled by persons carrying *S. aureus* then temperature-abused; ham; fermented sausages; cereal-filled pastries; cheese; milk; salads; peeled crustaceans, bivalve mollusks; mushrooms	Explosive vomiting and moderate diarrhea; symptoms usually resolve without treatment within 2 days of onset; death is rare
Vibrio cholerae non O1 and non O139	Sporadic	Raw bivalve mollusks, cross-contaminated cooked crustaceans	
Vibrio parahaemolyticus (*Vibrio parahaemolyticus* gastroenteritis)	Common in Japan; epidemic	Cooked crustaceans; raw bivalve mollusks; raw marine fish	Common where seafood is consumed raw
Small Round Structured Virus (SRSV), including Norwalk virus	Common	Raw bivalve mollusks, food handled by infected persons; bakery products	Symptoms are usually mild
Biogenic amines (e.g., histamine)	Infrequent	Scombroid fish, some cheeses	Biogenic amines are probably necessary for disease, serious for persons taking MAO

Chapter 9

Tightened, Reduced, and Investigational Sampling

9.1 Introduction
9.2 Application of Tightened Sampling and Investigational Sampling
9.3 Tightened Sampling Plans
9.4 Example of the Influence of Sampling Plan Stringency in Detecting Defective Lots
9.5 Selecting the Sampling Plan According to Purpose
9.6 Reduced Sampling
9.7 References

9.1 INTRODUCTION

Tightened sampling involves the use of more frequent sampling and/or more stringent sampling plans than normal. Reduced sampling involves less frequent sampling than normal. Investigational sampling consists of sampling to determine the cause of a problem.

Factors warranting tightened or reduced sampling may relate to the food product, the manufacturer, or the country of origin (Exhibit 9–1). In tightened sampling, the sampling plans discussed in earlier chapters are no longer applied. Tightened sampling usually requires increased sample size (n) with other adjustments to make the sampling plan more stringent. In a three-class plan with fixed m and M, stringency is increased by making c smaller or n larger. When two-class plans are involved and c is 0, the sample size n must be increased to obtain a more stringent plan, assuming m is fixed. In the case of reduced sampling, the frequency of sampling a particular food is reduced, but if a lot is sampled, the two- or three-class sampling plan originally specified is applied.

Investigational sampling is a term used to describe sampling that is related to a known or suspected problem. Investigational sampling differs from tightened sampling in that the cause of a problem is sought. Such problems can arise from failure of a lot (or a series of lots) to pass routine inspection or from some new information, such as field reports of illness or unexpected spoilage related to the product. Investigational sampling may be done to (1) confirm that a problem exists, (2) assist in describing the nature and extent of a problem, (3) provide information on possible sources of a problem, (4) help decide what to do with a product (e.g., determine whether the entire lot is unacceptable or if the contamination is limited to a certain portion), and (5) prevent the problem from reoccurring.

It is important to note the difference between investigational sampling and the type of sampling described elsewhere in this book. Attributes or variables sampling detects effects, whereas investigational sampling seeks to determine causes. The success of investigational sampling depends greatly on the expertise of the investigator, the investi-

Exhibit 9–1 Circumstances Warranting Tightened or Reduced Sampling of Food

Warranting increased frequency of sampling and/or more stringent sampling plan	Warranting reduced frequency of sampling
The food operation	
An audit indicates the operation does not have an adequate system of controls based on GHP* and HACCP†	The operation has an effective system of control based on GHP and HACCP
Records indicate a deviation at a CCP‡ in the HACCP plan has occurred	Records indicate the operation is under control
Information indicates the operation has used an ingredient from a source that has caused problems in other similar operations	
Food from the operation has recently been implicated in illness	Food from the operation has a favorable history of safety
The food	
The composition of the food differs from other foods of the same type, and an increase in a hazard is likely to occur under expected conditions of storage and distribution	The composition of the food differs from other foods of the same type, and the potential hazards will decrease or be eliminated during expected conditions of storage and distribution
Previous tests frequently unsatisfactory	Previous tests satisfactory
Routine tests for indicators have revealed a trend toward increased contamination	Routine tests for indicators show continuing control
The food has a history of being a source of food-borne illness	Rarely involved in foodborne illness
The food has been found to be a source of a newly emerging pathogen or new type of an existing pathogen	
Circumstances suggest involvement of this type of food in a recent outbreak	
A food that traditionally has been for the general public is to be directed toward a sensitive population	Not primarily intended for sensitive populations
New type of food or new formulation with reason to be concerned about a microbiological hazard	The parameters necessary for controlling foodborne illness are well known and widely applied
Examination results from different laboratories are in conflict, and disposition of the food is in question	
Country or region of origin	
Food control systems are in question	The food control systems are known to be equivalent for control of the food or ingredients in question
Endemic or epidemic situations exist that increase concern for consumers of the food	Endemic or epidemic situations do not exist that would increase concern for consumers of the food

*GHP-Good Hygienic Practices
†HACCP-Hazard Analysis Critical Control Point
‡CCP-Critical Control Points

gator's knowledge of the product and process involved in its manufacture, and the conditions of storage, distribution, and use. The attributes and variables plans previously discussed for lot acceptance are intended for routine application to lots or consignments presented for inspection at various stages prior to sale to the public. Two- and three-class attributes plans are used for regulatory, port-of-entry, and other receiver-oriented situations, where little information is available concerning the microbiological history of the lot. Variables plans can be used when the distribution of microorganisms (aerobic plate count, pathogens, indicators) is known. Variables plans may also be applicable to in-plant quality control, where these assumptions can be properly verified. In either case, the emphasis in lot acceptance sampling plans is to do the minimum amount of work necessary to obtain the degree of security that each plan can provide.

When a potential problem has been identified (e.g., recontamination of a product with *Salmonella*), further examination into the nature and extent of the problem is frequently required (i.e., investigational sampling). Such might be the case if a lot rejected under routine inspection is in dispute, or if the cause of a problem is being sought. For example, routine acceptance sampling may have indicated occasional recontamination in a plant from an increased rejection rate of lots. A larger random sample may then be taken from the rejected lots to confirm that a problem exists and to estimate the extent and distribution of the contamination. Emphasis may then shift to the processing plant where samples could be taken from various sites in an attempt to locate the source of contamination, e.g., certain processing equipment, a particular ingredient. It is necessary to identify the source so that appropriate measures can be taken to correct the problem. In another situation, analysis of rejected lots may show that contamination is associated with certain production times (e.g., first few hours, after breaks, following equipment repair). This information can be used to focus on the circumstances that may be causing contamination and how it may be prevented.

Note that these hypothetical investigations have used sampling at various stages in a food operation. Such sampling is rarely random. Biased sampling is much more efficient in an investigation because it takes advantage of prior knowledge of the operation, visual observations, and logic that may lead the investigator to sample those locations that are the most likely sources of the problem, e.g., a conveyor. Random sampling of easily accessible (and thus easily cleaned) sections of the equipment may ignore product accumulations in hollow gears that drive conveyors, which may escape proper cleaning and disinfection. A common approach to investigational sampling used by industry is to collect in-line samples of product at selected stages of processing with a bias toward those most likely to be contaminated or indicative of the source of a problem.

Biased sampling also is useful for suspected lots of food that are in storage or have been received at a location. For example, it may be evident from pack or container appearance that a dried product may have become wet due to container leakage during transport. In this case, it may be appropriate to take samples from the areas most likely to have become wet (e.g., the top packs on a pallet). Indications that a nonrandom problem may have occurred include transit pack damage, pallet collapse, or evidence of spoilage or discoloration in the outer packs suggesting abusive holding conditions. Other approaches to sampling, such as stratified, systematic, cluster, or various combinations of these, also may be used. Generally, once the purpose of sampling has been established, then the most efficient type of sampling applicable to this purpose is chosen. Such topics are beyond the scope of this book and are statistically complex, but an excellent discussion is found in Cochran (1977).

9.2 APPLICATION OF TIGHTENED SAMPLING AND INVESTIGATIONAL SAMPLING

Tightened and investigational sampling generally are applied to situations where there is an increased level of concern or perceived risk; for example, a process deviation has occurred, performance criteria have not been met, a product has failed to meet microbiological criteria, the food is from an operation with a history of inconsistent control, the food is from a region where there has been a recent increase in illness involving the same or similar type of food, or an environmental monitoring indicator has shown that the equipment used to produce or package the product did not meet acceptable hygienic criteria. Tighter sampling also may be warranted when there is insufficient knowledge about a particularly sensitive ingredient or a food is intended for a sensitive population and, thus, there is increased concern for making a correct decision as to its acceptability. For example, shipment from a new supplier or from a country in which the hygienic and manufacturing practices are not known may warrant increased sampling. These circumstances may confront both industry and regulatory agencies, and both may employ tightened sampling and investigational sampling. However, the objectives and approaches of industry and regulatory agencies may differ. Control authorities are primarily interested in protecting the consumer. Industry will have that same goal, but also must protect the economic interests of the company. Both seek to identify acceptable from unacceptable product, but it is far more important to industry to know why a problem may have occurred and how it can be corrected than solely to protect the consumer.

Control authorities may have less information available than industry about the source and production of a particular suspect lot, and therefore may be limited to the use of random sampling and tightened sampling. Industry, with much more information at its disposal, may choose to use an investigational approach, deliberately employing biased sampling. Biased sampling of finished product is often more efficient than random sampling when contamination might be associated with a time or sequence of production, such as the first product produced after a shutdown, shift change, ingredient switch, mechanical repairs, etc.

Occasionally, control authorities also may employ tightened sampling or investigational sampling. For example, if an outbreak occurs involving a certain product produced by several producers, the control authority will have a much broader focus than the individual manufacturers and may have more epidemiologic information at its disposal to help identify the source of the problem. In this case, the control authority may apply both biased sampling and tightened sampling, using random samples to determine which manufacturer is the source of the outbreak. Industry would be much more likely to employ tightened sampling to differentiate between acceptable and unacceptable lots of ingredients and/or finished product.

Tightened and investigational sampling can also be important in the design and application of Good Hygienic Practices (GHP) and Hazard Analysis Critical Control Point (HACCP). Tightened sampling, involving many more samples than would be practical for routine testing, can be used for process validation or as part of the HACCP plan verification. Investigational sampling may be required to identify areas in a plant where GHP needs to be changed to eliminate an organism of concern (e.g., more efficient disinfection techniques).

Failure to reach the critical limit (i.e., process criterion) for a Critical Control Point may necessitate tighter sampling of a lot. Experience with the food, the likely occurrence of a hazard, and the severity of the hazard(s) will influence the decision of whether sampling the lot or another option (e.g., reprocessing, destroying the lot) should be applied.

Suspicion that a food may contain a microbiological hazard may arise from a variety of sources of information:

- a hazard has been detected in other lots from the same operation.
- an audit of an operation has revealed questionable control.
- a consumer complaint has raised suspicion about the food.
- an ingredient recently has been implicated as a source of illness.
- an unfavorable trend in an indicator of pathogen contamination has been detected.

These and other circumstances (Exhibit 9–1) may lead to tightened sampling until accumulated evidence indicates this is no longer necessary.

9.3 TIGHTENED SAMPLING PLANS

One possible means to make a sampling plan more stringent is to increase the sample size (n) that is collected and analyzed. When two-class plans are involved (i.e., $c = 0$) and m is fixed, the number of analytical units (n) must be increased to obtain a more stringent plan. Another option for two-class sampling plans is not to change n but to increase the size of the analytical unit (e.g., from 25 g to 100 g) for testing. The spatial distribution and independence of the defectives throughout the lot will influence whether this option will increase confidence of detecting a defective sample unit within the lot and, thereby, lead to rejection of the food. When three-class plans with fixed m and M are involved, acceptance criteria can be made more stringent by making c smaller or n larger.

Typically, a sampling plan in tightened sampling will involve an increase in number of samples taken (i.e., increasing n). The two-class plans shown in Table 9–1 and 9–2 give some idea of how such plans might be affected by changing n. This is illustrated in another manner in Table 9–3 in which the probability of detecting one or more positive samples in batches with very low occurrence of defectives is given. Thus, if one unit in every 1,000 were contaminated, even a sample of $n = 500$ would result in a 61% chance of accepting the lot; that is, there is only a 39% chance of finding one or more contami-

Table 9–1 Probability of Acceptance (P_a) for Two-Class Sampling Plans with $n = 10$ to $n = 300$ and $c = 0$

Composition of lot		Number of sample units (n)						
% acceptable	% defective	10	20	30	50	100	200	300
99	1	0.90	0.82	0.74	0.61	0.37	0.13	0.05
98	2	0.82	0.67	0.55	0.39	0.13	0.02	<
97	3	0.74	0.54	0.40	0.22	0.05	<	
96	4	0.66	0.44	0.29	0.13	0.02		
95	5	0.60	0.36	0.21	0.08	0.01		
94	6	0.54	0.29	0.16	0.05	<*		
93	7	0.48	0.23	0.11	0.03			
92	8	0.43	0.19	0.08	0.02			
91	9	0.39	0.15	0.06	0.01			
90	10	0.35	0.12	0.04	0.01			

*< means $P_a < 0.005$.

Table 9–2 Probability of Acceptance (P_a) for Two-Class Sampling Plans with $n = 300$ to $n = 5000$ and $c = 0$

Composition of lot		Number of sample units (n)					
% acceptable	% defective	300	500	1000	2000	3000	5000
99.9	0.1	0.74	0.61	0.37	0.14	0.05	0.01
99.8	0.2	0.55	0.37	0.14	0.02	<	<
99.7	0.3	0.41	0.22	0.05	<		
99.6	0.4	0.30	0.13	0.02			
99.5	0.5	0.22	0.08	0.01			
99.4	0.6	0.16	0.05	<*			
99.3	0.7	0.12	0.03				
99.2	0.8	0.09	0.02				
99.1	0.9	0.07	0.01				
99.0	1.0	0.05	0.01				

*< means $P_a < 0.005$.

nated units with $n = 500$ when the contamination rate is 0.1%. Clearly, such a probability of detection/rejection would be unacceptable for highly toxic contaminants or severe microbiological hazards.

It is unlikely that the sampling plans with n between 300 and 5,000 would be used for microbiological testing because of laboratory effort and cost. It may be possible, however, to conduct simple nonmicrobiological tests on large numbers of sample units. For example, the options for testing a batch of canned food may be to conduct a simple test such as pH measurement, visual observation of can seams, examination for vacuum or swelling, or microscopic examination to identify potentially hazardous cans. While microbiological analysis involving conventional cultural media may be very effective in identifying contamination, a large number of examinations would not be practical. A measurement of product pH may be less likely to identify a container in which growth has taken place, but large numbers of analytical units can be examined. Because contamination may occur at a relatively low rate, a simple test that allows examination of a large number of units would be more effective.

Table 9–3 The Probability of Obtaining One or More Defectives in a Sample of n Sample Units with Proportion p of the Lot Defective

Number of sample units examined per sample, n	Probability of obtaining one or more defectives with the following proportions (p) of defective			
	0.01	0.001	0.0001	0.00001
200	0.87	0.18	0.02	0
1000	1.00	0.63	0.10	0.01
2000	1.00	0.86	0.18	0.02
3000	1.00	0.95	0.26	0.03
4000	1.00	0.98	0.33	0.04
5000	1.00	0.99	0.39	0.05

A common question is how many analytical units would need to be tested to provide some level of confidence that a lot will be rejected if it is hazardous to consumers? In general, the number of samples necessary will depend on the level of defective units that can be tolerated and the desired confidence to detect a defective lot. If, for example, a two-class attributes sampling plan is applied as a *zero tolerance* plan (i.e., $c = 0$), a defective lot would be detected as soon as at least one sample unit is defective. When sampling a given lot, the probability of finding at least one defective unit depends on the actual percentage of defectives in that lot and the number of sample units drawn. Hence, to select a sampling plan and derive the number of sample units (i.e., n), two decisions are first necessary: a decision on the percentage of defective units in a lot that defines a defective lot, and a decision on the required confidence level (e.g., 95%), i.e., on the desired probability of detecting at least one defective unit, if the lot is defective.

For such two-class plans, Table 9–4 provides guidance on the number of random sample units that would be needed to provide 90, 95, and 99% confidence that at least one defective sample unit will be detected in lots with defective levels ranging from 0.1 to 50% (Cannon & Roe, 1982). For example, if a confidence level of 95% is desired that a lot does not have 2% or greater defectives, then it would be necessary to collect 149 sample units from the lot. To increase the level of confidence to 99%, it would be necessary

Table 9–4 Number of Sample Units Required for 90, 95, and 99% Confidence of Detecting at Least One Defective Sample Unit in Lots with Defective Levels Ranging from 0.1 to 50%

Desired Confidence Level (%)	Definition of Defective Lot (%)	Number of Sample Units
90	50	4
	20	11
	10	22
	5	45
	2	114
	1	230
	0.1	2302
95	50	5
	20	14
	10	29
	5	59
	2	149
	1	299
	0.1	2995
99	50	7
	20	21
	10	44
	5	90
	2	228
	1	459
	0.1	4603

to collect 228 sample units. If the percentage of defectives that should be detected is 10%, then 44 sample units would be required for the same 99% level of confidence. As the proportion of defectives falls below about 1%, the intensity of sampling and testing increases to levels that are impractical in a microbiology laboratory.

In another form of tightened sampling, certain foods may be sampled at greater than normal frequency. The sampling plans discussed in Chapters 6–8 assume that every lot is sampled, and thus, frequency of sampling is not discussed. In practice, however, microbiological criteria and sampling plans are commonly established for a wide variety of foods, but very few lots are actually sampled. The established microbiological criteria define what is considered acceptable and are often applied on an as needed basis (see Table 4–3 and Exhibit 9–1). In practice, the frequency of testing a food can range from none, or rarely, to 100% of the lots. Increasing the application of an established sampling plan beyond that which is routinely applied would be an additional form of tightened sampling.

9.4 EXAMPLE OF THE INFLUENCE OF SAMPLING PLAN STRINGENCY IN DETECTING DEFECTIVE LOTS

Sampling plans of different stringency that are applied at different steps in the food chain or in an operation may reveal differing levels of control. For example, a drink mix containing nonfat dry milk (NFDM) was prepared by mixing dry ingredients. No kill step was applied. The ingredients were found to be negative for *Salmonella* using a case 14 sampling plan (see Chapter 8). Yet, the finished product was found to be positive using the same sampling plan. The ingredients were usually of a much greater quantity than the lot size of the product, such that a lot of NFDM tested at case 14 ($n = 30 \times 25$ g) would be used in several smaller lots of drink mix, also tested at case 14. The NFDM made up about 60% of the finished product. Thus, a greater quantity of NFDM was being tested as a component of the finished product than was tested to approve its use as an ingredient. This is illustrated in the following:

- Lot size of NFDM = 100,000 kg
  Each lot of NFDM sampled at case 14 ($n = 30$).
  Thus, 30×25 g = 750 g tested per 100,000 kg.
- Lot size of drink mix = 10,000 kg
  Each lot sampled at case 14 ($n = 30$).
  Thus, 30×25 g tested = 750 g per 10,000 kg.
- Drink mix = 60% NFDM or 6,000 kg NFDM per 10,000 kg lot of drink mix
  100,000 kg of NFDM was used to make 16.67 lots of diet drink.
  Since 750 g was tested per lot of drink mix, this meant that a total of 12,502.5 g was analyzed per 16.67 lots.
  Thus, 60% NFDM $\times$ 12,502.5 g of drink mix = 7,501.6 g of NFDM analyzed as a component of the drink mix.

In this example, 750 g of NFDM was sampled and tested for acceptance as an ingredient, but a total of 7,501.6 g was sampled and tested as a component of the finished drink mix. Thus, the NFDM was being tested at about a 10-fold greater stringency as a component of the finished product than when being tested for acceptance as an ingredient. Such experience should lead to a tightened sampling of the NFDM and/or a change of suppliers.

Thus, the potential exists for low level contamination to be missed by one sampling plan, but be detected by another more stringent sampling plan applied at another step in a food operation or elsewhere in the food chain. This example demonstrates that the choice of stringency of an attributes sampling plan applied to an ingredient must take into consideration lot size, use level, and microbiological criteria applied to the finished product. Further, if microbiological records are reviewed for the purpose of problem solving, negative results for ingredients should not eliminate them as a possible source of contamination until the balance between ingredient and finished product sampling plan stringency is considered.

9.5 SELECTING THE SAMPLING PLAN ACCORDING TO PURPOSE

Before choosing a sampling plan, the objective of the sampling should be clearly defined. Is the sampling intended to differentiate between acceptable and unacceptable product or to investigate and discover the cause of a problem? The stringency of the plan and the use of biased and unbiased sampling will depend greatly on the objective of the sampling.

One of the most difficult situations in choosing a sampling plan for either investigational or acceptance sampling is when there is zero tolerance for the attribute being tested, such as *Salmonella, Listeria monocytogenes,* or *E. coli* O157:H7. No sampling plan, short of 100% analysis of the food, can ensure complete absence of the defect. Thus, the investigator is faced with a situation where a decision must be made as to what plan stringency is adequate, but still practical to perform. Plan selection is much easier when a food safety objective or other established limit exists. The plan stringency needed to detect levels of the defect can then be determined. Sometimes the established limit may be too low to be practically sampled and tested for either investigational purposes or to differentiate between acceptable and suspect product. However, the investigator can at least determine the plan stringency that would be needed, even if it is not practical.

Even when the official regulatory position is "zero tolerance," there may be prescribed sampling plans that are widely used. Such is the case with *Salmonella* in the US. The US Food and Drug Administration has established a zero tolerance policy for *Salmonella* in most processed foods, yet employs sampling plans equivalent to International Commission on Microbiological Specifications for Foods cases 13–15 (see Chapter 8), referred to as categories I, II, and III in the FDA Bacteriological Analytical Manual (FDA, 1998). Even at the highest stringency ($n = 60$), a lot containing 2% defectives would test negative 30% of the time. It is not uncommon in the US to select a plan with a higher stringency to provide some security that the defect could be detected with greater sensitivity than if the normal attribute plan were used.

When selecting a sampling plan, the population to be sampled must be determined along with the potential sources of the problem. Consider for example a situation where three slicers are used to slice three varieties of meat. After slicing, equal portions of each variety are placed into a package and sealed. If one slicer has not been cleaned properly, it could be a source of contamination, particularly for the first half hour or more of slicing. A sampling plan that does not select samples from each slicer, or alternately does not include a sufficient number of samples to be relatively certain to include product from each slicer, will fail to detect the source of the problem. Clearly, if the sample results can identify a specific slicer, then a biased sample that ensures that the defective slicer is represented will be more efficient than a random sample of a sufficient number to be relatively certain each filler head is represented. This is especially true if the contamination may be time related, and one would like to have samples from each slicer for each time segment.

9.6 REDUCED SAMPLING

Reduced sampling involves sampling at a lesser frequency or, if circumstances warrant, not at all. When lots are sampled, however, the sampling plans described in Chapter 8 are recommended. Conditions that can lead to reduced sampling frequency may relate to the food or the source of the food (Exhibit 9–1). For example, reduced sampling is warranted when an audit of a food operation, as described in Chapter 4, leads to the conclusion that an adequate food safety management system is in place. Furthermore, foods of low risk (e.g., in which pathogens die or cannot multiply, or which have a long history of safety) could be sampled at a lesser frequency.

9.7 REFERENCES

Cannon, R. M. & Roe, R. T. (1982). *Livestock Disease Surveys: A Field Manual for Veterinarians.* Australian Bureau of Animal Health. Department of Primary Industry. Canberra, Australia: Australian Government Publishing Service.

Cochran, W. G. (1977). *Sampling Techniques,* 3rd edn. New York: John Wiley & Sons, Inc.

FDA (Food and Drug Administration). (1998). *Bacteriological Analytical Manual,* 8th edn., Revision A. Gaithersburg, MD: AOAC International.

Chapter 10

Experience in the Use of Two-Class Attributes Plans for Lot Acceptance

10.1 Introduction
10.2 The Concept of Zero Tolerance
10.3 The Need for Compromise
10.4 Application of Two-Class Sampling Plans for Pathogens Such as *Salmonella*
10.5 Problems in the Implementation of Stringent Sampling Plans
10.6 Relation to Commercial Practice
10.7 Discrepancies between Original and Retest Results
10.8 Summary
10.9 References

10.1 INTRODUCTION

The first portion of this chapter is slightly updated from Chapter 6, Book 2 of the International Commission on Microbiological Specifications for Foods (ICMSF, 1986). Very little of the text has been changed because the principles still hold true. Although the text is largely concerned with *Salmonella* in a variety of foods, the concepts also apply to other pathogens where presence/absence tests are applied and $c = 0$. The chapter from Book 2 is historically significant because when it was written, there was an increasing awareness and concern for *Salmonella* in a wide variety of foods, many of which were ready-to-eat. In response to this concern, a *zero tolerance* policy was applied by control authorities. Some companies buying ingredients from suppliers subsequently adopted this practice. As methods improved and the quantity of food analyzed increased, lower levels of contamination could be detected.

There has been an erroneous, but common, perception that *zero tolerance* equates to zero risk. This perception is not valid, as is evident in the following sections and elsewhere in this book. The concept of *zero tolerance* provoked questions about procedures to reduce the analytical workload associated with large numbers of samples and whether a positive lot could be retested. Some procedures are fallacious and do not provide the expected level of confidence. One procedure, compositing (pooling or bulking), proved highly successful and is now commonly applied for the analysis of *Salmonella* and certain other pathogens when research demonstrates no loss of test sensitivity.

Judgments about hazards from pathogens are based largely on microbiological analysis, but the results and conclusions are influenced by many factors, including the analytical method. For example, until after the late 1950s, methods of detecting *Salmonella* in food were inadequate because the method in use had been developed to analyze for the relatively large numbers of *Salmonella* in feces or blood of a diseased person. These methods could not detect the very low numbers of *Salmonella* expected to be present in foods. As more sensitive methods were developed, some foods previously thought to be generally free of *Salmonella* proved to be contaminated.

Today, highly sensitive methods have yet to be developed for *Campylobacter jejuni* and a variety of pathogens such as *Cyclospora cayetanensis* and viruses in foods. With each newly recognized or emerging pathogen, several years may elapse during which improvements are made in methodology, and the ecology of the pathogen becomes elucidated. Hence, opinions on frequency and levels of contamination by such pathogens need to be reviewed as more sensitive procedures become available.

Increasing the quantity of material examined has further improved the sensitivity of testing. Initially, for *Salmonella*, it was customary to examine samples of food weighing 1 or 10 g; but as methodology improved and the desire to detect defective lots increased, larger quantities of material were analyzed (e.g., 25 g or 50 g and occasionally more). The importance of this change will become evident in the following sections. A similar evolution occurred after about 1990 with *Listeria monocytogenes* and *Escherichia coli* O157:H7.

10.2 THE CONCEPT OF ZERO TOLERANCE

One objective of this book is to facilitate the adoption of realistic microbiological criteria. For example, some governments and companies have established criteria stating that certain pathogens "shall be absent," with no numerical tolerance being expressed. This practice is to be discouraged since it is not compatible with the use of food safety objectives and performance criteria. Microbiological criteria should include the parameters described in Chapter 5. Furthermore, at least four considerations challenge the ideal of *zero tolerance*:

1. No feasible sampling plan can ensure complete absence of a pathogen. Even when $c = 0$, it cannot be guaranteed that the lot is completely free of the organism, no matter how large the number of sample units, n. However, what can be stated is the probability of acceptance (P_a) for lots of various qualities, as a function of the given n and c. For example, when $n = 60$ and $c = 0$, then $P_a = 0.5$, meaning there is one chance in two that a lot will be accepted when 1% of the sample units are defective (Figure 10–1).
2. Plans in which $c = 0$ are not necessarily the most exacting. For example, if one sets a limit of 5% defectives in a lot, the plan $n = 95$, $c = 1$ will accept unacceptable lots less often, and acceptable lots more often than will the plan $n = 60$, $c = 0$. In other words, the former plan is the more discriminating, even though it admits the presence of the pathogen. Preference for $c = 0$ is influenced by the wish to emphasize that absence is the desired objective (although it cannot be guaranteed) and by the knowledge that pathogens such as *Salmonella* may be found by direct examination of certain foods and, having been demonstrated, the finding cannot be ignored. In addition, the work and cost associated with collecting, transporting, and analyzing 60 samples, even with compositing, is already a significant burden without increasing the number to 95 samples.
3. The sampling plans typically used for lot acceptance testing assume that defects (e.g., *Salmonella*) are randomly distributed through the lot. Section 10.7 and exam-

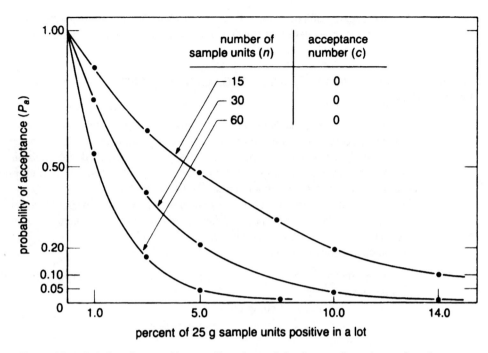

Figure 10–1 Relations between the sampling plan and the degree of security conferred—operating characteristic curves for three sampling plans proposed for *Salmonella*.

ples discussed elsewhere in the book demonstrate that this is often not the case. In addition, the plans assume that sample units are randomly drawn from the lot but, for practical reasons, this may be difficult, or impossible, to achieve.

4. It is not yet commercially possible to market some foods completely without pathogens. For example, most raw meats and poultry occasionally contain *Salmonella*. A sampling plan adjusted to this situation would seem more realistic and satisfactory than one based on an unattainable ideal of complete absence (i.e., zero tolerance).

10.3 THE NEED FOR COMPROMISE

The foregoing illustrates the difficulty of trying to make a reasonable compromise between a desire to eliminate pathogens completely from the food supply to protect consumers and current practicable methods of food production. In this respect, past decisions have often been arbitrary, and sometimes illogical. In most regions of the world, no restriction is enforced on the enormous quantities of poultry and carcass meats that are regularly contaminated with *Salmonella*, though these foods have been clearly shown by epidemiologic evidence to be major vehicles for outbreaks of salmonellosis. The risk entailed by regularly contaminated foods is probably as great as with an infective carrier whose detection would lead to his/her removal from a food operation. Sampling procedures are needed that remove the more highly contaminated lots and result in continuous improvement without completely disrupting the food supply.

To encourage a realistic choice of sampling plans for hazardous pathogens in foods, the problem is considered in further statistical detail with particular reference to *Salmonella*.

10.4 APPLICATION OF TWO-CLASS SAMPLING PLANS FOR PATHOGENS SUCH AS *SALMONELLA*

Any member of the genus *Salmonella* presents some degree of hazard to human health. Some, like *S. gallinarum* and *pullorum*, are of minor significance for humans; others, notably *S. typhi*, are severe hazards. But the majority, to which this discussion relates, represent a moderate direct hazard that, through cross-contamination and multiplication, can readily become widespread among a population via foods.

The numbers and prevalence of these organisms in food can, and should, be low. Foods presenting a minimal risk are examined for *Salmonella* only when there is unusual cause for concern. Under certain circumstances, sampling and testing may be needed to determine whether a lot of food is acceptable, such as when there is:

1. no knowledge of the manufacturing process or of earlier consignments from the manufacturer and
   —the food is of a kind commonly involved in salmonellosis or
   —the food is intended for a sensitive population.
2. special reason to suspect that the product from a particular factory may contain *Salmonella*.

Because most food control efforts should be directed to the areas of greatest risk, foods are assigned to categories according to the degree of risk they present. To each category is attached an appropriately stringent method of deciding the acceptability of the finished product. Case 10 (see Chapter 8) recognizes a hazard with reduced risk with normal preparation, such as exists when a food may contain *Salmonella* but will be consumed by ordinary consumers after cooking (e.g., red meats, poultry). Case 12 expresses a greater risk, where the conditions of use may favor spread and/or multiplication of the microorganism, and hence requires more stringent sampling. Case 11 is intermediate, reflecting no likely change in risk during normal conditions of storage, distribution, and preparation for serving. Because absence of *Salmonella* is the preferred criterion of acceptance ($c = 0$), the stringency of examination normally is increased by raising the number of sample units n. Table 10–1 shows recommended sampling plans for the various categories, with degrees of stringency related to hazard expressed by cases.

The ICMSF recommendations in Table 10–1 are similar in principle to those of the report of the *Salmonella* Committee of the National Academy of Sciences-National Research Council (NAS-NRC, 1969), in that both increase the stringency of the sampling plan with increase in the degree of risk caused by the conditions of food use. An obvious difference is that the proposals presented in Table 10–1 also involve less stringent plans for routine use. There are two reasons for this:

1. The sampling plans proposed by NAS-NRC were not intended for routine examinations, but for use when a question arises whether a product contains *Salmonella*. Routine inspections or audits will from time to time raise this question, and the NAS-NRC plans were intended for the special investigation it entails. For such investigations, the ICMSF agrees that plans are desirable in which the degree of stringency is as great as the NAS-NRC recommended (Table 10–1).

Table 10-1 Stringency of Sampling for *Salmonella* for Foods with Different Categories of Risk (All Two-Class Plans with c = 0)

		ICMSF		NAS/NRC*	
		n values[†]			
Hazard in use	Case	Normal routine	Special[‡]	Category	n values[†]
Reduced	10	5	15	B	13
Unchanged	11	10	30	C	29
Increased	12	20	60	D	60

*Based on the Report of the *Salmonella* Committee (NAS-NRC, 1969).
[†]n = number of sample units.
[‡]e.g., where the food will be eaten by susceptible consumers or for tightened inspection.

2. If a few highly pathogenic types, such as *S. typhi*, are regarded separately, the remaining *Salmonella* types that commonly occur in foods (to which Table 10-1 refers) usually involve mild to moderate disease risks. The NAS-NRC report does not make this distinction and considers all types of *Salmonella* together, implying a high degree of risk for all types. The ICMSF agrees that where circumstances (e.g., unusual susceptibility in the probable consumers) create a higher risk, sampling plans of higher stringency are desirable.

Figure 10-1 gives the operating characteristic curves that express the relationship between the degree of security conferred and the sampling plans proposed. For instance, consider the curve for the plan $n = 30$, $c = 0$. If the proportion of sample units defective (p) is 0.05 (i.e., 5% are defective), the probability of acceptance (P_a) is about 0.25 (i.e., 1 in 4), hence three out of four such lots would be rejected. Suppose the units are 25 g samples—a realistic supposition with a food such as dried or frozen egg. They are defective if examination shows them to contain any *Salmonella*. Hence, if 5% of the units are defective (i.e., 1 defective in 20), there is, on average, *Salmonella* contamination in 20×25 g = 0.5 kg of food. That is to say, the plan $n = 30$, $c = 0$, with analytical units of 25 g, will reject three out of four lots in which the prevalence of contamination is about two analytical units positive for *Salmonella* per kg.

To reduce laboratory effort while maintaining the stringency of the sampling plan, it would be helpful to combine either groups of analytical units or the enrichment cultures therefrom, using plans in which the presence of a single positive rejects the consignment. A laboratory investigation undertaken for ICMSF (Silliker & Gabis, 1973) indicated that different sized analytical units drawn from a single well-mixed sample unit of dried food give very similar results. A subsequent investigation (Gabis & Silliker, 1974) using foods of high moisture content (including eggs, poultry meat, meat, and meat products) showed comparable reliability in the use of composite analytical units; for example, examination of three composites of 20×25 g analytical units gave the same result as examination of all 60 analytical units individually (the total amount examined being 1500 g in both instances). This finding suggests that considerable reductions may be made in the cost of laboratory analyses by compositing (pooling) analytical units. Such compositing offers the prospect of large increases in the number of analytical units, n, with corresponding increase in the stringency of examination, without correspondingly increasing laboratory effort.

Numerous studies conducted since 1974 have continued to demonstrate the merits and limitations of compositing. Experience may vary for different food-pathogen combinations and analytical methods, indicating a need for caution and validation that compositing is acceptable. No reduction in the cost of collecting the necessary number of sample units can be expected, since analytical units to be composited would still have to be chosen randomly.

10.5 PROBLEMS IN THE IMPLEMENTATION OF STRINGENT SAMPLING PLANS

10.5.1 Conceivable Alternative Plans

As indicated above, for an infectious agent such as *Salmonella*, there are strong reasons for adopting sampling plans in which $c = 0$:

1. It is philosophically objectionable to some people to accept a lot containing a recognized pathogen, regardless of how the food is to be used in a manufacturing process or by the consumer.
2. There is the very great technical advantage of combining many analytical units for bacteriological examination in systems with $c = 0$, because a single positive decides the outcome.
3. The plan with $c = 1$ (or more) would always require the examination of (many) more analytical units, for equal probability of acceptance, than a plan with $c = 0$.
4. Even if the indicated probabilities of plans with $c = 0$ versus $c = 1$ are the same, their implications are not. For example, when a lot is accepted with the plan $n = 60$, $c = 0$, it is possible (though not certain) that the lot may not be contaminated at all. But using the plan $n = 95$, $c = 1$ and accepting one positive, it is known that the lot certainly is contaminated (though the probability that a lot with 5% of analytical units contaminated would be accepted is in both instances $P_a = 0.05$).

10.5.2 Fallacious Procedures

It might be suggested, for example, that if the plans $n = 60$, $c = 0$ and $n = 95$, $c = 1$ provide equivalent probability (for lots having, for example, 5% of units defective), an operator finding one positive in 60 analytical units might then proceed to examine another 35 (total $n = 95$) in the hope of clearing the lot if all the latter proved negative. But such a procedure is in reality a two-stage plan $n_1 = 60$, $c_1 = 1$, plus $n_2 = 35$, $c_2 = 0$, which has a greater probability of accepting an unsatisfactory lot than $n = 95$, $c = 1$. In fact, the probability P_a of this two-stage plan is 0.07, compared with 0.05 for the one-stage plan. Although in this example the difference is not great, there are situations where such two-stage procedures cause more serious error. Where two-stage sampling plans are actually being used, their operating characteristic curves should be computed and the resulting probabilities of acceptance evaluated.

A similar fallacy can arise where a plan requires a large number of analytical units, which would be unusually costly. Suppose the plan is $n = 95$, $c = 1$, but for economy, a group of only 20 units is tested initially. If one unit should fail in this initial group, an analyst might examine the remainder (in this case 75) with the idea that if a defective is not found in the second group, the first may be ignored. Nevertheless, the sampling plan that has been applied actually corresponds to $n_1 = 20$, $c_1 = 1$ and $n_2 = 75$, $c_2 = 0$. No justification exists for preferring the results from the second series. Similarly, if the plan were $n = 60$, $c = 0$ and an analyst felt uncertain about finding a defective unit in the first 20 samples, and the lot was accepted on the basis of the remaining 40 samples being negative, then the actual plan applied would be the two-stage sampling plan: $n_1 = 20$, $c_1 = 1$; $n_2 = 40$, $c_2 = 0$, which actually accepts more defective lots than would $n = 60$, $c = 0$.

10.6 RELATION TO COMMERCIAL PRACTICE

Stringent sampling plans like those just suggested represent an ideal not yet attainable for products such as poultry or carcass meats, which are found to be contaminated by *Salmonella* when specially sensitive methods of detection are used. At present, it is often recommended to use procedures such as rinsing or swabbing of the whole carcass, where in effect the analytical unit is most of the carcass surface and internal cavity, and a positive may correspond to levels of contamination on the order of one *Salmonella* cell per 2 kg for poultry or even one per 100 kg for carcass meats. Using such criteria, one finds slaughter lines producing contaminated carcasses with a frequency of 10% or more. For lots with 10% defectives, the relations shown in Section a of Table 10–2 are valid. The sampling plan $n = 60$, $c = 0$ would reject virtually all such material; $n = 5$, $c = 0$ would reject one-third of it (Table 10–2). Even the plan $n = 5$, $c = 1$ would reject one in twelve lots, a commercially severe rate of rejection. If, as a result of rejections on the basis of $n = 5$, $c = 1$, the quality of lots improved to the point where only 2% instead of 10% of carcasses were found to be contaminated, the relations would be as in Section b of Table 10–2. If 2% of carcasses are contaminated, the indicated average concentration of *Salmonella* cells is on the order of one per 50 kg for poultry, and one per 5000 kg or more for beef. If a 2% frequency of contamination was still thought to be too great, a similar pressure for further improvement (rejecting about 1 in 10 such contaminated lots) could be reintroduced by changing the sampling plan from $n = 5$, $c = 1$ to $n = 5$, $c = 0$.

The above illustrates how, by periodic adjustment of the stringency of the sampling plan, a steady pressure to improve could be applied without placing catastrophic restrictions upon the supply of the commodity at any time.

The sampling plan $n = 5$, $c = 1$ has P_a of about 0.65 for lots with $p = 0.24$ (Figure 10–2); that is, this plan would reject 1 out of 3 lots in which one-quarter of the units were

Table 10–2 Comparison of Some Operating Characteristics of Several Sampling Plans Relevant to Examination for Pathogens

	a. For lots with $p = 0.10$, i.e., 10% of samples contaminated			b. For lots with $p = 0.02$, i.e., 2% of samples contaminated		
Sampling plan	P_a*	P_r†	Approximate proportion of lots rejected	P_a*	P_r†	Approximate proportion of lots rejected
$n = 60$, $c = 0$	<0.005	0.995	199/200	0.30	0.70	2/3
$n = 10$, $c = 0$	0.35	0.65	2/3	0.82	0.16	1/6
$n = 5$, $c = 0$	0.59	0.41	1/3	0.90	0.10	1/10
$n = 3$, $c = 0$	0.73	0.27	1/4	0.94	0.06	1/16
$n = 5$, $c = 1$	0.92	0.08	1/12	1.00	<0.005	<1/200

*P_a probability of acceptance of lot
†P_r probability of rejection of lot

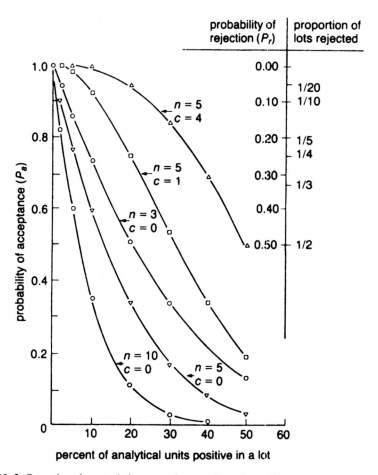

Figure 10–2 Operating characteristic curves for sampling plans with small numbers of analytical units.

contaminated. If these units were poultry carcasses weighing about 1 kg, this level of contamination would correspond to counts on the order of one *Salmonella* cell per 4 kg of food; while with beef carcasses, the figure would be on the order of one per 400 kg. This kind of situation is very different from the examination of analytical units on the order of 25 g (e.g., of egg). The plan $n = 60$, $c = 0$ rejects about one-third of lots in which 0.7% of analytical units are defective, which (these being 25 g analytical units) corresponds to a minimum level of contamination of about 1 cell per 3 kg. That is to say, the plan $n = 60$, $c = 0$, applied even to 25 g analytical units (e.g., of egg), actually provides less protection than the plan $n = 5$, $c = 1$ applied to whole-carcass samples of poultry of effective weight about 1 kg; even the latter plan would reject a substantial proportion of current production of poultry carcasses.

In such circumstances, the immediate application of the more stringent sampling plans to poultry, and still more to larger carcasses, seems at present unrealistic. The application of even so loose a plan as $n = 5$, $c = 1$ would immediately require a substantial improvement on current levels of contamination; and that improvement could then be continued by proceeding to plans of successively greater stringency. It is important to understand the difference between the sampling plans for some raw meats and those for other foods as regards *Salmonella*, and to note the effect of the size of the analytical unit on stringency of examination, as mentioned at the beginning of this chapter. (Editor's note: This assessment reflects the status of *Salmonella* in raw meat and poultry through the final revision of ICMSF Book 2 (ICMSF, 1986). Significant improvements in control of *Salmonella* have since been made by some processors through improved on-farm controls and procedures used for slaughtering and chilling. Information in this and previous sections should be considered when selecting a sampling plan for various food-pathogen combinations.)

10.7 DISCREPANCIES BETWEEN ORIGINAL AND RETEST RESULTS

The following is adapted from a newsletter prepared by Flowers & Curiale (1993). The text discusses two very common debates that arise in microbiological testing, namely:

1. One laboratory detects a pathogen in a lot and another laboratory does not.
2. Retesting a positive lot fails to confirm the initial result.

When a pathogen is detected in a food product by a two-class attributes plan, there is often a desire to retest retained samples or resample the lot in question to verify the finding. This is particularly true when there are economic and/or public health consequences associated with a positive result.

Often, retesting does not confirm the original positive result even when a much greater portion of product is analyzed in the retest. There may be a desire to believe the original result to be in error, perhaps due to contamination during sampling or analysis. Although laboratory or sampling error are plausible explanations, other explanations also should be considered such as nonrandom or heterogeneous distribution, low prevalence of contamination, and pathogen die-off between the original and repeat tests.

The discrepancy between original and retest results requires an understanding of how microorganisms may be distributed within a food product and the difference between prevalence and concentration of contamination. Prevalence refers to the frequency at which multiple samples from a given product test positive. Concentration refers to the number of cells present in a given amount of product. Consider the examples in Figure 10–3. Both lots, A and B, have the same prevalence but different concentrations. If lots A and B were each divided into 100 one-kg samples and each sample was tested individually with a method capable of detecting one cell kg^{-1}, both lots would likely result in 6 positives per 100 samples tested. However, lot B actually contains a 100-fold higher concentration, because each positive sample contains clumps of approximately 100 cells.

It is not uncommon in practice for microbial contamination to exist in clumps as represented in lot B. The number of cells per clump will vary with the nature of the product, source of contamination, and stability of the microbial contaminant in the product. Secondly, the distribution of the microorganisms in the product must be considered. If distribution is homogeneous, there is an equal opportunity for contamination to occur at any stage and time of the operation and any individual sample unit would be as likely to detect contamination as any other. If, on the other hand, contamination is limited to a certain segment of time during processing, then the defect would not be homogeneously distributed throughout the lot. If heterogeneous, then random sampling may not detect the organism.

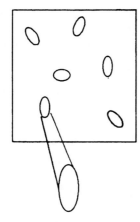 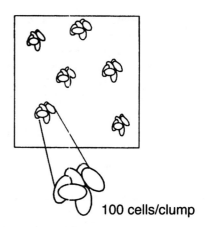

Lot A = 100 kg
Prevalence = 6 positive 1 kg sample units
Concentration = 6 cells/100 kg

Lot B = 100 kg
Prevalence = 6 positive 1 kg sample units
Concentration = 600 cells/100 kg

Figure 10–3 Illustration of two lots of product that upon analysis, may yield the same prevalence, but have 100-fold different concentrations.

Microbiologists commonly refer to random distribution of microorganisms as homogeneous and nonrandom distribution as heterogeneous. However, this is a misconception, as random refers to an irregular distribution that is still to be found if the material is "well homogenized." Figures 10–4a–c illustrate three types of nonrandom distribution in three lots of 20 consecutively produced boxes. In Lot A, the distribution is homogeneous but perfectly regular. Lots B and C represent heterogeneous distributions. In Lot B, contamination was greatest in the first sample, followed by decreasing levels of contamination in boxes 2, 3, and 4. This type of distribution of organisms commonly occurs when product is produced on contaminated equipment. The product flushes contamination out of the system upon start of production. Lot C is similar to B, but results from contamination being introduced into the system at some time during production. This type of contamination can originate from a variety of sources and causes, e.g., equipment failure followed by substitution of an unclean unit, contamination from line workers or maintenance personnel, contamination introduced from the process environment via aerosol created by cleaning in an adjacent area or contamination from outside the product stream falling into the product.

10.7.1 Nonrandom Distribution

Most microbiological sampling plans for lot acceptance testing assume contaminants are randomly distributed throughout a lot. In practice, microorganisms are often not randomly distributed, except in mixed liquid samples drawn from a container. Depending on when and how the contamination has occurred, the distribution of contaminants will vary considerably as to location and concentration within a batch. Consider the following examples:

Example 1: Assume, due to a sanitation failure, *Salmonella* has become established in a niche within one of 10 filler heads of a filling machine. During start-up, the first prod-

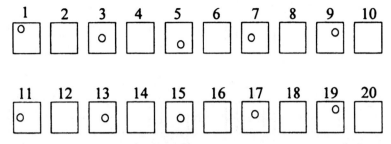

Figure 10–4a Illustration of regular (nonrandom) distribution throughout a production lot of 20 boxes (Lot A).

uct through this filler head will tend to flush out the contamination so the first product through that filler head will be more highly contaminated. Initially, one in every 10 containers would have *Salmonella*. If samples were collected every few minutes and a quantitative determination performed, the concentration of *Salmonella* in the product from the one filler head would decrease as illustrated as in Lot B of Figure 10–4b.

Analysis of random samples from the lot could by chance detect a positive sample. However, if a retest does not include one or more samples collected from the same filler head very early in production, the initial positive result will not be confirmed even if many samples are analyzed.

Example 2: A combination of condensation and product dust at the top of bucket elevators conveying dry product can provide sufficient moisture and nutrient to allow growth of *Salmonella* and, if introduced, a growth niche can occur during periods of no production. As production resumes, dry product conveyed near the top of the conveyor creates a drier environment. The residue can dry and slough off in clumps into the product stream. In this example, contamination is limited to a particular bucket, with product before or after the event being unaffected. As the product is subsequently transported, perhaps by air or screw conveyor, the clump breaks up and is diluted downstream. The effect is that *Salmonella* within the clump are distributed throughout the product stream much like a comet in space, with the highest level of contamination where the clump entered the product stream, followed by a dilution tail until the stream of product is not contaminated. This is illustrated as Lot C in Figure 10–4c.

In the above "comet-like" contamination, the first and last packages will be negative, but somewhere within the production lot is a series of contaminated packages of product.

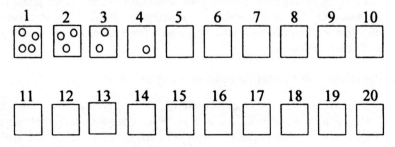

Figure 10–4b Illustration of heterogeneous (nonrandom) distribution throughout a production lot of 20 boxes (Lot B).

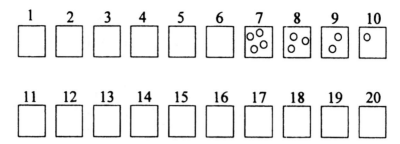

Figure 10–4c Illustration of heterogeneous (nonrandom) distribution throughout a production lot of 20 boxes (Lot C).

Again, depending on the sampling plan, this contamination may not be detected and if a positive sample is detected, extensive retesting may not confirm the positive result unless samples are collected from the affected portion of the lot.

In practice, "comet-like" contamination can occur with a wide variety of processed foods. For example, one commercial experience involved a dry-blended product that tested positive for *Salmonella* using a sampling plan of $n = 30$. The analytical units (25 g) were composited into two 375 g samples for analysis, for a total of 750 g. All the dry ingredients had been received with certificates indicating they had been sampled and tested negative for *Salmonella*. Since the blending operation was completely dry, there appeared to be little chance for contamination during mixing and packaging. Thus, the ingredients initially were suspected as the source of contamination and the remaining material was extensively resampled.

Eight pallets from an original shipment of 20 remained of a particularly suspect ingredient. Every 25 kg bag was sampled (375 g) and analyzed. Most bags from one pallet of this ingredient and a few bags from another tested positive for *Salmonella*; all others tested negative. Because the bags were sequentially numbered during filling, it was possible to determine that the initial contamination occurred during the middle of the day and decreased to below the detection limit within a couple of hours. Furthermore, one pallet that had been used contained bag numbers within the range of those found positive. The source was further confirmed when the *Salmonella* serotype isolated from the blended product matched the isolate from the ingredient. Obviously, had all three contaminated pallets been used, extensive resampling of the remaining pallets would have yielded negative results and the source of contamination would not have been determined.

Commercial experience with two-class attributes sampling plans for *Salmonella*, *L. monocytogenes*, and other infectious agents clearly demonstrates that, heterogeneous distribution of contamination can be a reason for discrepancies between initial test results and extensive retesting.

10.7.2 Low Prevalence of Contamination in a Lot or Batch

Microbiological criteria specify the quantity of product to be analyzed. Whether a pathogen such as *Salmonella* is detected also depends on its prevalence throughout the lot, its concentration in the analytical unit, and the sensitivity of the analytical method. When present at a high prevalence and concentration sufficient for detection by the test method, the initial analysis and subsequent retests will be positive. When present at a lower prevalence, not every test portion will contain the pathogen. Detection then depends on the probability of including a contaminated (i.e., positive) sample in the analysis. To confirm an initial positive result, a second positive sample must be selected. The probability of selecting a second positive sample in a row is much lower than the probability of selecting the first positive sample. Confirming an initial positive result when a pathogen is of low prevalence can be difficult, as described in the following.

In this example, a thoroughly blended product in a 500-L kettle contains one *Salmonella* cell per 3,750 g. If a two-class sampling plan is used that involves analysis of 375 g sample units, there is one chance in ten that each 375 g test sample will contain one *Salmonella* cell. If the first sample tests positive, then the second 375 g sample will likely test negative. The probability that two samples will test positive in a row is (1 in 10) × (1 in 10) = 1 in 100. Thus, there is only one chance in 100 that two consecutive samples will yield a positive result. Given this situation, a retest used to confirm an initial positive result will almost always increase the risk of accepting a contaminated lot. This is because both the initial and retest samples have equal probability of being positive when contamination is homogeneous. The probability that both will yield a positive result is the square of the probability of a single portion testing positive (Table 10–3). The lower the prevalence of contamination, the more difficult it will be to confirm. Confirmation will depend upon chance or testing until the prevalence of contamination is established. A very low prevalence of contamination is virtually impossible to confirm by resampling.

10.7.3 Low Prevalence of Contamination across Many Batches

Occasionally, test results indicate that contamination is occurring frequently, but at a very low level. In this example, a 500-L kettle is used to produce 20 lots per month of a homogeneous product and the lots are contaminated with one *Salmonella* cell per 3,750 g. Using a sampling plan involving $n = 15$ (25 g analytical units) and a composite of 375 g, only two lots will test positive. If sufficient retesting is conducted to establish that the prevalence of contamination in each of the two positive batches is one cell in 3,750 g, then it could be concluded that there was one chance in 100 that two positive lots would have been found. Since this is a low probability, one may suspect that the negative lots also may have been contaminated, but at low prevalence. A thorough analysis of several negative lots would determine whether the contamination is widespread. In any event, product histories should be charted and reviewed for sporadic instances of positive test results that indicate low level, widespread contamination.

Table 10–3 Probability of Detecting an Initial Positive and Then Finding a Second Positive When the Lot Is Retested

Probability of initial positive*	Probability of both initial and retest positive
1 in 2	1 in 4
1 in 5	1 in 25
1 in 10	1 in 100
1 in 20	1 in 400
1 in 50	1 in 2,500
1 in 100	1 in 10,000

*Probability of a positive = lot size in g × no. cells/size of test sample in g

10.7.4 Change in the Concentration of Contamination

In a contaminated ingredient, food, or food processing environment, the number of viable cells may increase, decrease, or remain the same with the passage of time. If a pathogen is multiplying, the concentration of contamination (cells g^{-1}) will increase and become easier to detect. However, if death is occurring, the probability of detection will decrease. An example of a survival curve for *Salmonella* introduced into a dry product is shown in Figure 10–5. During the first few days, the number of viable cells decreases rapidly. However, *Salmonella* can be detected because the concentration remains above the detection limit. A traditional *Salmonella* test requires about five days for a positive result. During this time, the concentration of viable cells has decreased to below the detection limit. If a retest is requested, there is an increased probability that the retest will not confirm the initial positive result. Rates for the initial rapid decrease and the subsequent slower decrease (Figure 10–5) vary by product, pathogen, and storage condition. Extensive experience with preparing inoculated samples for laboratory performance testing, evaluation of new methods, and challenge testing indicates that transferring pathogens such as *Salmonella* from a high moisture growth condition (e.g., broth cultures) into dry product results in a survival curve similar to that shown in Figure 10–5. A very similar curve would be expected for *Salmonella* that has multiplied in a plant environment and is then introduced into a dry food (e.g., the condensation/bucket elevator example). If product is sampled and tested the first few days after production, levels may be considerably higher than later when the concentration of survivors is lower. but more stable. Depending on the sensitivity of the method and the level of surviving cells after numbers stabilize, retesting may or may not confirm an original positive result.

10.8 SUMMARY

There are many possible explanations why a two-class attributes sampling plan may yield a positive result that is not confirmed by retesting. In practice, all of the phenom-

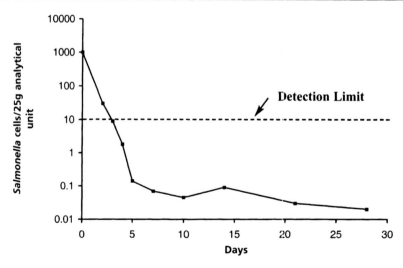

Figure 10–5 An example of a survival curve for *Salmonella* under dry conditions.

ena described above can be involved. Contamination may be heterogeneously and non-randomly distributed, at low prevalence, and unstable in the product. Disregarding an initial positive result because it is not confirmed by retesting should not be done unless there is clear reason to suspect laboratory error or contamination during sampling. Since it is very often difficult or impossible to confirm that results are erroneous due to laboratory or sampling error, it is extremely important that the laboratory performing qualitative tests has an effective quality assurance program to prevent laboratory error and to recognize contamination, should it occur.

10.9 REFERENCES

Flowers, R. S. & Curiale, M. S. (1993). Qualitative microbiology: discrepancies between original and retest results. *Scope* (A Technical Bulletin from Silliker Laboratories, Homewood, IL) 8, 1–5.

Gabis, D. A. & Silliker, J. H. (1974). ICMSF methods studies. II. Comparison of analytical schemes for detection of *Salmonella* in high-moisture foods. *Can J Microbiol* 20, 663–669.

ICMSF (International Commission on Microbiological Specifications for Foods) (1974). *Microorganisms in Foods 2. Sampling for Microbiological Analysis: Principles and Specific Applications.* Toronto: University of Toronto Press.

ICMSF (International Commission on Microbiological Specifications for Foods) (1986). *Microorganisms in Foods 2. Sampling for Microbiological Analysis: Principles and Specific Applications*, 2nd edn. Toronto: University of Toronto Press.

NAS-NRC (National Academy of Sciences-National Research Council) (1969). *An Evaluation of the* Salmonella *Problem.* Publication No. 1683, Washington, DC: NAS-NRC.

Silliker, J. H. & Gabis, D. A. (1973). ICMSF methods studies. I. Comparison of analytical schemes for detection of *Salmonella* in dried foods. *Can J Microbiol* 19, 475–479.

Chapter 11

Sampling To Assess Control of the Environment

11.1 Introduction
11.2 Principles of GHP
11.3 Post-Process Contamination
11.4 Establishment and Growth of Pathogens in the Food Processing Environment
11.5 Measures To Control Pathogens in the Food Processing Environment
11.6 Sampling the Processing Environment
11.7 References

11.1 INTRODUCTION

This chapter discusses the importance of microbiological testing to assess the risk of product contamination from the environment. Preventing contamination of ready-to-eat foods is emphasized. While limited to control of pathogens, the concepts also can be applied to microbial spoilage. Routine environmental sampling most likely will be applied in certain food processing plants and less likely at other steps along the food chain.

The microbiological safety of industrially manufactured foods requires effective design and implementation of Good Hygienic Practices (GHP) and Hazard Analysis Critical Control Point (HACCP) systems. GHP includes those prerequisite conditions that are necessary for pathogen control and implementation of an effective HACCP plan.

Food microbiologists have traditionally focused on the microbial ecology of raw materials and finished products. Such knowledge has been essential for limiting the number of pathogens in raw materials and for establishing critical limits at critical control points (CCPs) in an HACCP plan. It is, however, becoming clearer that some knowledge of the ecology of the food processing environment also is essential. Reports of foodborne outbreaks due to commercially manufactured foods reveal that many can be traced to weaknesses in GHP implementation. In some cases, post-process contamination has resulted from contact with contaminated equipment, such as during chilling, holding, conveying, or packaging.

In HACCP systems, CCPs are applied to prevent, eliminate, or reduce microbial hazards to acceptable levels. For many products, CCPs typically include a kill step (cooking, sterilization), but in minimally processed foods, CCPs may involve manipulation of one or more factors, such as low temperature storage, reduced a_w, and/or reduced pH, specifically designed to control growth of pathogens. Establishment of appropriate critical limits at specific processing steps ensures safety of the products, taking into account possible fluctuations during operations.

Experience indicates, however, that optimal application of a HACCP plan or a GHP program may not guarantee that contamination from the processing environment will not occur. Even the best surgical operations in hospitals carry the possibility of contamination that may be life threatening to the patient. Thus, it is possible to reduce, but not prevent, or eliminate, the likelihood of contamination when food is exposed to a processing environment (whether in an open or closed system). When this potential is recognized, certain manufacturers may decide to establish procedures to monitor the implementation of GHP and, where contamination can result in unsafe foods, to establish a routine sampling program to assess control of the environment.

11.2 PRINCIPLES OF GHP

The basic elements of GHP are described in Codex Alimentarius Commission's document *General Principles of Food Hygiene* (CAC, 1997). Specific recommendations for individual products or product groups (e.g., canned foods, aseptically processed and packaged low-acid foods, spices and dried aromatic foods, infant formulae, frozen foods, fruits, and vegetables) have been addressed in several documents (CAC, 1994, 1995).

Product-specific documents also have been issued by other international organizations to provide guidance on GHP implementation (IDF, 1992; IOCCC, 1993; ECFF, 1996; EC, 1997) and hygienic design of premises and processing equipment (EHEDG, 1997).

Some of the more important prerequisite programs that can minimize contamination from the processing environment are:

- layout of processing lines and control of the movement of personnel and mobile equipment to minimize cross-contamination from raw materials to finished product.
- equipment design and location for cleanability.
- cleaning and disinfecting procedures that are targeted toward the pathogens of concern to the process.
- scheduled preventive maintenance to minimize breakdowns during operation.
- appropriate waste removal.
- training and behavior of personnel relative to the target pathogens.

Visual and other sensory inspections are valuable tools to assess adherence to, and the effectiveness of, GHP. While such inspections can be performed routinely by trained specialists, all workers should be trained to be constantly vigilant and to report deviations from normality. Microbiological surveillance for pathogens or appropriate indicators provides supporting information to assess whether the environment is under control.

The potential for contamination should be considered in the design and implementation of the GHP program and, if contamination is likely to occur, some means for assessing its effectiveness should be adopted. Each manufacturing facility must consider control measures that are appropriate for the conditions of operation, the food, the process, and the pathogen(s) of concern to minimize in-process contamination (Cordier, 1997; Tompkin *et al.*, 1999).

11.3 POST-PROCESS CONTAMINATION

Despite strict control of all CCPs to ensure destruction of pathogens in raw materials, foods may subsequently become contaminated within the food operation. This typically results from two general circumstances:

1. adding a contaminated ingredient following a kill step; or
2. an environmental contaminant.

11.3.1 Contaminated Ingredients

Many cases could be cited where pathogens were introduced into ready-to-eat foods through contaminated ingredients, such as cocoa powder used to manufacture chocolate (*Salmonella*; Gästrin et al., 1972), paprika used to coat potato chips (*Salmonella*; Lehmacker et al., 1995), onions added to cheese curd (*Clostridium botulinum*; De Lagarde, 1974), sprouts added to sandwiches (*Escherichia coli* O157:H7; CDC, 1997), and hazelnut purée added to yogurt (*C. botulinum*; O'Mahoney et al., 1990).

Identifying high-risk ingredients during the hazard analysis when establishing a HACCP plan may involve epidemiologic or other published data, expert knowledge, and company databases on the incidence of pathogen(s) in the processing plant or in product. This should lead to the adoption of control measures, such as selection of suppliers, sampling plans for sensitive ingredients, or modification of recipes or processes. The addition of contaminated ingredients to ready-to-eat foods will not be further discussed. The principles described in the previous chapters are applicable to controlling this route of contamination.

11.3.2 Environmental Contamination

Foods cooked in the container in which they are sold and foods packaged using aseptic packaging systems are protected from contamination. Foods that are pasteurized, cooked, or undergo other processes that reduce pathogens and are subsequently exposed to the environment may acquire microorganisms before filling or packaging. Manufacturers of such foods take every reasonable precaution to prevent contamination from occurring after the food is cooked and before it is packaged, but it is impossible to prevent completely.

Some ready-to-eat foods are not cooked, but receive mild treatments that may alter their microflora, or they acquire microflora from the processing environment. For example, a range of fish products (e.g., pickled, gravad, or cold-smoked) receive very mild treatments, yet they will much more likely carry a microflora that reflects the microorganisms associated with the post-processing equipment and environment.

Ready-to-eat foods typically have been subjected to a wide range of processing conditions, some of which are intended to eliminate pathogens. Subsequent handling, storing, conveying, sorting and packaging, etc., create opportunities for contamination. Even the environment to which the foods are exposed can be a potential source of pathogens. Pathogens such as *Salmonella* or *Listeria monocytogenes*, and also spore-formers, can be introduced into a food operation by a number of vectors. Conditions permitting, they may become established and multiply, particularly in sites in the processing environment that are difficult to clean and disinfect. If this occurs, during production the pathogens may be transferred to product contact surfaces in the vicinity of the site and pass downstream with the flow of the product. In recent years, evidence has accumulated indicating that some pathogens can become established in the processing environment and persist for long periods, even years.

The probability of contamination of the food product increases with prevalence; the concentration of the pathogen is also influenced by the type of process, the hygiene procedures, and whether those procedures will spread the pathogen if they are not properly performed (Holah et al., 1993). Severe failure of one or more steps in GHP may lead to high levels of pathogens in the processing environment and to product contamination (Langfeldt et al., 1988; Steuer, 1992).

11.3.3 Examples of Outbreaks Due to Recontamination

Numerous outbreaks of illness due to the consumption of processed foods that have been recontaminated have been reported and described (Table 11-1). The number of implicated foods from industrial establishments, however, is relatively small in comparison to the total, with most outbreaks being related to foods prepared in food service operations, catering operations, homes, etc. (Beckers et al., 1985; ICMSF, 1988; Motarjemi & Käferstein, 1997). The size of the outbreaks associated with industrially manufactured foods, however, can be very large due to the quantity of product and distribution throughout a country or region. In addition, some outbreaks have been known to last several months before they are recognized and the source of the manufactured food is identified. The causes and origin of the contamination are seldom published, but sufficient data are available to demonstrate the importance of post-process contamination.

The following examples illustrate the importance of adequate GHP procedures and the establishment of sites or niches supporting microbial growth and survival.

Salmonellosis due to consumption of chocolate presents a typical example. In an outbreak due to *S. eastbourne* (Craven et al., 1975), inadequate separation between areas for storage and handling of raw and roasted cocoa beans led to contamination of roasted beans. Analysis of environmental samples identified the outbreak strain at different points, in particular on the wheels of a forklift used in both the raw and roasted bean areas.

In a large listeriosis outbreak in Switzerland due to the consumption of a soft cheese, Vacherin Mont d'Or, the *L. monocytogenes* serotype causing the outbreak was traced back to several ripening centers collecting otherwise sound cheeses from different small cheese makers. In particular, the serotype causing the outbreak was found on the trays used for cheese storage during maturation (Bille, 1990).

11.4 ESTABLISHMENT AND GROWTH OF PATHOGENS IN THE FOOD PROCESSING ENVIRONMENT

Effective GHP procedures targeted toward the pathogens of concern require knowledge of the incidence, distribution, fate, and behavior of the pathogens in food processing environments. A fundamental question that must be considered for each manufacturing facility is whether contamination can occur due to microorganisms that are merely transient in the premises or pathogens that become established in the premises.

11.4.1 Transient Versus Established Microorganisms

Transient microorganisms are introduced into a food plant through raw materials, humans, packaging supplies, etc., and do not become established in the environment and multiply. Transients, however, can contribute significantly to the types and numbers of microorganisms on the resulting food. Each raw agricultural commodity carries a wide variety of microorganisms and, possibly, certain pathogens (ICMSF, 1998). In general, industrial operations convert raw commodities into further processed products under con-

Table 11-1 Examples of Outbreaks Attributed to Environmental Contamination

Product	Pathogen	Comments	Reference
Ice cream	S. enteritidis	Pasteurized ice cream mix in tanker truck previously used for transporting raw liquid eggs	Hennessy et al. (1996)
Infant formulae	S. ealing	Contamination from the processing environment, insulation material of the drying tower	Rowe et al. (1987)
Soft cheese	S. berta	Cheese ripening in buckets previously used for chicken carcasses	Ellis et al. (1998)
Cooked sliced ham	S. typhimurium	Cooked ham placed into containers previously used for curing raw pork	Llewellyn et al. (1998)
Chocolate	S. napoli	Possibly contaminated water used in double-walled pipes, tanks, and other equipment	Gill et al. (1983)
Chocolate	S. eastbourne	Contamination from the processing environment	Craven et al. (1975)
Butter	L. monocytogenes	Contamination from the processing environment	Lyytikainen et al. (2000)
Hot dogs	L. monocytogenes	Contamination from the processing environment	Anonymous (1999)
Canned salmon	C. botulinum	Contamination from the processing environment, cooling water	Anonymous (1984); Stersky et al. (1980)
Lasagne	S. aureus	Growth of S. aureus in the processing equipment, improper cleaning	Woolaway et al. (1986); Aureli et al. (1987)
Different foods	E. coli O157:H7	Contaminated meat grinder and equipment at retail level	Banatvala et al. (1996)
Chocolate milk	Y. enterocolitica	Probably during manual mixing of pasteurized milk and chocolate syrup	Black et al. (1978)

continues

Table 11-1 continued

Product	Pathogen	Comments	Reference
Canned meat	S. typhi	Use of nonpotable water for can cooling	Ash et al. (1964); Stersky et al. (1980)
Crabmeat	S. aureus	Contamination during manual picking of cooked meat	Bryan (1980)
Canned mushrooms	S. aureus	Possible growth of S. aureus in the brine bath before canning	Hardt-English et al. (1990)
Flavored yogurt	E. coli O157:H7	Pump previously used for raw milk	Morgan et al. (1993)
Pastry	S. enteritidis PT4	Equipment previously used for raw eggs or insufficiently cleaned piping and nozzles used for cream	Evans et al. (1996)
Yeasts	S. münchen	Contamination from the processing environment	Joseph et al. (1991)
Pasteurized milk	S. typhimurium	Possibly cross-connection between raw and pasteurized milk	Lecos (1986)
Pasteurized milk	E. coli O157:H7	Contamination from pipes and rubber seals of the bottling machine	Upton & Coia (1994)
Mexican type cheese	L. monocytogenes	Contamination from the processing environment	Linnan et al. (1988)

ditions that minimize cross-contamination during processing. The cleaning and disinfecting procedures normally are adequate to control the transient flora so that each day of operation is separate from the previous day. For example, the salmonellae serotypes present in one group of animals on day 1 may differ from those on day 2, and the resulting raw meat or poultry from each day will reflect the difference. Upon cooking, the salmonellae are killed and, should the products become contaminated, the serotypes will reflect those in the post-cooking environment. This concept applies generally to most commodities that receive a kill step during processing.

Resident microorganisms are introduced into the environment, become established, multiply, and persist for days or years. Normal cleaning and disinfecting procedures control their numbers but may not eliminate them from the environment. Commercial experience suggests the pathogens can be categorized as follows:

- extensive history of establishment—nontyphoid *Salmonella, L. monocytogenes*
- some history or the potential exists for establishment—*Staphylococcus aureus, E. coli O157:H7, Yersinia enterocolitica, Bacillus cereus, C. botulinum, C. perfringens*
- no history of establishment—*S. typhi*, shigellae, *Campylobacter jejuni*, viruses, parasites

Temperature is an important determinant of whether a pathogen can become established as a member of the resident flora of a food plant. If the environmental temperature is at, or near, the lower limit for growth, the pathogen would most likely remain transitory. For example, salmonellae and *E. coli* O157:H7 would not be expected to become established in a beef cutting, trimming, grinding, and packaging facility with a room temperature of 10 °C or below. The second most important factor is available moisture. Food processing environments that are *continuously* dry (e.g., no condensation or leaking roofs, no wet cleaning) would prevent establishment and growth of all the pathogens listed above.

Pathogens primarily of human origin (e.g., *S. typhi*, shigellae) have not been reported to become established in a modern manufacturing facility. Pathogens requiring a human host or living cells (e.g., viruses, parasites) cannot multiply in a food processing environment. The high temperature and microaerophilic conditions required for growth and sensitivity to dehydration preclude *C. jejuni* from becoming established in the processing environment.

Evidence for the establishment of spore-forming pathogens is sparse, but commercial experience with spore-forming spoilage organisms in a variety of foods is evidence for this possibility. For example, examination of spoiled canned foods in which spore-formers were acquired as post-retorting contaminants of leaking cans supports this possibility (Davidson *et al.*, 1981; Matsuda *et al.*, 1985; Lake *et al.*, 1985a). A survey for mesophilic anaerobic spore-formers in three canning plants found significant numbers of these spores on various equipment and can conveyor tracks (Lake *et al.*, 1985b). The recovery of spore-formers from spoiled low acid canned foods must be interpreted with caution (Segner, 1979). Post-process contamination of canned tuna fish and salmon with *C. botulinum* type E from within the canning factory has occurred (Johnston *et al.*, 1963; Dack, 1964; Denny, 1982; Anonymous, 1984). There is evidence that *B. cereus* can adhere to dairy pipelines and lead to contamination of milk and milk products (Pirttijärvi *et al.*, 1998). Furthermore, *B. cereus* isolates from one section of a dairy process differed from those of another, depending on the prevailing temperature (e.g., > 30 °C versus 2–4 °C).

While the potential theoretically exists for the establishment of *Y. enterocolitica*, this has not been demonstrated. For example, in pork slaughtering and cutting operations, the primary source has been found to be the animals being processed (i.e., tonsil area of pigs and, thus, the intestinal tract and feces) and not *Y. enterocolitica* that have become established in the slaughtering environment (Nesbakken *et al.*, 1994; Fredriksson-Ahomaa *et al.*, 2000).

Despite the ambient temperatures during slaughtering, available evidence indicates that salmonellae and *E. coli* strains do not become established in the slaughtering environment (Magwood *et al.*, 1967; Bryan *et al.*, 1968; Cherrington *et al.*, 1988). Instead, the isolates detected on carcasses reflect those of the incoming animals.

A few examples will illustrate the importance of pathogens that do become established in the food processing environment. During the 1960s, nonfat dry milk was implicated as a source of salmonellosis (CDC, 1966). Subsequent investigations revealed that salmonellae could be detected in dried milk products and environmental samples from many different manufacturing plants (Ray *et al.*, 1971; Blackburn & Ellis, 1973). In Ray *et al.* (1971), the frequency of isolation was highest in the first product after start of operation (24.9%). The frequency then decreased during the production run (first acceptable bag of product, 9.0%; middle bag, 2.5%; last bag, 4.1%). Also, salmonellae were detected in 21.1% of the environmental samples from nine plants, with none of the plants testing negative. Such data led to considerable improvements in the processing conditions and the use of environmental sampling programs. More recently, an outbreak has been attributed to milk powder containing *Salmonella mbandanka*. Investigational sampling of the processing plant established a link between contamination of external areas of the buildings (roof) and the internal processing environment. Multiplication of the pathogen was shown to have occurred in a scrubber used to wash exhaust air and was an ideal source of contamination within the factory (Langfeldt *et al.*, 1988).

A major outbreak of salmonellosis attributed to a toasted oat breakfast cereal is another example of the importance of post-process contamination. While this contamination was originally thought to have involved the application of a solution of vitamins, this did not prove to be the case. Instead, the data indicated the pathogen had become established in the environment because it was recovered from the equipment on the production line, floor, and exhaust system. The contamination was limited to only one of 12 production lines in the manufacturing plant. Further evidence that *S. agona* had become a resident along that particular production line was recovery of *S. agona* from product produced on March 27 and recovery of the same strain from the equipment when the plant was investigated after May 28. It was estimated that as many as 16,000 cases of salmonellosis occurred before the problem was detected and the product was removed from the market (Breuer, 1999). These and similar experiences in other food operations (e.g., dried eggs, chocolate, peanut butter, soy protein, gelatin, cooked diced poultry, meat and bone meal, fish meal) demonstrate the ability of salmonellae to become established in processing environments that operate at ambient temperatures.

The number of reports of long-term persistence of specific DNA types of *L. monocytogenes* in food establishments continues to grow. One type was detected in an ice cream plant over a seven-year period (Miettinen *et al.*, 1999). In a cold-smoked fish operation, a specific randomly amplified polymorphic DNA (RAPD) type was isolated from products over a three-year period. This specific type was found in the slicing environment, whereas a variety of others were detected in the raw fish handling area (Fonnesbech-Vogel *et al.*, 2001b). The outbreak due to the soft cheese, Vacherin Mont d'Or, occurred between 1983 and 1987, indicating a period of long persistence (Bille, 1990). Other examples involving a variety of environments and foods have been reported (Cox *et al.*, 1989;

Lawrence & Gilmour, 1995; Salvat *et al.*, 1995; Nesbakken *et al.*, 1996; Loncarevic *et al.*, 1996; Unnerstad *et al.*, 1996; Ericsson *et al.*, 1997; Autio *et al.*, 1999; Fonnesbech-Vogel, 2001; Norton *et al.*, 2001).

S. aureus was found to colonize the rubber fingers of feather picking (plucking) machines and serve as the primary source for the strains found on raw chicken carcasses (Gibbs *et al.*, 1978; Notermans *et al.*, 1982; Adams & Mead, 1983; Dodd *et al.*, 1988; Mead & Dodd, 1990; Mead, 1992). The presence of a unique strain of enterotoxigenic (type B) *S. aureus*, phage type 94/96 with resistance to colistin, ampicillin, and penicillin, was recovered from Swiss cheese implicated in an outbreak. This strain was detected in cheese produced in one factory over a period of at least two years. Since the milk was pasteurized, this unique strain was introduced after pasteurization. The source was reportedly "contamination of the starter culture, probably through one of the company employees" (Todd *et al.*, 1981). In another outbreak involving 2,112 suspect vats of cheddar, Kuminost, and Monterey cheese, 59 vats were found to contain enterotoxin A. They had been produced between May and November during a time when process records showed the plant was having difficulty with slow and inactive starters. All the vats were phosphatase negative, indicating that the milk was contaminated after pasteurization (Zehren & Zehren, 1968). A survey of a whey powder plant concluded that *S. aureus* is infrequently found in the environment and is transient in nature, based on changing phage patterns. *S. saprophyticus* and *S. xylosus* were frequently isolated from the environment, each being associated with certain locations (Kleiss *et al.*, 1994).

It is very helpful to understand the difference between transient and resident microorganisms when designing a system to control in-process contamination. Furthermore, this distinction is essential when investigating and correcting a contamination problem (see section 11.6.8 and Chapter 9). A determination of whether pathogens are transient or resident can be made using a variety of traditional or more sophisticated DNA-based typing methods. Recovering the same isolates over a period of time is evidence that the pathogen is a resident of the environment.

The remainder of this chapter briefly considers factors that should be considered to minimize contamination. Some are applicable to both transient and resident pathogens.

11.5 MEASURES TO CONTROL PATHOGENS IN THE FOOD PROCESSING ENVIRONMENT

11.5.1 Minimizing Entry of Pathogens

It is not possible to prevent the introduction of pathogens into food processing facilities. This realization means GHP procedures should be targeted to control those pathogens most likely to be of concern for the food and conditions of operation. Among the sources of pathogens in food processing areas are:

- *Raw agricultural commodities* such as raw cocoa beans, raw milk, raw meat, poultry, fish or seafood, vegetables, fruit, nuts, and spices represent important sources of pathogens to food processing facilities. Physical separation of raw materials through plant design and layout is necessary to minimize entry of pathogens into processed product areas.
- *Food handlers and maintenance personnel* can be a source of food contamination. The effectiveness of gloves and other measures to minimize recontamination has been evaluated; the evidence suggests that training personnel in hygiene is more

important than the obligation to wear gloves (Fendler *et al.*, 1998a, b). Despite this finding, both hand washing and disposable gloves are common in many manufacturing facilities.

- *Personal clothing* and, in particular, shoes can transfer pathogens from one area to another. While preventive measures can be adopted, such as changing shoes or boots, it may be more effective to design the plant layout to direct the flow of personnel. The use of footbaths to control pathogens on shoes, boots, and equipment remains highly controversial. Some have found that maintaining clean, dry floors is more effective. Maintenance and management personnel and control authorities whose responsibilities require them to routinely visit and inspect all areas of the plant jeopardize control.
- *Air and water.* The role of air as a source of direct contamination of products is probably overestimated and not normally as important as other sources of contamination. Air quality is important in unique situations such as in the operation of aseptic packaging, air chilling, or drying and air conveying systems. The filters for compressed air can become a source of contamination if not properly maintained. Aerosols created during cleaning can disperse microorganisms throughout the processing environment. Water as it enters a manufacturing facility should be potable and in abundant supply. Water can become contaminated through use and serve as a vehicle for transmitting pathogens (e.g., water for washing, conveying, chilling) if not controlled.
- *Insects and other pests* such as flies, cockroaches, or rodents can serve as vectors for pathogens. Direct cross-contamination is assumed in one report where the spread of *Salmonella* in a household was investigated (Michanie *et al.*, 1987). In food processing facilities, however, insects rarely cause direct contamination, but can act as vehicles for transmission (Kopanic *et al.*, 1994; Olsen & Hammack, 2000; Urban & Broce, 2000).
- *Transport equipment,* such as racks, carts, trolleys, forklifts, or similar equipment, can be important vectors for transferring microorganisms throughout a facility. Some facilities color-code certain equipment and limit its use to specific areas.

11.5.2 Minimizing the Establishment of Pathogens in the Environment

Transport of pathogens into facilities is the first step in the process of pathogen contamination, but it is the establishment and multiplication that increase the risk that food will become contaminated. Establishment in the factory environment may be linked to one or more of the following factors:

- "macro" design
- "micro" design
- adherence and colonization of microorganisms.

11.5.2.1 Macro and Micro Design

Food processing rooms (e.g., walls, ceilings, floors, windows, doors, conduits or trays for electrical cable, overhead pipes) should be designed, constructed, and installed to minimize accumulation of dust and other material that can serve as sites potentially harboring microorganisms. They should be designed and installed to allow for efficient cleaning. Additional information is available (ICMSF, 1988; CAC, 1997; EHEDG, 1997).

Surfaces of equipment, floors, and walls may appear smooth to the naked human eye, but in reality microscopic examination of materials, such as stainless steel, rubber, or Teflon, shows characteristic rough structures where microorganisms can hide and that are conducive to adherence. Also, the microstructure of the materials of floors and walls is often rough or porous and may deteriorate due to exposure to water, heat, chemicals, and mechanical stress to form cracks and crevices.

Models are being developed that consider the hygiene aspects of food processing (Zwietering & Hasting, 1997a, b).

11.5.2.2 Adherence and Colonization: Biofilms and Niches

A range of factors contribute to microbial colonization of processing environments. When introduced into processing facilities, free or planktonic microorganisms may be on a carrier (e.g., dust, droplets, food particles) and, given the opportunity, tend to rapidly attach to equipment or environmental surfaces. This initial attachment phase is followed by adhesion. Adhesion depends on factors such as the type of surface (e.g., stainless steel, rubber), the type of microorganism and its physiological status, the physico-chemical status of the surface, and the existing microflora. Exposure to acid conditions or mild heat seems to predispose cells and trigger the phenomena of attachment and adhesion. Individual cells adhering to solid surfaces could already be classified as biofilms (nascent biofilm). This status confers ecological advantages such as increased resistance against dehydration, heat, and the effect of cleaning and disinfection agents (Norwood & Gilmour, 1999).

Biofilms. The most extreme forms of biofilms are found in systems with constant exposure to water, such as in pipes, flumes, air-cooling units, drains, and floors. The more developed and complex (mature) the biofilm, the stronger the protective effect against environmental stresses, the protection being enhanced by the presence of biopolymers and other material.

The formation of biofilms can occur in almost any environment with sufficient moisture and nutrients. Biofilms have been studied mainly on surfaces in contact with a liquid phase, such as pipes, cooling towers, and plate heat exchangers. Data from food processing environments, however, are limited (Taylor & Holah, 1996; Mettler & Carpentier, 1998). Detailed information on the formation, structure, and biochemistry of biofilms can be found in Wimpenny (1995) and Kumar & Anand (1999).

The microbial composition of biofilms depends on the food being processed and the operating conditions. In a survey of 17 different food environments, Gibson *et al.* (1995) found that pseudomonads and Enterobacteriaceae, in particular, were common colonizers of surfaces. The microflora in fish processing environments differs depending on the type of product being processed. Thus, the environment of smokehouses is dominated by pseudomonads, Enterobacteriaceae, and some yeasts (Bagge *et al.*, 2001). Similar types are found in processing facilities for semi-preserved fish, whereas factories manufacturing lumpfish roe harbor a flora consisting of *Vibrio* spp. and pseudomonads (Bagge *et al.*, 2001). Notably, these organisms can be easily recovered from many areas of the plant after cleaning and disinfection, indicating their ability to resist these regimens.

External stresses such as heat, acidity, and water activity can lead to increased resistance to adverse environmental conditions. In most of the cases, however, research on this topic has been restricted to the food matrixes and multi-hurdle systems. Reports of specific problems in food processing environments are very limited. Most have focused on resistance to disinfectants, particularly in biofilms (Carpentier & Cerf, 1993). Others have been devoted to the effect of environmental factors, such as moisture, drought, temperature, etc., on microbial survival (Gundermann & Johannssen, 1970; Gundermann & Glück, 1971; Gundermann, 1972; Dickgiesser, 1978) and their behavior, for example, in aerosols (Marthi et al., 1990; Walter et al., 1990).

Niches. A niche is a site harboring microorganisms that, typically, is impossible to reach and clean with normal cleaning and sanitizing procedures. The processing environment may appear visually clean and acceptable. Examples include hollow rollers on conveyors, cracked tubular support rods on equipment, the space between close fitting metal-to-metal or metal-to-plastic parts, worn or cracked rubber seals around doors, saturated insulation, interfaces between floors and equipment, cracks, and crevices. Niches tend to be continually wet and hold residues of food and other material. Microorganisms, including pathogens, can become established and multiply within these sites, which then serve as a reservoir from which the pathogen is dispersed during operation and contaminates food contact surfaces and food. In a controlled environment, the niche usually affects only the food along one processing or packaging line and not the product on adjacent lines. Microbiological testing is necessary to detect a niche.

11.6 SAMPLING THE PROCESSING ENVIRONMENT

Environmental sampling is used to:

- assess the risk of product contamination
- establish a baseline for when the facility is considered under control
- assess whether the environment is under control
- investigate a source of contamination so corrective actions can be implemented.

11.6.1 Assessing the Risk of Product Contamination (Investigational Phase)

Biased, investigational sampling techniques (Chapter 9) are most appropriate for assessing whether a product can become contaminated with a pathogen. The purpose of this sampling is to find a target pathogen if it is present. The selection of samples will reflect the experience of the investigator and the process being investigated. In general, in-process food samples and/or sponge samples may be collected. Food samples should be collected from stages throughout the process that could permit contamination. Consideration should be given to collecting samples at different production times (e.g., first product, middle of production, end of production, following a break in production, after mechanical repairs, after a change in ingredients or packaging material). Product residues (e.g., shavings from slicing machines, sweepings from the floor of a dried milk plant) also can be collected since they serve as a type of composite sample from the process.

Environmental samples can be collected from product contact surfaces and the immediate surrounding. Sponges (Silliker & Gabis, 1975) or large cotton pads are suitable for this purpose. Since the intent is to detect the presence of a pathogen, if present, large areas should be sampled without regard to dimension. Environmental samples should also be collected at different times as just described for product samples.

Certain samples represent historical information measuring accumulation over time. Examples include residues from hollow bodies, cracks, and crevices; joints and seals between floor and equipment; water residues from siphons; drains; dust from vacuum cleaners; brooms, brushes, and mops; air filters; and wet or old insulation material.

In addition to analyzing for pathogens, samples may be tested for indicators (*E. coli*, listeriae) relevant to the pathogen(s) of concern. All the data should be organized in a manner that a baseline can be established for what is considered normal when all GHP procedures are in control.

11.6.2 Establishing a Routine Environmental Sampling Program

Routine environmental sampling programs are usually focused on one pathogen and/or indicator organism and involve a limited sampling regime. The purpose of the routine sampling program is to detect increased risk of product contamination before it actually occurs. Data from the investigational phase are used to select sampling sites, times, frequencies, and types of samples that will most effectively fulfill that purpose. Some of the basic process control concepts described in Chapter 13 may be applicable in the design and implementation of an environmental sampling program. The use of trend analysis, in particular, is useful. Statistical analysis is seldom of value, however, if the program involves presence/absence testing for a pathogen, such as in a cooked product area. In this case, any positive sample is a warning that requires further investigation.

11.6.3 Sampling Locations: The Zone Concept

Some manufacturers design their environmental sampling programs around zones having different levels of risk (Figure 11-1). For example, the zone of highest risk (zone 1) includes product contact surfaces over or through which product passes during processing. The next zone (zone 2) consists of equipment and other items that are in close proximity to the product flow and may indirectly lead to product contamination. The third zone may include items or areas that are less likely to lead to product contamination, but nevertheless may hinder efforts to control pathogens and from which pathogens could potentially be transferred to zones 1 or 2. The fourth zone is outside the processing area and if not maintained at an acceptable level of cleanliness could increase the risk of introducing pathogens into zones 1–3. The concept of zones with differing levels of risk can be used to select sites for a routine environmental sampling program. In addition, the concept can be used as a teaching aid for plant personnel. It is important to recognize that the zone concept has not been standardized and the items designated for each zone differ among food operations, depending on accumulated data and experience.

The purpose of routine environmental sampling is to verify that the GHP procedures are controlling the risk of product contamination, with sampling locations selected according to the risk of product contamination. In the case of heat-processed foods, for example, attention should be on the zones after cooking and where the products are exposed (e.g., where cooked ham is sliced and packaged). Emphasis should be placed on sites in zone 1 and supplemented with sites from zone 2, as needed. Figure 11-2 illustrates possible sampling sites in four different food operations. Equipment located between when the food is smoked/heated and protected by wrapping or packaging is of greatest concern (designated by a shaded area). Samples from product contact surfaces of equipment (i.e., zone 1) that could be included in a routine sampling program have been identified. In addition, samples from floors or other sites in close proximity to the flow of the product (i.e., zone 2) have been indicated.

The types of samples collected for a routine sampling program should be determined from the investigational sampling data and experience as the program is implemented. Environmental sponge samples are more commonly collected for analysis than are product samples. Examples of sampling sites for the four food operations in Figure 11-2 are

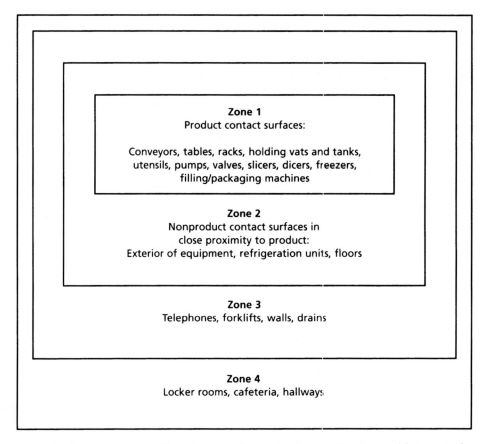

Figure 11–1 Zone concept to illustrate areas of highest risk (zone 1) to lowest risk (zone 4) for product contamination.

provided in Table 11–2. Environmental samples are intended to detect a microbial indicator or pathogen, if present. This may lead the sampler to collect from a large area from one piece of equipment and from a small area on another, relying on experience and previous results for guidance.

The sampling program could include in-process product samples where this may provide additional benefit over sponge sampling. The type of material sampled will depend on the type of product and processing line.

11.6.4 Number, Frequency, and Time of Sampling

Environmental sampling protocols are not statistically designed sampling plans. Instead, they are based on experience and knowledge of the sites most likely to detect a failure in GHP. The knowledge base continues to increase over time, enabling adjustments to further improve the system's sensitivity without increasing analytical costs unnecessarily.

The number of samples and frequency of sampling are normally determined by knowledge of the operation and its variability. Knowing when to collect samples may be more important than increasing the number or frequency of samplings. For example, it may be

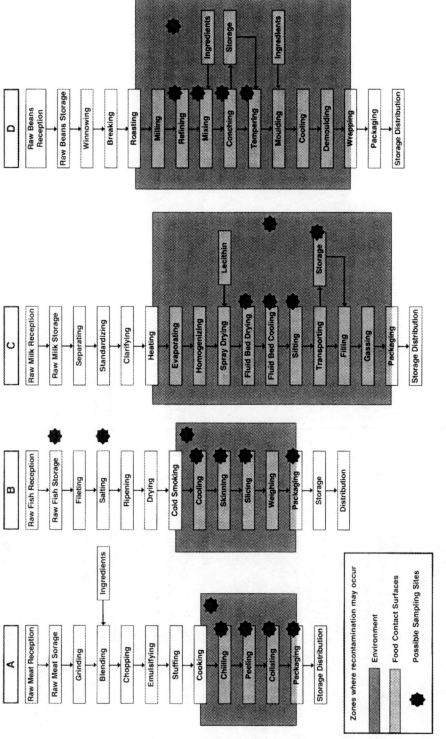

Figure 11–2 Flow diagram for 4 production lines and potential sampling sites in the environment and on product contact surfaces. A: Frankfurter sausages; B: Cold smoked salmon; C: Milk powder; D: Chocolate

Table 11–2 Examples of Possible Sampling Sites in the Environment of Four Processing Systems

	Frankfurters	Cold smoked salmon	Milk powder	Chocolate
	Food contact surfaces (Sponge samples)			
Equipment	Brine chiller	Transport racks	After dryer	Mills
	Peeler table	Trays	After cooler	Refiner
	Casing removal system	Tables	Pipes	Conches
	Hopper after peeler	Skinner	Conveyor belts	Storage tanks
	Incline conveyor	Brine injector/tank for brining	Tote bins	Pipes
	Collator	Slicer	Silos	Mixers
	Containers/tubs/trays	Scale		Molds
	Final conveyor	Conveyor belts		Tempering equipment
	Packaging machine	Packaging machine		Transport trolleys
	Scale			Trays
Utensils	Knives	Knives	Brushes	Scrapers
	Tools	Tools	Scrapers	Paddles
		Scrapers		Spatulas
	Environmental sponge samples (Indirect contact but close to product)			
Equipment	Overhead piping	Overhead piping	Lids	Lids
	On/off switches/buttons	On/off switches/buttons	Covers	Covers
	Legs of equipment	Legs of equipment	External surfaces of silos	Piping above equipment
	Sides of equipment	Sides of equipment		Internal surface cooling tunnel
	Bottom of equipment	Bottom of equipment		
	Floor by packaging line	Floor near packaging line		
	Floor of holding cooler	Floor of chill room		
	Environmental sponge samples (Indirect contact)			
Environment	Floors	Floors	Floors	Floors
	Walls	Walls	Walls	Walls
	Refrigeration units	Refrigeration units	Pipes	Pipes
	Drains	Drains	Ducts	Ducts
	Doors	Doors	Overhead structures	Overhead structures
				Drains

known that sponge samples collected during operation will yield more useful information about whether the environment is under control than samples collected from cleaned equipment before the start of operation, which may provide misleading results. In another situation, the first product from the process may represent the highest risk. Other important factors to consider include ease of sampling, whether the integrity of the product being processed will be jeopardized and, in particular, safety for the person collecting the samples.

Environmental sampling plans normally involve an established, routine sampling plan with a defined minimum number of samples. The number, frequency, timing, and sampling sites may be increased when evidence indicates increased risk of contamination. Tables 11–3a,b,c, and d list examples of sampling sites, numbers and frequencies for the four operations described previously. In addition, the examples indicate a possible increased frequency when data indicate such a need.

When judged necessary, special samples can be collected, such as during construction, during major changes to equipment, or after unforeseen events such as heavy storms that could have caused contamination of the processing environment (e.g., leaking roof).

To minimize the analytical workload and cost, samples may be composited (i.e., bulked) before analysis (e.g., from the same sampling point over one week or from different sampling points along a processing line). Compositing should only be done when the quality of the information obtained is known to be unaffected.

11.6.5 Sample Collection

Meaningful information is gained only if appropriate tools are used to collect samples. Sampling materials (e.g., sponges or cotton pads, utensils, cups, bags) must be sterile to avoid contamination. Proper labeling and description of the samples are essential. Before analysis, samples should be stored under appropriate conditions according to the characteristics of the sample (see Chapter 12) so that numbers of the target organism(s) neither increase nor decrease.

The sampling tools need to be adapted to the type of sample to allow for efficient sampling. Spatulas and scrapers of different size are used to collect residues from surfaces, holes, crevices, etc. Paintbrushes can be used to collect dusty samples on surfaces of equipment or on the infrastructure (e.g., electric cables, control panels, etc). Cotton plugs, sponges, or gauze pads are useful for liquid or moist residues as well as surfaces with limited amounts of residues. Sponges must not contain antimicrobial substances (Daley *et al.*, 1995). Other sampling tools such as pipettes may be required, depending on the situation (i.e., flexibility is needed to adapt the means of taking samples to local situations).

The sampling tools must not introduce other nonmicrobiological hazards into the processing system. In most manufacturing plants, for example, a strict policy exists that glass cannot be introduced into the processing environment.

11.6.6 Sample Analysis

Samples taken for microbiological surveillance are analyzed for specific microorganisms, i.e., pathogens of concern (e.g., *Salmonella, L. monocytogenes*) and for indicators, such as *E. coli*. Usually, traditional microbiological methods are used, although some more rapid methods are gaining acceptance.

A routine testing program for a pathogen may be supplemented with tests for hygiene indicators. Pathogen testing is usually qualitative (presence/absence). Such analyses are, however, time consuming, costly, and cumbersome. Therefore, the potential presence of pathogens may be assessed using indicators. Quantitative determination of Enterobacteriaceae, for example, allows the assessment of moisture control in environments intended to remain dry. The use of biochemical reactions to detect adenosine triphosphate (Flickinger, 1996) or residual proteins (Patrick & Bayliss, 1997) enlarges the concept of hygiene indicators. Although not directed toward the detection of viable microorganisms, the presence of residues of the manufactured products indicates the potential for growth and multiplication of microorganisms and thus for recontamination.

Of particular interest for environmental monitoring are sub-typing methods, which are used for pathogens and increasingly for other microorganisms. While biochemical methods are of limited use, serological, biological (phage-typing), and molecular methods, such as pulsed field gel electrophoresis, RAPD, ribotyping, and others, find more and more applications. These methods, and especially the molecular ones, are very powerful tools to trace microorganisms within processing facilities and establish links between isolates. Details on the technical aspects of these methods have been provided (Farber, 1996; Olive & Bean, 1999).

11.6.7 Managing the Data from Environmental Sampling

An environmental sampling program is useful only when data acquired are organized and frequently reviewed. In addition to reviewing the most recent data, it is also helpful to review data for the past quarter or year to detect weaknesses and trends that may not be otherwise evident. Normal, routine levels of sampling can be continued as long as the data are within the limits of acceptance. Detection of increased levels of indicators, such as Enterobacteriaceae, may not trigger immediate corrective actions but should serve as a warning. The response will depend on the type of product manufactured.

11.6.8 Investigational Sampling To Determine the Source of Contamination

When a trend or other information indicates an increased risk of contamination, the reason should be determined. This procedure involves a combination of increased sampling and biased, investigational sampling leading to collection of the data necessary to identify the source of contamination and to implement corrective actions. Tables 11–3a,b,c, and d provides examples of increased levels of sampling that could be used for the four processes introduced in Figure 11–2.

During investigational sampling to detect the source of contamination, it is important to assemble and review historical environmental sampling data that may reveal trends or suggest a particular cause. Of particular interest are data for finished product, ingredients, and the processing environment. These data may not include tests for the organism in question, but correlating data may provide insight into changes in the microbiology of the ingredients, process, or the environment. In addition, production records may associate the problem with specific events, such as construction, remodeling, mechanical repairs, or tests of new equipment or product formulations.

In addition, specialists with appropriate experience can conduct visual inspections of the cleaning and disinfecting procedures, the status of the equipment and operating conditions, employee practices and movement of equipment, product, and personnel.

The goal of the intensified sampling is to detect the contamination and its cause. Investigational sampling may identify biofilm formation, a niche supporting microbial growth in the equipment, or other causes. When pursuing a source such as a niche, it is

Table 11–3a Example of an Environmental Sampling Plan for *L. monocytogenes* in a Frankfurter Operation

	Frankfurters	
	Normal level	*Increased level*
Environmental sponge samples		
Floor in peeler area	1 x week	1–3 x day
Floor in vicinity of collating and packaging line	1 x week	1–3 x day
Equipment sponge samples, etc.		
Brine chill solution	1 x week	1–3 x day
Peeling table	1 x week	1–3 x day
Hopper/incline conveyor after peeler	1 x week	1–3 x day
Collator	1 x week	1–3 x day
Conveyor before packaging	1 x week	1–3 x day
Packaging machine	1 x week	1–3 x day
Finished product		
Finished product	1 x biweekly	1–3 x day

usually necessary to dismantle completely any suspected equipment. As this is being done, samples should be collected to confirm, if possible, that the source has been found. In such situations, care must be taken not to disseminate pathogens dislodged from their niche into the environment by careless handling of material and samples.

The increased frequency of sampling is normally not restricted to samples of the same type taken at the usual location. In reality, the source of the pathogen needs to be investigated to determine whether it is a transient or resident strain in the factory environment. Tracing, source detection, or troubleshooting can be a long and tedious process requiring repeated sampling campaigns, where attention is focused on sites in addition to those

Table 11–3b Example of a Sampling Plan for *L. monocytogenes* in a Smoked Fish Plant

	Cold smoked salmon	
	Normal level	*Increased level*
Environmental sponge samples		
Floor of chill room	1 x week	1–3 x day
Floor in vicinity of slicing line	1 x week	1–3 x day
Equipment sponge samples		
Brine injector/brine tank	1 x week	1–3 x day
Racks exiting chiller	1 x week	1–3 x day
Conveyor/table	1 x week	1–3 x day
Skinner	1 x week	1–3 x day
Slicer	1 x week	1–3 x day
Scale	1 x week	1–3 x day
Packaging machine	1 x week	1–3 x day
Finished product		
Finished product	1 x biweekly	1–3 x day

Table 11-3c Example of a Sampling Plan for *Salmonella* in a Dried Milk Powder Plant

	Milk powder	
	Normal level	Increased level
Environmental sponge samples		
Area zone drying tower	1 x week	Several points
Area lecithination room	1 x week	Several points
Area zone after dryer	1 x week	Several points
Area zone storage silos	1 x week	Several points
Area zone packaging	1 x week	Several points
Equipment sponge samples		
Star valve	1 x week	1 x day
Filters	1 x month	Several points
Cyclone residues	1 x week	1 x day
Sieves after cooler	1 x day	Several points
Silos	1 x week	1 x day
Packaging machine	First product	10 x day
Finished product		
Finished product	3 x batch	10 x batch

included in routine sampling. Preconceived ideas about the source should be viewed with caution and have been known to prolong the effort of detecting the true source. Sampling may be expanded to adjacent zones of lower concern to obtain a more complete picture of the extent of the contamination. Dismantling equipment, removing filters, etc. may be necessary to access sites that could harbor a niche.

Table 11-3d Example of a Sampling Plan for *Salmonella* in a Chocolate Plant

	Chocolate	
	Normal level	Increased level
Environmental sponge samples		
Area zone after roaster	1 x week	Several points
Area zone refining	1 x week	Several points
Area zone mixer	1 x week	Several points
Area storage rework	1 x week	Several points
Area storage ingredients	1 x week	Several points
Equipment sponge samples		
Mixer	1 x week	1 x day
Conches	1 x day (rotation)	1 x day (all)
Tempering equipment	1 x day (rotation)	1 x day (all)
Storage tanks	1 x day (rotation)	1 x day (all)
Molding	1 x week	Several points
Rework vessel	1 x week	1 x day
Finished product		
Finished product	3 x batch	10 x batch

Two examples of investigational sampling in problem solving are provided below.*

Listeria monocytogenes in *processed, cold-filled cheese*. In this example, *L. monocytogenes* was isolated from processed cold-filled cheese. Upon investigating the source, 80 finished product samples from a one-month period, as well as over 200 ingredient samples and 1230 environmental samples, were analyzed. *L. monocytogenes* was found in six finished product samples, three ingredient samples, and one environmental sample. The only ingredient found to be contaminated was butter from a particular supplier. The environmental sample that proved positive was the underside of an elevated electrical cable stand to a chopper.

All 14 isolates were of the same serotype (1/2 b), multilocus electrophoresis type (ET202), ribotype (DD0941), and phagetype (non-typable). The probability that all 14 isolates would, by chance, have the same pattern is extremely remote and indicates a common source of contamination. Most likely, the butter was the source of contamination for the processed cheese and the single environmental positive resulted from spilled butter or cheese during manufacture.

Since only one of the 1230 environmental sponge samples was positive, it can be concluded that the unique isolate was not a resident of the facility producing processed, cold-filled cheese.

This study illustrates how serotyping, electrophoresis, ribotyping, and phage typing can be used in investigations to detect the source of contamination.

Staphylococcus aureus in fermented sausage. High numbers (e.g. $> 10^5$ g^{-1}) of *S. aureus* detected during a routine verification at the end of fermentation for a dry fermented sausage could have been due to a variety of reasons. Initially, product produced before and after the suspect lot was analyzed to determine whether the problem was continuing or an isolated problem. Furthermore, individual lots produced during the period in question were sub-lotted into smaller batches by time of production. Review of the processing records for the incriminated lots helped to determine the origin of the problem. In addition, equipment and environmental sampling was conducted.

Since *S. aureus* numbers decline during the drying/aging step that follows fermentation, quantitative analysis for *S. aureus* might yield misleading information. Thus, the product was tested for heat stable thermonuclease as an indicator of high numbers of *S. aureus*. The samples were collected from the outer few mm of the sausages since this is where growth of *S. aureus* would occur. The accumulated information indicated that the products most likely to contain detectable levels of heat stable thermonuclease were the first batches produced after weekend shutdowns. Sponge samples from equipment initially failed to detect a source of contamination; however, the production records clearly suggested growth of *S. aureus* in the processing equipment during shutdown as the probable cause. Reexamination of the equipment revealed growth in an area of the equipment that was difficult to reach and clean (and sample). The investigation concluded that *S. aureus* multiplication occurred in a niche in the equip-

---

*Unpublished case studies.

ment during shutdown on weekends. Meat moving through the unclean equipment during start of production on Monday mornings became contaminated with relatively high numbers of *S. aureus*. As production continued, subsequent batches were contaminated with decreasing numbers of *S. aureus*. The lower levels did not reach unacceptable levels during fermentation.

11.7 REFERENCES

Adams, B. W. & Mead, G. C. (1983). Incidence and properties of *Staphylococcus aureus* associated with turkeys during processing and further processing operations. *J Hygiene* 91, 479–490.

Anonymous (1984). Botulism risk from post-processing contamination of commercially canned foods in metal containers. *J Food Prot* 46, 801–816.

Anonymous (1999). Update, multistate outbreak of listeriosis—United States, 1998-1999. *Morbidity & Mortality Wkly Rpts* 47, 1117–1118.

Ash, I., McKendrick, G. D. W., Robertson, M. H. et al. (1964). Outbreak of typhoid fever connected with corned beef. *Br Med J* (i), 1474–1478.

Aureli, P., Fenicia, L., Gianfranceschi M. et al. (1987). Staphylococcal food poisoning caused by contaminated lasagne. *Arch Lebensmittelhyg* 38, 159–165.

Autio, T., Hielm, S., Miettinen, M. et al. (1999). Sources of *Listeria monocytogenes* contamination in a cold-smoked rainbow trout processing plant detected by pulsed-field gel electrophoresis typing. *Appl Env Microbiol* 65, 150–155.

Bagge, D., Ng, Y. Y., Hjelm, M. et al. (2001). The microflora of process equipment in three different fish processing industries. *Appl Env Microbiol* 67, 2319–2325.

Banatvala, N., Magnano, A. R., Carter, M. L. et al. (1996). Meat grinders and molecular epidemiology: two supermarket outbreaks of *Escherichia coli* O157:H7 infection. *J Infect Diseases* 173, 480–483.

Beckers, H. J., Daniels-Bosman, M. S. M., Ament, A. et al. (1985). Two outbreaks of salmonellosis caused by *Salmonella indiana*. A survey of the European Summit outbreak and its consequences. *Int J Food Microbiol* 2, 185–195.

Bille, J. (1990). Epidemiology of human listeriosis in Europe, with special reference to the Swiss outbreak. In *Foodborne Listeriosis*, pp. 71–74. Edited by A. J. Miller, J. L. Smith & J. G. A. Somkuti. Amsterdam: Elsevier.

Black, R. E., Jackson, R. J., Tsai, T. et al. (1978). Epidemic *Yersinia enterocolitica* infection due to contaminated chocolate milk. *New Engl J Med* 298, 76–79.

Blackburn, B. O. & Ellis, E. M. (1973). Lactose-fermenting *Salmonella* from dried milk and milk-drying plants. *Appl Microbiol* 26, 672–674.

Breuer, T. (1999). CDC investigations: The May 1998 outbreak of *Salmonella agona* linked to cereal. *Cereal Foods World* 44, 185–186.

Bryan, F. L. (1980). Epidemiology of foodborne diseases transmitted by fish, shellfish, and marine crustaceans in the United States, 1970–1978. *J Food Prot* 43, 859–876.

Bryan, F. L., Ayres, J. C. & Kraft, A. A. (1968). Contributory sources of salmonellae on turkey products. *Am J Epidemiol* 87, 578–591.

CAC (Codex Alimentarius Commission) (1994). *Foods for Special Dietary Uses (Including Foods for Infants and Children*, Vol. 4. FAO/WHO Food Standards Programme. Rome: Codex Alimentarius Commission.

CAC (Codex Alimentarius Commission) (1995). *General Requirements (Food Hygiene)*, Vol. 1B. FAO/WHO Food Standards Programme. Rome: Codex Alimentarius Commission.

CAC (Codex Alimentarius Commission) (1997). *Joint FAO/WHO Food Standards Programme, Codex Committee on Food Hygiene. Food Hygiene, Supplement to Volume 1B-1997. Recommended International Code of Practice, General Principles of Food Hygiene*. CAC/RCP 1-1969, Rev. 3.

Carpentier, B. & Cerf, O. (1993). Biofilms and their consequences with particular reference to hygiene in the food industry. *J Appl Bacteriol* 75, 499–511.

CDC (Centers for Disease Control and Prevention) (1966). Salmonellosis associated with nonfat dried milk. *MMWR Morb Mortal Wkly Rep* 15, 385–386.

CDC (Centers for Disease Control and Prevention) (1997). Outbreaks of *Escherichia coli* O157:H7 infection associated with eating alfalfa sprouts—Michigan and Virginia, June–July 1997. *MMWR Morb Mortal Wkly Rep* 46, 741–744.

Cherrington, C. A., Board, R. G. & Hinton, M. (1988). Persistence of *Escherichia coli* in a poultry processing plant. *Letters Appl Microbiol* 7, 141–143.

Cordier, J. L. (1997). Das Hygienekonzept als Voraussetzung für die erfolgreiche Selbstkontrolle. *Mitt Gebiet Lebensmittelunters Hyg* 88, 9–17.

Cox, L. J., Kleiss, T., Cordier, J. L. *et al.* (1989). *Listeria* spp. in food processing, non-food and domestic environments. *Food Microbiol* 6, 49–61.

Craven, P. C., Mackel, D. C., Baine, W. B. *et al.* (1975). International outbreak of *Salmonella eastbourne* infection traced to contaminated chocolate. *Lancet* i, 788–793.

Dack, G. M. (1964). Characteristics of botulism outbreaks in the United States. In *Botulism, Proceedings of a Symposium*, pp. 33–40. Edited by K. H. Lewis & K. Cassel. Cincinnati, OH: Robert A. Taft Sanitary Engineering Center.

Daley, E. F., Pagotto, F. & Farber, J. M. (1995). The inhibitory properties of various sponges on *Listeria* spp. *Letters Appl Microbiol* 20, 195–198.

Davidson, P. M., Pflug, I. J. & Smith, G.M. (1981). Microbiological analysis of food product in swelled cans of low-acid foods collected from supermarkets. *J Food Prot* 44, 686–691.

De Lagarde, E. A. (1974). *Boletin Informativo del Centro Panamericano de Zoonosis*, Vol. 1. Buenos Aires, Argentina: Centro Panamericano de Zoonosis.

Denny, C. B. (1982). Industry's response to problem solving in botulism prevention. *Food Technol* 36, 116–118.

Dickgiesser, N. (1978). Untersuchungen über das Verhalten Gram-positiver and Gram-negativer Bakterien in trockenem und feuchtem Milieu. *Zbl Bakt Hyg 1. Abt* 167, 48–62.

Dodd, C. E. R., Chaffey, B. J. & Waites, W. M. (1988). Plasmid profiles as indicators of the source of contamination of *Staphylococcus aureus* endemic within poultry processing plants. *Appl Environ Microbiol* 54, 1541–1549.

EC (European Commission) (1997). *Harmonization of Safety Criteria for Minimally Processed Foods*. Inventory Report FAIR Concerted Action FAIR CT96-1020. Brussels: Directorate-General XII, Science, Research and Developement. (http://134.58.114.110/onderzoek/harmony/)

ECFF (European Chilled Food Federation) (1996). *Guidelines for the Hygienic Manufacture of Chilled Foods*. London: Chilled Food Association.

EHEDG (European Hygienic Equipment Design Group) (1997). EHEDG guidelines and test methods. *Trends Food Sci Technol* 8, 1–90.

Ellis, A., Preston, M., Borczyk, A. *et al.* (1998). A community outbreak of *Salmonella berta* associated with a soft cheese product. *Epidemiol Infect* 120, 29–35.

Ericsson, H., Ecklow, A., Danielsson-Tham M.-L. *et al.* (1997). An outbreak of listeriosis suspected to have been caused by rainbow trout. *J Clin Microbiol* 35, 2904–2907.

Evans, H. R., Tromans, J. P., Dexter, E. L. S. *et al.* (1996). Consecutive *Salmonella* outbreaks traced to the same bakery. *Epidemiol Infect* 116, 161–167.

Farber, J. M. (1996). An introduction to the hows and whys of molecular typing. *J Food Prot* 59, 1091–1101.

Fendler, E. J., Dolan, M. J. & Williams, R.A. (1998a). Handwashing and gloving for food protection. Part I: Examination of the evidence. *Dairy Food Environ San* 18, 814–823.

Fendler, E. J., Dolan, M. J., Williams, R. A. *et al.* (1998b). Handwashing and gloving for food protection. Part II. Effectiveness. *Dairy Food Environ San* 18, 824–829.

Flickinger, B. (1996). Plant sanitation comes to light—an evaluation of ATP-bioluminescence systems for hygiene monitoring. *Food Qual* June/July, 1–15.

Fonnesbech-Vogel, B., Jørgensen, L. V., Ojeniyi, B. *et al.* (2001a). Diversity of *Listeria monocytogenes* isolates found in cold-smoked salmon from different smokehouses assessed by randomly amplified polymorphic DNA analyses. *Int J Food Microbiol* 65, 83–92.

Fonnesbech-Vogel, B., Huss, H. H., Ojeniyi, B. *et al.* (2001b). Elucidation of *Listeria monocytogenes* contamination routes in cold-smoked salmon processing plants detected by DNA-based typing methods. *Appl Environ Microbiol* 67, 2586–2595.

Fredriksson-Ahomaa, A., Korte, T. & Korkeala, H. (2000). Contamination of carcasses, offals, and the environment with YADA-positive *Yersinia enterocolitica* in a pig slaughterhouse. *J Food Prot* 63, 31–35.

Gästrin, B., Kaempe, A., Nystroem, K. G. et al. (1972). *Salmonella durham* epidemi spridd genom kakaopulver. *Laekartidingen* 69, 5335–5338.

Gibbs, P. A., Patterson, J. T. & Thompson, J. K. (1978). The distribution of *Staphylococcus aureus* in a poultry processing plant. *J Appl Bacteriol* 44, 401–410.

Gibson, H., Taylor, J. H., Hall, K. H. et al. (1995). *Biofilms and Their Detection in the Food Industry*. R&D Report No. 1. Chipping Campden, Gloucestershire, UK: Campden & Chorleywood Food Research Association.

Gill, O. N., Sockett, P. N., Bartlett, C. L. T. et al. (1983). Outbreak of *Salmonella napoli* infection caused by contaminated chocolate bars. *Lancet* i, 574–577.

Gundermann, K. O. (1972). Untersuchungen zur Lebensdauer von Bakterienstämmen in Staub unter dem Einfluss unterschiedlicher Luftfeuchtigkeit. *Zbl Bakt Hyg 1. Abt* 156, 422–429.

Gundermann, K. O. & Glück, S. (1971). Untersuchungen zur Ueberlebensdauer von Bakterien auf Oberflächen und der Möglichkeiten ihrer Beeinflussung. *Arch Hyg* 154, 480–487.

Gundermann, K. O. & Johannssen, H. (1970). Untersuchungen zur Ueberlebensdauer von Bakterien auf Oberflächen und der Möglichkeiten ihrer Beeinflussung. *Arch Hyg* 151 102–109.

Hardt-English, P., York, G., Stier, R. et al. (1990). Staphylococcal food poisoning outbreaks caused by canned mushrooms from China. *Food Technol* 44, 74–78.

Hennessy, T. W., Hedberg, C. W., Slutsker, L. et al. (1996). A national outbreak of *Salmonella enteritidis* infections from ice cream. *New Engl J Med* 334, 1281–1286.

Holah, J. T., Taylor, J. H. & Holder, S. (1993). *The Spread of* Listeria *by Cleaning Systems*, Part II. Technical Memorandum No. 673. Chipping Campden, Gloucestershire, UK: Campden and Chorleywood Food Research Association.

ICMSF (International Commission on Microbiological Specifications for Foods) (1988). *Microorganisms in Foods 4. Application of the Hazard Analysis Critical Control Point (HACCP) System To Ensure Microbiological Safety and Quality.* Oxford: Blackwell Scientific Publications, Ltd.

ICMSF (International Commission on Microbiological Specifications for Foods) (1998). *Microorganisms in Foods 6: Microbial Ecology of Food Commodities.* Gaithersburg, MD: Aspen Publishers, Inc.

IDF (International Dairy Federation) (1992). Hygiene management in dairy plants. *Bull IDF* 276, 1–68.

IOCCC (International Organization of Cocoa, Chocolate and Confectionery) (1993). *Guidelines for the Establishment of Good Hygiene Practices.* Brussels: IOCCC.

Johnston, R. W., Feldman, J. & Sullivan, R. (1963). Botulism from canned tuna fish. *Publ Health Rpts* 78, 561–564.

Joseph, C. A., Mitchell, E. M., Cowden, J. M. et al. (1991). A national outbreak of salmonellosis from yeast flavoured products. *Communic Disease Rpt*, 1 Review, R16–R19.

Kleiss, T., van Schothorst, M., Cordier, J. L. et al. (1994). Staphylococci in a whey powder plant environment: an ecological survey as a contribution to HACCP studies. *Food Control* 5, 196–199.

Kopanic, R. J., Sheldon, B. W. & Wright, C. G. (1994). Cockroaches as vectors of *Salmonella*: laboratory and field trials. *J Food Prot* 57, 125–132.

Kumar, C. G. & Anand, S. K. (1999). Significance of microbial biofilms in food industry: a review. *Int J Food Microbiol* 42, 9–27.

Lake, D. E., Graves, R. R., Lesnieski, R. S. et al. (1985a). Post-processing spoilage of low-acid canned foods by mesophilic anaerobic sporeformers. *J Food Prot* 48, 221–226.

Lake, D. E., Lesnieski, R. S., Graves, R. R. et al. (1985b). Enumeration and isolation of mesophilic anaerobic sporeformers from cannery post-processing equipment. *J Food Prot* 48, 794–798.

Langfeldt, N., Heeschen, W. & Hahn, G. (1988). Zum Vorkommen von Salmonellen in Milchpulver: Untersuchungen zur Kontamination durch Analyse kritischer Punkte. *Kieler Milchwirtschaftl Forschungsber* 40, 81–90.

Lawrence, L. M. & Gilmour, A. (1995). Characterization of *Listeria monocytogenes* isolated from poultry products and from the poultry-processing environment by random amplification of polymorphic DNA and multilocus enzyme electrophoresis. *Appl Environ Microbiol* 61, 2139–2144.

Lecos, C. (1986). Of microbes and milk: Probing America's worst *Salmonella* outbreak. *Dairy Food Environ San* 6, 136–140.

Lehmacker, A., Bockemühl, J. & Aleksic, S. (1995). Nationwide outbreak of human salmonellosis in Germany due to contaminated paprika and paprika-powdered potato chips. *Epidemiol Infect* 115, 501–511.

Linnan, M. J., Mascola, L., Lou, X. D. et al. (1988). Epidemic listeriosis associated with Mexican style cheese. *New Engl J Med* 319, 823–828.

Llewellyn, L. J., Evans, M. R. & Palmer, S. R. (1998). Use of sequential case-control studies to investigate a community *Salmonella* outbreak in Wales. *J Epidemiol Comm Health* 52, 272–276.

Loncarevic, S., Tham, W. & Danielsson, M-L. (1996). The clones of *Listeria monocytogenes* detected in food depend on the method used. *Lett Appl Microbiol* 22, 381–384.

Lyytikainen, O., Autio, T., Maijala, R. et al. (2000). An outbreak of *Listeria monocytogenes* serotype 3a infection from butter in Finland. *J Infect Diseases* 181, 1838–1841.

Magwood, S. E., Rigby, C. & Fung, P. H. J. (1967). Salmonella contamination of the product and environment of selected Canadian chicken processing plants. *Can J Comp Med Vet Sci* 88–91.

Marthi, B., Fieland, V. P., Walter, M. V. et al. (1990). Survival of bacteria during aerosolization. *Appl Environ Microbiol* 56, 3463–3467.

Matsuda, N., Komaki, M., Ichikawa, R. et al. (1985). Cause of microbial spoilage of canned foods analyzed during 1968–1980. *Nippon Shokuhin Kogyo Gakkaishi* 32, 444–449.

Mead, G. C. (1992). Colonization of poultry processing equipment with staphylococci, an overview. In *Prevention and Control of Potentially Pathogenic Microorganisms in Poultry and Poultry Meat Processing, Proceedings 10. The Attachment of Bacteria to the Gut*, pp. 29–33. Edited by A. Pusztai, M. H. Hinton & R. W. A. W. Mulder. Beekbergen, The Netherlands: DLO Spelderholt Centre for Poultry Research and Information Services.

Mead, G. C. & Dodd, C. E. R. (1990). Incidence, origin and significance of staphylococci on processed poultry. *J Appl Bacteriol Symp Suppl*, 81S–91S.

Mettler, E. & Carpentier, B. (1998). Variation over time of microbial load and physicochemical properties of floor materials after cleaning in food industry premises. *J Food Prot* 61, 57–65.

Michanie, S., Bryan, F. L., Alvarez, P. et al. (1987). Critical control points for foods prepared in households in which babies had salmonellosis. *Int J Food Microbiol* 5, 337–354.

Miettinen, M. K., Björkroth, K. & Korkeala, H. J. (1999). Characterization of *Listeria monocytogenes* from an ice cream plant by serotyping and pulsed field gel electrophoresis. *Int J Food Microbiol* 46, 187–192.

Morgan, D., Newman, C. P., Hutchinson, D. N. et al. (1993). Verotoxin-producing *Escherichia coli* O157 infections associated with the consumption of yogurt. *Epidemiol Infect* 111, 181–187.

Motarjemi, Y. & Käferstein, F. K. (1997). Global estimation of foodborne diseases. *World Health Stats Quarterly* 50, 5–11.

Nesbakken, T., Kapperud, G. & Caugant, D. A. (1996). Pathways of *Listeria monocytogenes* contamination in the meat processing industry. *Int J Food Microbiol* 31, 161–171.

Nesbakken, T., Nerbrink, E., Røtterud, O.-J. & Borch, E. (1994). Reduction of *Yersinia enterocolitica* and *Listeria* spp. on pig carcasses by enclosure of the rectum during slaughter. *Int J Food Microbiol* 23, 197–208.

Norton, D. M., McCamey, M. A., Gall, K. L. et al. (2001). Molecular studies on the ecology of *Listeria monocytogenes* in the smoked fish processing industry and implications for control strategies. *Appl Environ Microbiol* 67, 198–205.

Norwood, D. A. & Gilmour, A. (1999). Adherence of *Listeria monocytogenes* strains to stainless steel coupons. *J Appl Microbiol* 86, 576–582.

Notermans, S., Tips, P., Rost, J. A. & van Leeuwen, W. J. (1982). *Staphylococcus aureus* in pluimveeslachtlijnen. *Tijdschr Diergeneesk* 107, 889–895.

Olive, D. M. & Bean, P. (1999). Principles and applications of methods for DNA-based typing of microbial organisms. *J Clin Microbiol* 37, 1661–1669.

Olsen, A. R. & Hammack, T. S. (2000). Isolation of *Salmonella* spp. from the housefly, *Musca domestica L.*, and the damp fly, *Hydrotea aenescens*, at caged layer houses. *J Food Prot* 63, 958–960.

O'Mahoney, M., Mitchell, E., Gilbert, R. J. et al. (1990). An outbreak of foodborne botulism associated with contaminated hazelnut yogurt. *Epidemiol Infect* 104, 389–395.

Patrick, M. & Bayliss, C. L. (1997). *Evaluation of the Konica Swab & Check*. Technical Notes No. 123. Leatherhead, Surrey, UK: Leatherhead Food Research Association.

Pirttijärvi, T. S. M., Ahonen, L. M., Mannuksela, L. M. et al. (1998). *Bacillus cereus* in a whey process. *Int J Food Microbiol*, 44, 31–41.

Ray, B., Jezeski, J. J. & Busta, F. F. (1971). Isolation of salmonellae from naturally contaminated dried milk products. *J Milk Food Technol* 34, 389–393.

Rowe, B., Hutchinson, D. N., Gilbert, R. J. et al. (1987). *Salmonella ealing* infections associated with consumption of infant dried milk. *Lancet* i, 900–903.

Salvat, G., Toquin, M. T., Michel, Y. & Colin, P. (1995). Control of *Listeria monocytogenes* in the delicatessen industries: the lessons of a listeriosis outbreak in France. *Int J Food Microbiol* 25, 75–81.

Segner, W. P. (1979). Mesophilic aerobic sporeforming bacteria in the spoilage of low-acid canned foods. *Food Technol* 33, 55–59, 80.

Silliker, J. H. & Gabis, D. A. (1975). A cellulose sponge sampling technique for surfaces. *J Milk Food Technol* 38, 504.

Stersky, A., Todd, E. & Pivnick, H. (1980). Food poisoning associated with post-process leakage (PPL) in canned foods. *J Food Prot* 43, 465–476.

Steuer, W. (1992). Hygienische Probleme bei der Teigwarenherstellung. *Mitt Gebiet Lebensmittelunters Hyg* 83, 151–157.

Taylor, J. H. & Holah, T. (1996). A comparative evaluation with respect to the bacterial cleanability of a range of wall and floor surface materials used in the food industry. *J Appl Bacteriol* 81, 262–266.

Todd, E., Szabo, R., Robern, H. et al. (1981). Variation in counts, enterotoxin levels and TNase in Swiss-type cheese contaminated with *Staphylococcus aureus*. *J Food Prot* 44, 839–348.

Tompkin, R. B., Scott, V. N., Bernard, D. T. et al. (1999). Guidelines to prevent post-processing contamination from *Listeria monocytogenes*. *Dairy Food Environ San* 19, 551–562.

Unnerstad, H., Bannerman, E., Bille, J. et al. (1996). Prolonged contamination of a dairy with *Listeria monocytogenes*. *Neth Milk Dairy J* 50, 493–499.

Upton, P. & Coia, J. E. (1994). Outbreak of *Escherichia coli* O157 infection associated with pasteurised milk supply. *Lancet* 334, 1015.

Urban, J. E. & Broce, A. (2000). Killing of flies in electrocuting insects traps releases bacteria and viruses. *Current Microbiol* 41, 267–270.

Walter, M. V., Marthi, B., Fieland, V. et al. (1990). Effect of aerosolization on subsequent bacterial survival. *Appl Env Microbiol* 56, 3468–3472.

Wimpenny, J. (1995). Biofilms: Structure and organisation. *Microbiol Eco' Health Dis* 8, 305–308.

Woolaway, M. C., Bartlett, C. L. R., Weneke, A. A. et al. (1986). International outbreak of staphylococcal food poisoning caused by contaminated lasagne. *J Hyg* 46, 67–73.

Zehren, V. L. & Zehren, V. F. (1968). Relation of acid development during cheesemaking to development of staphylococcal enterotoxin A. *J Dairy Sci* 51, 645–649.

Zwietering, M. H. & Hasting, A. P. M. (1997a). Modelling the hygienic processing of foods—a global process overview. *Food and Bioproducts Processing* 75 (C3), 159–167.

Zwietering, M. H. & Hasting, A. P. M. (1997b). Modelling the hygienic processing of foods—influence of individual process stages. *Food and Bioproducts Processing* 75 (C3), 168–173.

Chapter 12

Sampling, Sample Handling, and Sample Analysis

12.1	Introduction	**12.5**	Sample Analysis
12.2	Collection of Sample Units	**12.6**	Recovery of Injured Cells
12.3	Intermediate Storage and Transportation	**12.7**	Errors Associated with Methods and Performance of Laboratories
12.4	Reception of Samples	**12.8**	References

12.1 INTRODUCTION

Many different types of samples are collected and submitted to the laboratory for analysis. Some are sample units from lots or consignments of foods or ingredients for lot acceptance determination. Others may be for investigational purposes to assess control of the environment, investigate the source of a problem, or validate a process. Some may have legal implications relative to a lawsuit or for regulatory compliance. This chapter briefly discusses some of the major factors that should be considered when collecting sample units, shipping them to a laboratory, preparing them for analysis, and analyzing them.

A sample unit is a small portion of the lot in which the distribution of both numbers and types of microorganisms should be representative of the whole lot. This requirement is only completely fulfilled when the food and processing conditions result in a homogeneous distribution of microorganisms. This only occurs in well-agitated fluid products. In general, the distribution of microorganisms is presumed to be random.

When cells are evenly distributed within the solid or liquid food, i.e., with a variance (σ) squared equal to the mean (μ) ($\sigma^2 = \mu$), the distribution approximates to a Poisson distribution. More frequently, cells occur in clumps or aggregates and their spatial distribution is erratic, with the squared variance larger than the mean ($\sigma^2 > \mu$), representing, for example, a negative binomial distribution (Jarvis, 1989).

The presence and distribution of pathogens, indicators, and/or spoilage microorganisms in finished products depend on many factors, including the level and type of the initial microflora of raw materials, the structure of those raw materials, their quantities and ratio in recipes, and the processing conditions. Formulating and processing the food will also present certain microorganisms with the opportunity to multiply as discrete colonies in certain of the parts of the food (Brocklehurst *et al.*, 1995; Hills *et al.*, 1997), while others will be inhibited or even die. Post-process contamination, particularly when sporadic, may lead to the random presence of very low levels of pathogens.

Representative samples can be collected from liquids, such as milk, in bulk if the product is thoroughly mixed before sampling. While adequate mixing can be achieved in small volumes, the efficiency of mixing decreases as the bulk volume increases. In the case of dry foods, the difficulty in obtaining representative samples increases as the size of particles increases, e.g., from fine powders, such as milk powder or flour, to coarser particles, such as cereal grains, up to large particles or pieces, such as nuts or raisins. Similar difficulties occur with meat, from ground (minced) meat, to trimmed pieces, large cuts, and whole carcasses, and with multi-component foods, such as prepared dishes with several ingredients.

Similar limitations apply to samples from the processing line. Sampling where residues accumulate increases the probability of detecting deviations. For environmental samples, which should play an important role as an early warning of the ingress of, and colonization by, pathogens, the situation is even more complex. Factors including the complexity of the environment, the presence of niches where multiplication may occur, interactions with competitive microorganisms, and changing environmental conditions, such as relative humidity, temperature, exposure to disinfectants, and so on, have an important impact on the distribution of pathogens such as salmonellae and *Listeria monocytogenes*. Another important factor is the small quantity of sample often available. This reinforces the need to have a well-designed sampling plan and well-trained personnel to collect the samples.

The distribution of microorganisms is also important when an analytical unit is withdrawn from the sample unit. Homogenization minimizes the heterogeneous distribution of microorganisms in the analytical sample. As demonstrated by Kilsby & Pugh (1981) using meat, the degree of homogenization of microorganisms increases with more finely chopped meat. The variance of counts decreased from 0.334 to 0.075 for whole pieces to minced and finely chopped meat, respectively, and the likelihood of detecting salmonellae also increased.

Using the most appropriate analytical standard methods or validated alternative methods is prerequisite to obtaining reliable results. Several essential practical aspects, such as the need to adapt methods to certain food matrices, to take account of the possible presence of injured microorganisms, and to apply good laboratory practices, etc., have an important impact on the successful detection or enumeration of the microorganisms sought.

The aim of this chapter is to review the steps from sample collection to the analysis of the analytical unit, and to highlight potential problems that could affect the final result.

12.2 COLLECTION OF SAMPLE UNITS

The purpose of sampling a food is to collect a representative sample to obtain information on its microbiological status. Samples are collected throughout the whole food chain, serving different purposes depending on the information required. The needs of different parties, such as regulatory authorities, food manufacturers, or researchers, may be quite different. The reasons for sampling in the food chain, whether at the farm, at the processing facility, in a warehouse, at retail, or at the consumer's home, need to be considered. This facilitates selection of samples, correct interpretation of results, and drawing valid conclusions.

Sampling may be done for investigative purposes, to obtain information and increase the knowledge of the occurrence of microorganisms in food products, or to assess their behavior during processing and storage. Regulatory authorities sample at predetermined points in the food chain to judge whether imported or commercially available foods comply with current legislation. Intensive sampling by the same authorities would be carried out during an outbreak to identify its source. In commerce, sampling is performed to confirm compliance with specifications between suppliers and customers. Samples taken at intermediate steps of the processing line, from surfaces of equipment in contact with the product or from the processing environment, allow the food processor to verify adherence to Good Hygienic Practices (GHP) and the efficacy of preventive measures.

12.2.1 Steps in the Sampling Procedure

Initial sampling represents the step where physical samples are taken, according to a preestablished sampling plan, at specified sampling points. *Sample handling* encompasses all subsequent steps including the description, recording, and labeling of the samples, their intermediate storage before dispatch, their transport to the analytical laboratory, and their reception, registration, and storage up to the preparation of the analytical unit(s) before analysis. All steps up to the analysis must be controlled to ensure the quality of the sample units and to allow reliable traceability. The guidelines provided in this chapter are considered good practice and should help to optimize results.

12.2.2 Containers

Unless sample units comprise products in their original packaging, such as sachets, tins, boxes, bottles, or cups, the samples drawn must be placed in clean and sterile containers. The containers used should be of appropriate size and have a sufficiently wide opening to allow clean and easy transfer of the samples. Depending on the sample unit, different containers may be used, such as plastic bottles, jars, or bags, metallic cans, or boxes. Glass containers, however, should not be used in manufacturing facilities where breakage could lead to the presence of glass in the products.

Using sterile disposable containers is the most convenient, but for reasons of cost this may not be possible. If reusable containers are used, they should be thoroughly washed, rinsed, dried, and sterilized before further use. Depending on the material, containers are sterilized either at 121 °C for 15 min in an autoclave, or for at least 1 h at 160 °C in a hot air oven. Details of the different options for the sterilization of containers can be found in APHA (1992).

It is important to prevent samples from leaking during handling and, in particular, during transport. Screw caps are the most appropriate closing devices for jars and bottles, and plastic bags must be sealed securely. If rubber or cork is used for caps or closures, it must be confirmed that no adverse interactions with the samples can occur.

12.2.3 Sampling Utensils and Devices

Utensils used to collect the sample units should be sterile and adapted to permit the most appropriate and representative sampling according to the type of food. Sterile scissors, knives, saws, can openers, and other tools can be used to open boxes, packs, sacks, or cans. Samples of the actual food product can then be collected using sterile scoops, spoons, triers, probes, drill bits, forks, cork-borers, pipettes, or swabs, depending on the type of food sampled. Examples and further details can be found in FAO (1990).

For environmental sampling, utensils need to be adapted to the sampling site, with no limitations to their type. Sponges or swabs are most suitable to sample food contact surfaces, floors, or walls. Scrapers, scoops, spatulas, or brushes are suited to collecting residues in cracks, crevices, or underneath equipment. Pipettes (plastic) can be used to collect water in drains, and air sampling devices can be used to monitor the air.

Containers and utensils should be packaged individually and be presterilized, thereby avoiding the need for disinfection between samples. Disposable plastic gloves are recommended when collecting certain samples (e.g., sponge samples from equipment), a new glove being used for each sample. Sterilization of utensils by means of a torch should be avoided and may even be dangerous (e.g., risking explosions in dusty environments).

Sampling ports or taps directly installed on equipment, such as tanks, are usually sterilized in situ by means of steam. They require particularly careful attention (design, cleaning, and sterilization procedures) to avoid contamination of the product.

12.2.4 Sampling Procedures

The method of sampling should be adjusted to fit the purpose. For lot acceptance determinations, the intent is to collect sample units that are representative of the lot. To assess control of the processing environment, a routine procedure should have been developed including sample sites and method of sampling. When investigating the source of a problem, the sampling sites, foods, ingredients, times, frequencies, etc. are not predetermined and are left to the discretion of the sampler. While it is desirable to make every effort to collect samples that fulfill the intended purpose, the safety of the individual collecting the sample, and others who may be affected, takes precedence. Under dangerous circumstances, sample collection should not be attempted. Likewise, if the method of sampling can jeopardize the food and render it injurious to consumers, the sampling procedure should be changed or the sample should not be collected.

Samples should be collected by persons previously trained and instructed, using appropriate methods and, above all, aseptic techniques. The timing and sampling point in the food chain or a food process may be crucial for data interpretation. When regulatory authorities collect samples at a port-of-entry, timing is determined by the arrival of trucks or ships. Samples collected by authorities to verify compliance with legal requirements may be collected during storage, distribution, retail, or food service. Samples collected at different steps in the processing line, such as after production and packaging and before and after cleaning, allow food manufacturers to verify adherence to GHP and the efficacy of Hazard Analysis Critical Control Point (HACCP) plans.

For packaged products, sampling is straightforward and need only follow preestablished sampling plans. For ultra high temperature products, for example, routine samples are taken at defined intervals during processing, while *event samples* are taken after startup, after stops in production, or after changes of paper rolls and strips to monitor and assess the impact of such events (Cordier, 1990). In other circumstances, packed units may be taken from pallets according to a predetermined scheme.

If the packaging material needs to be opened to collect the sample unit, appropriate precautions must be taken to avoid contamination. External surfaces must be cleaned to remove dust and soil and, if multiple layers of packaging are used (e.g., bags of flour, sugar, or other dry products), the outer layer(s) can be removed. This allows access to cleaner surfaces, which may then be disinfected, if necessary, before opening or cutting.

Sufficient quantities should be collected to allow additional analyses to be performed as necessary, or for unforeseen inquiries.

If packages show signs of "blowing" (swelling), extra care should be exercised to avoid dissemination of contaminated material. In such circumstances, the whole package should be sent to the laboratory.

Ideally, the food should be mixed before collecting the sample units, but this may not be practical or possible. For liquid samples, such as milk, ice cream mixes, or beverages stored in vats or tanks equipped with stirrers, it is usually relatively easy to obtain representative samples. If there is no evidence of recent stirring and mixing, thorough mixing with a sterile ladle is recommended before collecting the sample.

In some foods, however, sampling specific components, sections, or layers of multicomponent products is desirable. Particular care should be exercised to achieve a good separation. Examples of special considerations are provided in Table 12–1.

Water (potable water, processing water) should be sampled from taps or built-in sampling devices after flushing sufficient water through those taps or built-in devices to obtain a representative sample. A sterile ladle or other means can be used for water or brine solutions in open systems. Addition of a neutralizing solution is necessary to neutralize disinfectant residues, if present.

Air sampling in processing environments may be performed to assess the microbiological status of air. Many techniques have been described, from simple passive collection using sedimentation plates to active air samplers based on impaction or impingement. Those methods and available equipment are reviewed by Henningson & Ahlberg (1994).

Special consideration should be given to sampling during investigation of foodborne outbreaks since the sampling can differ considerably from routine inspections. Using professional judgment, samples of food or equipment surfaces should be collected at the most appropriate locations where contamination, survival, or growth might have occurred (Bryan, 1999).

12.2.5 Sample Labeling

Sample units must be clearly labeled and identified to permit good traceability. This can be achieved by writing descriptive terms or numbering each sample unit directly on the container or a firmly attached label, making sure that the ink cannot be washed off.

In addition, a sampling report should be prepared with relevant details stating, for example, the time of sampling, sample site, particular observations on packaging, etc. The comments may vary widely depending on the purpose of sampling and what is to be learned from the analysis. Thus, the report should indicate the reason for sampling and, if known by the person collecting the samples, the types of analyses to be performed. If necessary, for example, for legal reasons or in case of dispute, the report should be signed by the person responsible for the sample collection as well as by representatives of the parties concerned. In such instances, sample containers may be sealed with an official seal that makes tampering impossible.

12.3 INTERMEDIATE STORAGE AND TRANSPORTATION

Collected sample units should be transported to the laboratory as promptly as possible. However, in certain situations, intermediate storage cannot be avoided before samples are dispatched, e.g., when daily dispatch is impossible. For environmental sampling, samples may be collected at regular intervals from specific sites over a period of time and, when assembled, dispatched. When samples are stored, the storage conditions must

Table 12-1 Special Considerations When Sampling Food Commodities

Phase	State	Form	Food Ingredient	Recommendations for sampling
Solid	Dry	Fine powder	Dry milk, cocoa powder	From center of container unless problems suspected at the surface
		Large part	RTE* cereals, dry pasta	Select representative components and blend before drawing sample; may be rehydrated directly
		Blocks	Dates, sugar, cheese	If hard, use sterile hammer and chisel (also mortar and pestle) to draw sample; try to draw samples from several locations
		Multicomponent	Soup mix, spices	Entire unit or selected components drawn for sample
Solid	Moist	Particles	Ground meat	Multiple portions drawn from one or more packages or units
		Chunk	Tote of beef primal cuts	Selected pieces or liquid weep (drip) from large tote
		Blocks	Whole poultry, cheese	Whole bird rinsed in diluent or media; try to draw samples
		Carcass	Beef	Prescribed suspected area swabbed or sponged for testing
		Sausage	Sliced, cooked	Rinse or blend
			Fermented sausage	Selected areas for selected analyses, e.g., test for *S. aureus* near the surface, and enterohemorrhagic *E. coli* in the center
		Multicomponent	Cream dessert, stew	Selected areas for selected analyses, e.g., test for *S. aureus* near or in cream filling; test for *C. perfringens* in liquid portion of the stew
Solid	Frozen	Particles	IQF† peas/beans	Frozen sample either "stomached" or rinsed with diluent
		Blocks	Frozen egg	Collected with (sterile) funnel and drill
		Multicomponent	Pizza	Sector sampled while frozen unless specific components tested
Liquid		Single phase	Water	Membrane filter kept in conditions that avoid death or multiplication
		Multiple nonmiscible phases	Salad dressing	Well-mixed representative sample, or specific layer
		Homogeneous multiple phases	High calorie supplement	Representative weighed sample

*RTE—ready-to-eat
† IQF—instant quick frozen

prevent changes, i.e., microbial growth or death. Storing dry or shelf-stable products or samples poses relatively few problems, but wet and perishable samples are more delicate and need to be refrigerated or even frozen. The storage conditions considered most appropriate must be established to ensure that the target flora remains unchanged. Loss of viability during frozen storage and subsequent thawing can be a particular problem (e.g., *Campylobacter jejuni* vegetative cells of *Clostridium perfringens*).

For transportation, all samples must be packed to avoid breakage and spillage. Where necessary, containers should be protected by additional packaging material. Finished products, raw materials, and environmental samples should be packaged separately or shipped separately to avoid possible problems of cross-contamination.

Samples of perishable chilled or frozen products should be shipped in insulated containers to maintain the appropriate temperature with a refrigerant, such as ice, dry ice, or freezer packs. For particularly sensitive or important samples, temperature recorders or indicators should be included to record in-transit temperatures that could affect the analytical results.

12.4 RECEPTION OF SAMPLES

When received at the analytical laboratory, sample units should be inspected visually for damage or spillage, temperature checked as necessary, and the samples cross-checked against the sample report. Information that could influence the analytical results should be noted on the sample report. Today laboratory information management systems are widely used, allowing for easy and complete registration, thus ensuring good traceability.

12.5 SAMPLE ANALYSIS

12.5.1 Withdrawing Analytical Units

Samples should be analyzed as promptly as possible and, if not, stored under conditions that do not permit either death or multiplication of the target microflora.

Paramount in the process of withdrawing sample units is to use aseptic technique to prevent contamination of the analytical unit. The first step of the preparation for microbiological analysis is to withdraw a representative analytical unit from each sample unit. Samples must be mixed thoroughly to obtain a representative analytical unit. Liquid, semi-liquid, and, to some extent, powders in containers with a headspace, can be mixed by inverting or shaking the container. Analytical units should be removed as soon as possible after mixing. The external parts of packages or containers are first disinfected using chlorinated or iodinated solutions. Opening of packages is performed taking all necessary measures to prevent contamination. Sterile utensils are used to open packages, mix, and withdraw the necessary aliquots.

12.5.2 Dilution and Homogenization of the Analytical Unit

Weighing the analytical unit into the appropriate diluent (first dilution) must be performed aseptically. With dusty products, it is recommended to work under a hood or in a laminar flow cabinet to avoid dispersion of dust that might be contaminated with pathogens.

Most analytical methods recommend weighing the analytical unit with a precision of ± 0.1 g of the desired weight. The size of the analytical unit has a significant effect on the coefficient of variation of the sample weight. The larger the analytical unit, the lower the coefficient of variation and the better the accuracy. For practical reasons, however, analytical units of 10 g up to 50 g are typical for quantitative methods and are diluted 1:10. For qualitative analyses, 25 g analytical units are frequently analyzed individually or as composites of 100 g, 200 g, or more.

For residue collected from the environment, only small quantities, commonly 1 g or less, may be available. Nevertheless, such samples can provide very valuable information.

Some products, such as liquids, semi-liquids, and paste-like samples, are normally mixed and dispersed rather easily. For products such as margarine, butter, chocolate, etc., gentle warming (to about 40 °C) improves dispersion and dissolution of the food matrix and facilitates the release of microorganisms. Other types of samples, in particular solids, require special treatments using homogenizers or blenders. Diluted homogenized sample units can then be further processed according to the protocols of specific analytical methods. Detailed information on sample dilution and homogenization is provided in various documents, such as ISO documents (1999).

Preparation of the first dilution, either for qualitative (enrichment) or quantitative (direct) analyses, can have a marked impact on the microbial population in the sample. This is often due to changes in the physico-chemical characteristics of the suspension as compared to the food matrix. Where the food does not confer protection, the composition of the diluent is important to prevent or minimize shock/injury to the microorganisms that can result from rapid changes in the ionic concentrations. This has long been recognized and is discussed, for example, by Keller et al. (1974).

Use of stomachers or blenders usually has no impact on the viability of microorganisms. An exception is high-speed blenders that may damage certain cells, for example anaerobes, due to exposure to oxygen (Ray, 1979).

Dry foods must be rehydrated slowly to avoid die-off due to osmotic shock (e.g., freeze-dried cells) (Ray et al., 1971). This was confirmed by van Schothorst et al. (1979), who showed that the conditions of rehydration of foods can have a major impact on the recovery of salmonellae. Slow rehydration, e.g., using a soaking procedure, resulted in higher recovery rates than shaking.

In foods, the ratio between dead, injured, and healthy cells varies, as does the extent of injury to individual cells. Numerous studies have been devoted to this subject and reviews are available (Ray et al., 1971; Mackey, 2000).

12.6 RECOVERY OF INJURED CELLS

Injury of microbial cells can have a significant impact on the final results and must be taken into account during analysis. Injured or damaged microbial cells should be allowed time to repair that damage before growth is initiated. This "time to repair" is seen as an increased lag phase and affects the incubation time of both solid and liquid media. Time spans recommended in standard methods are generally those optimal to recover the maximum number of microorganisms. Reducing this incubation period can lead to a significant reduction in counts, or in the case of qualitative methods, to false negatives. This was shown for *Salmonella*, for example, when preenrichment for 6 h gave lower recoveries than preenrichment for 24 h. Extending preenrichment to 48 h did not increase the number of *Salmonella*, and numbers were sometimes lower than that at 24 h (D'Aoust et al., 1992; Hammack et al., 1993).

Food components can markedly affect recovery of both injured and healthy cells, as is shown by detection of *Salmonella* in different food matrices. Rapid rehydration of dehydrated products, such as milk powder or feeds, reduces the recovery rate (Van Schothorst *et al.*, 1979; D'Aoust & Sewell, 1986). Dehydrated culinary products, such as soups or concentrated bouillons, and raw materials, such as spices, onions, etc., require higher dilution rates (1:20 up to 1:100) or the addition of substances to neutralize the inhibitory effects of salt and certain food components (Andrews *et al.*, 1995). Bacteriostatic and/or bactericidal substances present in cocoa may inhibit growth of *Salmonella* during the preenrichment (Zapatka *et al.*, 1977). Inhibition is neutralized by adding casein or nonfat dry milk to the preenrichment broth (Poelma *et al.*, 1981; IOCCC, 1990). In fatty foods, surfactants such as Tween 80 improve recovery (D'Aoust *et al.*, 1982). Hydrocolloids may affect and inhibit recovery of salmonellae by thickening and changing the pH of the preenrichment broth. Using appropriate dilutions, adding enzymes to reduce viscosity, and adjusting pH have been shown to improve handling and recovery (Amaguaña *et al.*, 1996, 1998). Gel formation of gelatin during incubation will affect recovery, and dilution (1:20) or using papain to reduce the viscosity are recommended (Jay et al, 1997; Amaguaña *et al.*, 1998).

Environmental samples can be heavily contaminated and addition of malachite green to the preenrichment broth to inhibit competitive microorganisms has been shown to enhance recovery of salmonellae (Van Schothorst & Renaud, 1985).

Optimization of preenrichment conditions is therefore essential for the detection of salmonellae and should be considered for other pathogens. Increasing the number and type of selective enrichment broths and selective plating media will not improve detection of injured and healthy cells if they are not permitted to recover and grow during preenrichment.

Selective agents in direct plating media can have an adverse effect on particular microorganisms (Ray, 1979). This is of particular concern if low levels need to be detected. Selective agents, such as sodium lauryl sulphate, Oxgall, brilliant green, bile salts, sodium desoxycholate, crystal violet, and others, are used in different media to detect Enterobacteriaceae, coliforms, or *Escherichia coli*. Comparative studies with basal agar supplemented with one of these selective components have shown inhibition rates between 0 and 99% for *E. coli* (Ray, 1979). This has been confirmed by others using different experimental approaches, underlining the need for careful choice of media in the microbiological analysis of foods.

Numerous attempts have been made to overcome this problem either through choice of selective agent, introducing a resuscitation step in the procedure, use of a nonselective medium, or addition of betaine or pyruvate to enhance recovery of injured or stressed cells (Ray, 1979; Johnson & Busta, 1984; Mackey *et al.*, 1994; Marthi & Lightfoot, 1990).

12.7 ERRORS ASSOCIATED WITH METHODS AND PERFORMANCE OF LABORATORIES

The errors associated with quantitative methods, such as colony count techniques, differ from those for qualitative methods, such as presence/absence tests. Errors affecting the quality of data obtained by analytical laboratories have been discussed in detail (Jarvis, 1989). Quality of results is characterized by the accuracy of the method, i.e., the ability to provide results equal to, or close to, the real value. The repeatability (r) of a method reflects the difference between two single results obtained when the same sample is analyzed by the same analyst under identical analytical conditions. The reproducibility (R), on the other hand, represents the difference obtained between two laboratories. Examples of r and R values for different analytes are provided in Table 12–2.

Over recent years, considerable efforts have been made in the accreditation of laboratories. This will have a positive outcome, provided accreditation is not just considered as a system or a commercial asset. If properly implemented, it should be used for continual

Table 12–2 Examples of Repeatability (r) and Reproducibility (R) for Different Analytes in Mashed Potatoes, Rehydrated with Cultures. (Data provided by the Silliker Laboratories).

Estimation of laboratory variance (*Repeatability*) using the average variation of homogeneity samples for proficiency test materials. Each set examined by the same technician (20 tests).

Method	$s^{2\dagger}$	s	CV‡ (%)	Repeatability (r)
Aerobic mesophilic counts (n=80)	0.0126	0.1122	2.46	0.317
C. perfringens spores (n=40)	0.103	0.321	7.29	0.908
Coliforms (n=80)	0.057	0.239	6.42	0.677
Coliforms (MPN)* (n=80)	0.109	0.331	9.29	0.935
E. coli (MPN) (n=40)	0.150	0.387	9.73	1.096
Yeasts (n=40)	0.023	0.152	3.31	0.425
Molds (n=40)	0.022	0.150	4.16	0.424

Estimation of the variance among laboratories (*Reproducibility*) using the average variation from 13 proficiency test sets (1995–1998).

Method	s^2	s	CV (%)	Reproducibility (R)
Aerobic mesophilic counts (n=1669)	0.068	0.261	5.4	0.37
C. perfringens spores (n=660)	0.353	0.594	15.2	1.68
Coliforms (n=1439)	0.138	0.371	10.0	1.05
Coliforms (MPN) (n=1066)	0.191	0.437	11.4	1.24
E. coli (MPN) (n=706)	0.435	0.659	17.8	1.87
Yeasts (n=701)	0.129	0.359	8.7	1.02
Molds (n=735)	0.084	0.290	8.9	0.82

†s–standard deviation
‡CV–coefficient of variation
*MPN–most probable number

improvement of analytical performance and laboratory procedures, such as the quality of the media used and their preparation and control of incubator and water bath temperatures. Skills and training of personnel, standardization of laboratory practices, and so on, should also be addressed.

Participation of laboratories in proficiency tests organized and offered by national, professional, or commercial organizations and requested by accreditation bodies also represents an opportunity for improvements (Black & Craven, 1990; Peterz, 1992; Berg et al., 1994). Such proficiency tests facilitate benchmarking performance of the laboratory and identification of weaknesses that need improvement. It must, however, be realized that the type of samples provided for proficiency testing has limitations related to the preparation and the viability of microbial population. Consequently, check samples for proficiency testing are not available for all food matrices. The concentration of pathogens is frequently relatively high and a competitive flora is not always included in check samples. Such samples do not, therefore, accurately assess the laboratory's ability to detect very low numbers of injured cells that may occur in actual food samples. The use of reference materials containing very low levels of injured cells, such as the *Salmonella* product developed in the Netherlands (Foundation, 2001), may be more useful in assessing the laboratory performance or reliability of a method. Reference materials have been developed for various microorganisms (Peterz & Steneryd, 1993; In't Veld et al., 1995).

Simplified or alternative methods are often used to cope with the large number of analyses and to obtain results more quickly. This is legitimate and can accommodate a sudden influx of samples, e.g., environmental, to detect a source of contamination. This is more effective than applying more cumbersome standard methods that limit the number of samples that can be analyzed.

It is extremely important to underline the need to use alternative methods that have been validated. This not only allows more results to be obtained sooner, but also guarantees reliability of results. A number of validation procedures exist, ranging from a simple peer review of alternative methods by an expert panel to extremely thorough procedures based on extensive comparative and collaborative studies (Andrews, 1996; Lombard et al., 1996; Rentenaar, 1996; Scotter & Wood, 1996). The development of a technical protocol within Microval (Rentenaar, 1996) represents an example of how to standardize validation procedures and to develop an International Organization for Standardization/Comit Europen de Normalisation standard that will be issued during 2001.

12.8 REFERENCES

Amaguaña, R. M., Sherrod, P. S. & Hammack, T. S. (1996). Usefulness of cellulase in recovery of *Salmonella* spp. from guar gum. *J Assoc Off Anal Chem Int* 81, 853–857.

Amaguaña, R. M., Hammack, T. S. & Andrews, W. H. (1998). Methods for the recovery of *Salmonella* spp. from carboxymethylcellulose gum, gum ghatti and gelatin. *J Assoc Off Anal Chem Int* 81, 721–726.

Andrews, W. H. (1996). AOAC International's three validation programs for methods used in the microbiological analysis of foods. *Trends Food Sci Technol* 7, 147–151.

Andrews, W. H., June, G. A., Sherrod, P. S. et al. (1995). Salmonella. In *FDA Bacteriological Analytical Manual*, 8th edn., pp. 5.01–5.20. Gaithersburg, MD: AOAC International.

APHA (American Public Health Association) (1992). *Compendium of Methods for the Microbiological Examination of Foods*, 3rd edn. Edited by C. Vanderzant & D. F. Splittstoesser. Washington, DC: American Public Health Association.

Berg, C., Dahms, S., Hildebrandt, G. et al. (1994). Microbiological collaborative studies for quality control in food laboratories: reference material and evaluation of analyst's errors. *Int J Food Microbiol* 24, 41–52.

Black, R. G. & Craven, H. M. (1990). Program for evaluation of dairy laboratories testing proficiency. *Austral J Dairy Technol* 45, 86–92.

Brocklehurst, T., Parker, M., Gunning, P. et al. (1995). Growth of food-borne pathogenic bacteria in oil-in-water emulsions: II. Effect of emulsion structure on growth parameters and form of growth. *J Appl Bacteriol* 78, 609–615.

Bryan, F. L. (1999). *Procedures To Investigate Foodborne Illness*, 5th edn. Committee on Communicable Diseases Affecting Man. Des Moines, Iowa: International Association of Milk, Food, and Environmental Sanitarians Inc.

Cordier, J. L. (1990). Quality assurance and quality monitoring of UHT processed foods. *J Soc Dairy Technol* 43, 42–45.

D'Aoust, J. Y., Maishment, C., Stotland, P. et al. (1982). Surfactants for the effective recovery of *Salmonella* in fatty foods. *J Food Prot* 45, 249–252.

D'Aoust, J. Y., Sewell, A. M. & Warburton, D. W. (1992). A comparison of standard cultural methods for the detection of foodborne *Salmonella*. *Int J Food Microbiol* 16, 41–50.

D'Aoust, J. Y. & Sewell, A. M. (1986) Detection of *Salmonella* by the enzyme immunoassay (EIA) technique. *J Food Sci* 51, 484–488, 507.

FAO (Food and Agriculture Organization) (1990). *Manuals of Food Quality Control. 5. Food Inspection*. Rome: Food and Agriculture Organization of the United Nations.

Foundation for the Advancement of Public Environmental Protection (SVM) (2001). Bilthoven, NL. (http://www.svm.rivm.nl/index.html)

Hammack, T. S., Satchell, F. B., Andrews, W. H. et al. (1993). Abbreviated preenrichment period for recovery of *Salmonella* spp. from selected low-moisture dairy foods. *J Food Prot* 56, 201–204.

Henningson, E. W. & Ahlberg, M. S. (1994). Evaluation of aerosol samplers: a review. *J Aerosol Sci* 25, 1459–1492.

Hills, B. P., Manning, C. E., Ridge, Y. et al. (1997). Water availability and the survival of *Salmonella typhimurium* in porous systems. *Int J Food Microbiol* 36, 187–198.

IOCCC (International Office of Cocoa, Chocolate, and Sugar Confectionery) (1990). Detection of *Salmonella*. In *Microbiological Examination of Chocolate and Other Cocoa Products*. Brussels: IOCCC.

In't Veld, P. H., Notermans, S. H. W. & Van den Berg, M. (1995). Potential use of microbiological reference material for the evaluation of detection methods for *Listeria monocytogenes* and the effect of competitors: a collaborative study. *Food Microbiol* 12, 125–134.

ISO (International Standardisation Organisation) (1999). *Microbiology of Food Products—General Rules for the Preparation of Samples for Microbiological Analysis*. ISO/WD 6887-1 to 6887-5.

Jarvis, B. (1989). *Statistical Aspects of the Microbiological Analysis of Food. Progress in Industrial Microbiology*, Vol. 21. Amsterdam: Elsevier.

Jay, S., Grau F. H., Smith K. et al. (1997). *Salmonella*. In *Foodborne Microorganisms of Public Health Significance*, 5th edn., pp. 169–229. North Sydney, NSW: Australian Institute of Food Science and Technology.

Johnson, K. M. & Busta, F. F. (1984). Detection and enumeration of injured bacterial spores in processed foods. In *The Revival of Injured Microbes*, pp. 241–256. Society for Applied Bacteriology Symposium Series No. 12. Edited by F. A. Skinner, A. D. Russell & M. H. E. Andrew. London: Academic Press.

Keller, P., Sklan, D. & Gordin, S. (1974). Effect of diluent on bacterial counts in milk and milk products. *J Dairy Sci* 57, 127–128.

Kilsby, D. C. & Pugh, M. E. (1981). The relevance of the distribution of micro-organisms within batches of food to the control of microbiological hazards from foods. *J Appl Bacteriol* 51, 345–354.

Lombard, B., Gomy, C. & Catteau, M. (1996). Microbiological analysis of foods in France: standardized methods and validated methods. *Food Control* 7, 5–11.

Mackey, B. M. (2000). Injured bacteria. In *The Microbiological Safety and Quality of Food*, pp. 315–341. Edited by B. M. Lund, T. C. Baird-Parker & G. W. Gould. Gaithersburg. MD: Aspen Publishers, Inc.

Mackey, B. M., Boogard, E., Hayes, C. M. et al. (1994). Recovery of heat-injured *Listeria monocytogenes*. *Int J Food Microbiol* 22, 227–237.

Marthi, B. & Lightfoot, B. (1990). Effect of betaine on enumeration of airborne bacteria. *Appl Environ Microbiol* 56, 1286–1289.

Peterz, M. (1992). Laboratory performance in a food microbiology proficiency testing scheme. *J Appl Bacteriol* 73, 210–216.

Peterz, M. & Steneryd, A. C. (1993). Freeze-dried mixed cultures as reference samples in quantitative and qualitative microbiological examinations of food. *J Appl Bacteriol* 74, 143–148.

Poelma, P. L., Andrews, W. H. & Wilson, C. R. (1981). Pre-enrichment broths for recovery of *Salmonella* from milk chocolate and edible casein: collaborative study. *J Assoc Off Anal Chem* 64, 893–898.

Ray, B. (1979). Methods to detect stressed microorganisms. *J Food Prot* 42, 346–355.

Ray, B., Jezeski, J. J. & Busta, F. F. (1971). Effect of rehydration on recovery, repair and growth of injured freeze-dried *Salmonella anatum*. *Appl Microbiol* 22, 184–189.

Rentenaar, I. M. F. (1996). Microval, a challenging Eureka project. *Food Control* 7, 31–36.

Scotter, S. & Wood, R. (1996). Validation and acceptance of modern methods for the microbiological analysis of foods in the UK. *Food Control* 7, 47–51.

Van Schothorst, M. & Renaud, A. M. (1985). Malachite green pre-enrichment medium for improved *Salmonella* isolation from heavily contaminated samples. *J Appl Bacteriol* 59, 223–230.

Van Schothorst, M., Van Leusden, F. M., De Gier, E. et al. (1979). Influence of reconstitution on isolation of *Salmonella* from dried milk. *J Food Prot* 42, 936–937.

Zapatka, F. A., Varney, G. W. & Sinskey, A. J. (1977). Neutralization of bactericidal effect of cocoa powder on salmonellae by casein. *J Appl Bacteriol* 42, 21–25.

Chapter 13

Process Control

13.1 Introduction
13.2 Knowledge of the Degree of Variability and the Factors That Influence Variability
13.3 Process Capability Study
13.4 Control during Production: Monitoring and Verifying a Single Lot of Food
13.5 Control during Production: Organizing Data from Across Multiple Lots of Food To Maintain or Improve Control
13.6 Use of Process Control Testing as a Regulatory Tool
13.7 Investigating and Learning from Previously Unrecognized Factors or Unforeseen Events
13.8 References

13.1 INTRODUCTION

This chapter discusses sampling and testing to assess whether food operations are under control (i.e., correct procedures are being followed and criteria are being met) and using the data to make the adjustments necessary to maintain control. Food operations must be controlled to produce foods of consistent quality and safety. A controlled process requires being proactive and informed of the factors that influence variability. Process control technology can be applied to the manufacture of a single lot of food produced on one day, or to multiple lots produced over days or years, and to both batch and continuous processes.

As described in Chapters 3, 4, and 5, the process or product criteria that are used to assess whether a process is under control can be based on a food safety objective or a performance criterion that is specific to the food, the process, and the microbial hazard(s). Product criteria can be standards, specifications, or whatever a control authority or processor considers necessary to ensure that a process is controlled and the food is safe when used as intended.

Since about 1990, there has been increasing interest in the food industry in quality enhancement, through programs that stress the use of structured quality systems, such as International Organization for Standardization (ISO) certification. The concept of continuous improvement has led to greater use of data in a more organized manner to improve quality and production efficiencies. This interest in quality systems occurred during a period when Hazard Analysis Critical Control Point (HACCP) planning was being more widely adopted throughout the industry. HACCP systems are, in essence, that portion of an establishment's overall process control system that focuses on food safety. Collectively, these programs have led to a greater awareness of the need and value of process control technology. Although food safety is emphasized below, the concepts described can be applied to ensuring microbiological quality.

Microbiological testing is conducted at various points along the food chain for a number of different purposes. Some of these are summarized in Table 1–2 and Exhibit 5–1. In Chapters 4 and 5, microbiological testing for evaluation of lots or consignments of food (raw materials and end products) in commerce is discussed. However, due to the time required for most microbiological analyses and the relative insensitivity of even the most stringent sampling plans, microbiological testing is of limited value for monitoring in quality and safety assurance programs (Chapters 6, 7, 9; NRC, 1985; ICMSF, 1988; NACMCF, 1997). More rapid tests must necessarily be used and will typically involve sensorial, chemical, and/or physical measurements such as time, temperature, acidity, pH, moisture, a_w, flow rates, and so on. The principles of process control technology described in this chapter can be applied to all such measurements. Thus, while the emphasis of the book is on microbiological testing, it should be kept in mind that other measurements will more commonly be used in process control systems (see Chapter 4).

HACCP is often referred to as a preventive system. However, from a statistical standpoint, HACCP would be more appropriately described as a means of minimizing the variability of a system. A statistical approach to safety can be effectively applied in HACCP systems. Control chart methods (see sections 13.3.1 and 13.4) provide an objective and statistically valid means to assess ongoing processes, and as such are particularly applicable to monitoring. Lot acceptance sampling can be used in verification and, to a very limited extent, in monitoring. These two statistical techniques for controlling quality and safety are well developed and documented in many textbooks, manuals, and periodicals (e.g., ASTM, 1951; BSI; Duncan, 1986; Grant & Leavenworth, 1972; ISO/TC 69; Massart et al., 1978). The statistics of lot acceptance plans have been dealt with in detail in Chapters 6 and 7. Control chart methods with applications to food technology are treated in detail by Kramer & Twigg (1982).

This chapter is intended to serve as a brief introduction to the use of statistical methods for monitoring and verification. For more information on process-oriented control systems in the food industry and the statistical tools used for that purpose, other texts are recommended (Steiner, 1984; Hubbard, 1990; DeVor et al., 1992; Smith, 1998; Juran et al., 1999; Ledolter & Burrill, 1999).

Controlled food operations require:

1. *Knowledge of the significant hazards*
   The principles of HACCP should be considered for all food operations. At a minimum, a hazard analysis should be conducted. If the conclusion drawn is that significant hazards are not likely to occur, a HACCP plan would not be established.
2. *Knowledge of the factors that are necessary for control*
   If significant microbial hazards have been identified, it is necessary to investigate the conditions of processing that must be controlled to prevent, eliminate, or reduce their occurrence to acceptable levels and then to establish control measures. This will clarify the importance of certain steps in the process (i.e., critical control points (CCPs)) and establish specific Good Hygienic Practices (GHP) for controlling the identified hazards.
3. *Knowledge of the extent of variability and factors that influence variability*
   Understanding the factors that influence variability is an essential element of a process control system. Most food operators understand the factors that influence the cost of producing a food and strive to control each factor according to its relative impact on cost and profit. This same concept can be applied to producing safe

foods. Processes with a high degree of variability, particularly when that variability is not recognized or understood, are more likely to produce unacceptable, and possibly hazardous, food. Each process is unique, owing to differences in plant layout, equipment design and performance, equipment maintenance and cleanability, personnel, type of food being produced, and other factors. The conditions that influence variability at CCPs must be understood, as well as the extent of variability that can occur. The information should then be used to determine how this variability might be controlled within an acceptable range.

4. *Establishing criteria for the factors that must be controlled*

   The information from item 3 above should be used to establish operating parameters that take account of variability and ensure that critical limits are met. Through continuous improvement, variability can be reduced and result in improved safety, quality, and process efficiency. With limits having been established at critical steps in the operation, procedures must be developed to monitor those limits to ensure they are met during operation.

5. *Establishing monitoring procedures*

   A wide variety of measurements, such as sensorial, physical, chemical, and microbiological, are used for monitoring food processes. The method of choice will be the simplest, easiest, cheapest, and safest available that can provide, in a timely manner, the information needed to adjust the process and maintain control. Ideally, the measurements will be continuous with adjustments being made automatically. The measurements may include processing parameters (e.g., temperature, humidity, pH), food collected at different stages in processing, finished product, and/or environmental samples. A permanent record should be created for subsequent verification.

6. *Organizing and interpreting data*

   Considerable quantities of data are generated to evaluate current production, and if organized correctly, the data have great value. The data can be used to determine longer-term trends and to facilitate continuous process improvement. With the availability of powerful computers, data can be organized into databases that allow rapid interrogation and provide a cumulative record.

7. *Using the data to improve control and measure change*

   The value of an effective process control system is most evident when data are organized and used to further increase knowledge about a process and the factors affecting variability. The longer-term goal should be to use the data to reduce variability and achieve more tightly controlled processes. Properly organized, the data can be used to measure the effect of modifications to equipment and other factors in the process. In addition, the data can be used to detect trends over multiple lots of food that may indicate a gradual loss of control. When data are organized and routinely reviewed, adjustments can be made to maintain control during production of one lot or across multiple lots of food.

8. *Responding to the data*

   In the best situations, after a process has been modified, data will indicate reduced variability and improved control. This reaffirms the operator's understanding of the factors affecting variability and how the process may be further improved. Occasionally, trends may be unfavorable and indicate the need to determine which factor has changed and requires correcting. Many examples could be cited where failure to recognize change and to respond effectively has led to unsafe food and foodborne illness.

9. *Investigating and learning from previously unrecognized factors or unforeseen events*

   Occasionally, an unexpected change occurs in a process that results in loss of control. Upon investigation, the cause is determined and controls are implemented to prevent a similar event from occurring again. This could occur through equipment malfunction and lead to a new preventive maintenance procedure to avoid future malfunctions. Another possibility could involve weather changes (e.g., higher humidity in summer compared with winter and its effect on processes that involve drying) or power outages from storms or shortages of electricity in a region. Isolated events may result in no change to an existing process control system as long as the monitoring procedures and confidence in control are not affected, but more frequent occurrences may require a prepared plan of action.

Items 1 and 2 above are well known as the first two principles of HACCP (ICMSF, 1988; CAC, 1997; NACMCF, 1997) and are not discussed here.

13.2 KNOWLEDGE OF THE DEGREE OF VARIABILITY AND THE FACTORS THAT INFLUENCE VARIABILITY

13.2.1 Establishing a Baseline for a Process

Food manufacturers collect a variety of samples for microbiological analysis, including samples from equipment and the processing environment, ingredients, in-line food samples, and finished product. The selection of samples and choice of analyses are influenced by the type of food and food operation. The samples should generate data that can help the operator assess the degree of control and prevent problems. Since one or more days may elapse between sampling and obtaining a result from a microbiological analysis, the data provide a history of past performance.

While not always clearly articulated, the purpose of this sampling is often to provide a "microbiological history" of the food product and the processing conditions (Buchanan, 2000). By acquiring data over time on the microbial population, manufacturers establish a baseline for the level of control that is attainable when GHP procedures and the HACCP system are under control. Once established, subsequent analyses that differ from the baseline indicate a deviation from the norm due to a change in operating conditions. The baseline data also can be used to establish microbial specifications.

The extent of microbiological testing is typically limited and is not intended to provide assurance of any specific batch or lot of food. If sufficient data have been accumulated over multiple lots, statistical analysis can greatly enhance the usefulness of the data. This type of analysis, sometimes referred to as cross-batch testing, is similar to data analysis for a single batch, except that the data are collected over time and multiple batches are analyzed cumulatively. An underlying assumption is that when a process is under control, the between-batch variability is small.

13.2.2 Types of Microbiological Data for Baseline

Microbiological data can be collected from at least five sources that differ by location or time of sampling within the process (i.e., ingredients, in-line samples, end product, equipment/environmental samples, and shelf-life samples).

Ingredient data can be viewed as a type of in-line sample. Periodic microbiological testing of ingredients can be used to verify that a step in the process that is not directly under the control of the manufacturer is, in fact, under control. The microbiological limits for ingredients are typically defined through purchase specifications. Thus, periodic microbiological testing of ingredients by the purchaser is a means of verifying that the supplier is meeting the specifications. Ingredient testing is particularly important when microbial levels could be so large as to overwhelm the controls designed into the food safety system or when a sensitive ingredient is used in a process that has no kill step (e.g., blending dry ingredients for powdered infant formula or chocolate manufacture).

In-line sampling consists of collecting samples of the food at different steps in the process. The data provide information needed to understand the effects of each step on the microbial population. The time at which in-line samples are collected may be important (e.g., at start, middle, or end of production). Sampling before and after a critical step in a process can be used to validate the effectiveness and variability associated with control measures. Although end product testing is common, this may not provide the most useful information. Analysis of the finished product provides an integrated measure of all the steps that contributed to the total population, but if the product does not meet a specification, the results do not identify the cause of the problem. In-line samples may be necessary to identify the cause.

End product testing is generally not necessary on a routine basis when records indicate a process is in control. The merits and limitations of end product testing as a measure of process control are discussed in Chapters 4 through 7 and elsewhere in this book.

Equipment and environmental tests are used to measure the effectiveness of GHP within a food operation. Visual inspection is the most common method of assessing whether equipment has been adequately cleaned/sanitized between production runs. Many operators routinely supplement the visual inspection with sponge or swab samples collected from the equipment as a means of verification. The sampling sites and frequency of sampling are determined by experience and whether problems are detected. In other operations, an inspector decides when and where to collect microbiological samples, usually only when there is uncertainty or it is desirable to confirm an observation.

The value of environmental sampling for microbial control is discussed in Chapter 11. The correlation between the microbiological status of the environment and a food is highly dependent on the characteristics of the food manufacturing system. For example, an enclosed processing system has fewer opportunities to become contaminated from the surrounding environment; so the relationship between environmental samples and the product would be tenuous. Conversely, food production systems with a great deal of manipulation and exposure to the plant environment are more susceptible to environmental contamination and there is more likely a stronger correlation between microbes detected in environmental samples and the microbiological quality or safety of such a food.

Shelf-life sampling is a specialized form of end product testing that involves holding finished product for longer than the time specified on the package and then examining it for specific attributes. Shelf-life testing is most often used to establish code-dating practices based on quality attributes (e.g., time to spoilage). However, shelf-life testing can be used to evaluate whether certain pathogens can increase, not change, or decrease during the normal expected conditions of storage and handling. If the pathogen is initially present at levels for which detection by standard microbiological analytical methods is unlikely, holding the product under "market or use conditions" can offer one means of estimating consumers' potential exposure at the time of consumption.

13.2.3 Determining the Causes of Variability

Variability can occur for many reasons. There may be inherent differences in an ingredient from one lot to the next. Solid foods may vary in dimension or weight. Liquids and semi-liquids may differ in viscosity and other attributes. The variety, species, or age of an ingredient can introduce differences in composition, texture, and other properties of the food that is being processed. Equipment maintenance and performance can be a significant influence on variability. Facilities having the same equipment will likely experience different degrees of variability depending on equipment age, model, maintenance, etc. In certain regions, season and weather can significantly affect humidity and drying conditions, such as for dry cured hams, fermented sausages, and legumes and cereals during harvesting and subsequent storage. The composition of milk (e.g., fat content) varies with season and must be taken into account when manufacturing cheese. The factors having a significant impact on variability must be determined for each food operation and a decision made whether those factors are sufficiently important to be controlled. The food safety system must take account of variability when establishing limits and monitoring procedures and be sufficiently conservative to ensure that the food being produced will be acceptable and safe.

13.2.4 Desired Patterns of Variability in Foods and Food Processing Parameters

There are two distinct patterns of variability in food characteristics and in the control parameters of food processing systems. The first applies to intended characteristics, attributes, or process parameters, such as pH, temperature, or even the concentration of intentionally added microorganisms in fermented products. In this pattern, there is a *target value* and some degree of *variation* above and below that value. These intended characteristics are controlled by keeping the mean value for the process near the target value and minimizing the spread above and below the mean.

Intended characteristics or attributes of a food or a process are usually controlled in terms of the normal distribution, which is derived from a mathematical theorem called the "Law of Large Numbers." That theorem states, in simple terms, that a collection of mean values from a large number of samples, all from the same original distribution, will show the variation described by the normal distribution. Since the normal distribution applies to mean values, a great deal of control theory is based on the distributions of the mean values of multiple measurements. This has proven both reliable and valuable.

The means of even a small number of samples of a distribution are often nearly normally distributed. For many intended food characteristics or control parameters, such as pH and temperature, the distribution of individual values is essentially normal even when the number of samples being considered is relatively small, e.g., $n = 5$. In fact, for many characteristics or attributes in nature, such as the pH of an individual sample or the population density of bacteria in a growth medium, an individual sample can be viewed as the outcome of many small samplings (i.e., means). Therefore, if one can represent the attribute or characteristic in the correct units, the distribution of individual results will approximate to a normal distribution. In the case of bacteria cultured in an environment that supports non-steady state growth, such as occurs in a food, the increases in population density typically occur exponentially. Declines under adverse conditions are also roughly exponential. It has long been observed that "concentrations" of microorganisms have a log-normal distribution and consequently have a normal distribution when concentrations are expressed as logarithmic values.

The second pattern of variability applies to unintended and undesirable characteristics, such as concentrations of spoilage microorganisms or specific pathogens. In this instance, the goal is to keep the level as low as possible based on technological, economic, or public health considerations. In the case of infectious pathogens, the target value is often the absence of the biological agent by a specific microbiological assay in a given weight of product. However, even when the target value is absence in a specified quantity of food, there is some distribution of values inherent in the target. The rare chance occurrence of a pathogen is not necessarily indicative of the process being out of control. Typically, target values for indicators (e.g., coliforms) are as low as can be achieved with GHP and HACCP and, when indicator organisms are present, they are within levels considered acceptable for the process and food. Control of unintended characteristics such as microbiological contamination is usually achieved using two targets: a limit on the *fraction of product that shows contamination* and an upper limit on the *concentration of contaminants* that occurs in that fraction.

13.3 PROCESS CAPABILITY STUDY

Confidence in the performance and variability of individual processes, or entire systems, is based on data from past performance, often expressed as microbiological baselines. The use of historical, product development, and/or initial control chart data to set warning and action control limits is termed a process capability study (Bothe, 1997). Such studies can be considered part of the process validation activities that establish the efficacy of a new process or process step to control a microbial hazard. Like the HACCP plan itself, a process capability study is valid only for the process for which it was conducted. A change in the process that might alter product safety would require a new process capability study. Ideally, the process validation would be of sufficient duration to quantify all factors (e.g., seasonality, alternate sources of raw materials) that could influence process performance. However, this is seldom achieved, so subsequent verification activities are designed to acquire additional data to enhance the initial process capability study.

A process capability study has two parts:

- collecting data that show the distribution of microbial measurements when a process is operating as intended; and
- using the data to derive one or more limits on sequential patterns of data.

Whether for attribute (i.e., presence/absence) data or microbial numbers, the setting of microbiological limits will ultimately be a matter of judgment, experience, and considering the risk and consequences of failure. A process capability study of an attribute associated with a CCP often would be useful for establishing the critical limit, e.g., temperature of pasteurization. Similarly, a process capability study that examines the frequency and extent of contamination by a specific pathogen or indicator microorganism for a process could provide the basis for establishing criteria for use in verification. Several countries have acquired national microbiological baseline data on the levels of microorganisms in various food products. Those data can be viewed as industrywide process capability studies that provide a measure of the mean and variability of the foods produced by the industry. Such studies can be used to establish microbiological criteria and then, subsequently, to measure the effectiveness of industry performance and regulatory control policies and practices.

13.3.1 Establishing Criteria and Monitoring Procedures

Some of the general principles that determine which process evaluation techniques are most appropriate are introduced below.

Statistical process control (SPC) allows the processor to control the mean and minimize the variability of each important control parameter. If the process has achieved a predictable repeatability, the mean and the variability should remain relatively constant over time. In HACCP, SPC can be used to determine the capability of a process to maintain its mean and variability in relation to both the monitoring of CCPs and verification analyses. Data from HACCP records provide a temporal history of how well the system has been controlled. Such data are usually more useful when displayed graphically on charts called *control charts* (Figure 13-1 through Figure 13-5). First introduced in 1924 by Shewhart, control charts are the primary tool for visualizing, comparing, and analyzing process control data. Accordingly, statistical concepts related to SPC will be discussed mainly in terms of these charts.

The stringency of a process control system is established by *control values* for parameters in the process or product that require intervention to maintain control. In HACCP, the control values associated with CCPs are critical limits used to separate acceptability from unacceptability (CAC, 1997). Product produced during a process deviation would be unacceptable to release until it could be determined that the loss of control did not result in food of unacceptable, or uncertain, quality and safety.

The stringency of a food safety system is a relative attribute, and establishment of a control value is a decision that requires consideration of risk and the consequences of a system failure. If a control value is set in such a way that even minor changes in the mean or the variability will violate the established value, then the system is very stringent. Conversely, if substantial deviation from the mean or a substantial increase in the variance is tolerable without an intervention, then the system is not stringent. Control values are set in different ways with different types of control charts, so examples of each will be discussed as various control charts are introduced. However, a discussion of some general concepts is often helpful in providing a framework within which the individual approaches can be interpreted.

13.3.2 Types of Variability and Error

Two sources of variation are associated with food processes. The first is referred to as "common causes" of variation inherent in a process when it is operating as intended, i.e., in control. Reducing the common causes of variation would improve the process. The second source of variation is referred to as "special causes" (or assignable causes) of variation. This variability can occur when one or more steps in a process are no longer operating as intended (i.e., out of control). In this instance, identification and correction of the special cause of variation returns the system to the original degree of variability within which it was designed to operate.

Differentiating common and special causes of variation is usually achieved by comparing series of measurements that indicate whether the products meet the specifications previously established when the process was under control. In their simplest form, the results of such a set of measurements fall into four categories:

1. The measurement correctly identifies the system as in control.
2. The measurement correctly identifies the system as out of control.
3. The measurement incorrectly identifies the system as out of control.
4. The measurement incorrectly identifies the system as in control.

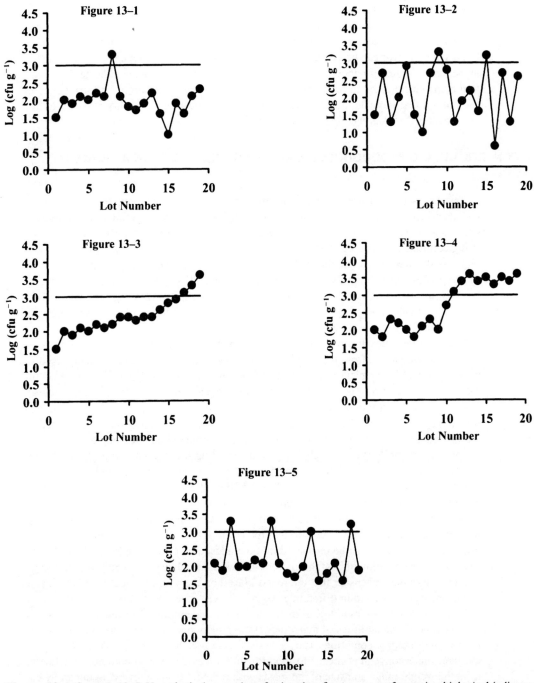

Figures 13–1 through 13–5 Hypothetical examples of using data from an assay for a microbiological indicator to verify the effectiveness of a food safety system. Examples depicted include the system under control (13–1), lack of control due to excess variability (13–2), loss of control due to gradual (13–3) and abrupt (13–4) process failures, and loss of control due to a reoccurring, transitory failure (13–5). The solid horizontal line depicts a hypothetical microbiological criterion above which a sample is considered only marginally acceptable. A criterion based on the presence of more than a specified number of marginal samples within a specified period of time would be the basis for determining if a process is out of control and requires corrective actions.

Categories 3 and 4 above are referred to as Type I and Type II errors, respectively, when the measurements lead to incorrect interpretations. A Type I error (category 3 above) indicates that a process is out of control when, in fact, it is still operating as intended. Such a Type I error can be viewed as a false alarm. Conversely, a Type II error (or missed signal error—category above) occurs when a series of measurements indicate that a system is in control when in fact it has an assignable source of variation. The stringency of the criteria that indicate the emergence of a special cause of variability is based on the consequences of making Type I versus Type II errors.

13.4 CONTROL DURING PRODUCTION: MONITORING AND VERIFYING A SINGLE LOT OF FOOD

Two types of data are collected during HACCP implementation, from monitoring CCPs and from verification. Control chart methodology is ideally suited to monitoring CCPs and other control points in an objective and consistent fashion. Decisions reached using control charts can be made with a measured degree of confidence. If the risk associated with a hazard is high, the control chart can be constructed to minimize the chances of the process going out of control. If the measurement is instantaneous, e.g., temperature, control is active, rather than passive, with control charts following the progress of a lot of food while it is being manufactured, detecting impending problems, and allowing adjustments to be made before control is lost.

The need for quick feedback as a food is being processed precludes the use of microbiological tests in control chart applications. Thus, control charts are normally used to record physical or chemical measurements. Control charts are easily produced when automatic in-line measuring is used (e.g., recording temperature during milk pasteurization). Control can be built into the system when a critical limit is not met, such as when milk is diverted by an automatic flow diversion valve to a balance tank, or holding tank, for repasteurization after the equipment has been adjusted to the required operating temperature.

Another example of control charting is in recording the internal temperature of beef roasts during cooking. To achieve a performance criterion of a 6 $\log_{10}$ reduction of salmonellae, it would be necessary to cook to an internal end point temperature of 62.8 °C and remain at this temperature or higher for a minimum of 4 minutes. Other time-temperature combinations could be used to meet the same performance criterion (6 $\log_{10}$ reduction of salmonellae). To ensure that the entire lot meets the performance criterion, internal temperature must be measured in the coldest area of the largest roasts in the lot. In addition, distribution of heat throughout the oven should be periodically verified (e.g., quarterly, monthly) to ensure that the roasts placed on the bottom, middle, and top layers of the cooking racks and in different areas of the oven meet the criterion.

Because continuous monitoring of the internal temperature of a number of roasts is even beyond the capabilities of many processors, many operators establish process criteria for the cooking conditions (i.e., oven temperature, humidity, time) as a simpler, cheaper, more convenient means to monitor the process. It is also common practice to limit the weight range and size of the roasts within the lot to avoid overcooking and yield loss among lighter weight roasts. Frequent measurements of the oven temperature can be collected automatically from thermocouples strategically placed throughout the oven. The data for the oven can be recorded automatically onto a computer-based chart logger that provides a visual record as the lot is being cooked. The oven operator can examine the chart to determine whether adjustments are necessary to meet the critical limit of

62.8 °C. When experience indicates that the roasts should be at, or above, the required temperature, a temperature probe is inserted into selected roasts to verify the criterion has been met. The lot would then be held for four minutes at that temperature before chilling would begin. Some operators may take account of the fact that the internal temperature will continue to rise several degrees even with no further application of heat.

An alternative approach to meeting the performance criterion could be based on the total lethality for salmonellae that occurs during the heating and chilling cycle. After placing thermocouples into the roasts and documenting the time the internal temperature is at 57, 58, 59, 60 °C, etc. during heating and chilling, the cooking procedure is calculated from the incremental lethality that occurs at the times and temperatures above 57 °C, the sum of which would meet the 6 $\log_{10}$ reduction performance standard for salmonellae. This approach should be continuously monitored with thermocouples placed in the largest roasts to provide a temperature profile that can be used for verification.

It is important to note that biased sampling is used in the selection of roasts to be monitored. Information about the largest roasts and, in some cases, the location of the roasts within an oven, is used to ensure all the roasts meet the criterion. If any of the selected roasts have not met the established criteria, then the entire lot must continue to be cooked. While information about variability is very important when establishing the cooking and monitoring procedures, information about both the mean and variability will determine the quality of the product and profitability of the process. In this example, and in many other food processes, a statistical evaluation is not used to decide when each individual lot is cooked. Statistical evaluation, however, can be an important tool in the design and validation of a process.

13.5 CONTROL DURING PRODUCTION: ORGANIZING DATA FROM ACROSS MULTIPLE LOTS OF FOOD TO MAINTAIN OR IMPROVE CONTROL

This section primarily considers the organization and interpretation of data collected from multiple lots produced over days, months, or even years to enhance control of production and to provide the information necessary for continuous improvement. However, the information in this section also applies to individual lots of food.

The emergence of process control programs such as HACCP has led to a substantial shift in the intent of microbiological testing programs. While testing of individual lots of some food still occurs, increasingly industry and control authorities are focusing their testing programs on verifying that food control systems are effective. While many of the microbiological assays employed for these two approaches are virtually identical, the statistical tools and assumptions that help in interpreting the results of monitoring and verification differ from those for batch testing. Particularly for verification testing, data from multiple batches, often over extended periods of time, are evaluated. Thus, unlike batch testing, this aspect of verification requires consideration of both within-batch and between-batch variability.

It is important to reemphasize that the purpose of such testing is not approval for release of batches or characterization of particular lots of product. In the case of HACCP systems, individual batches are characterized by monitoring CCPs, not verification testing. Instead, the purpose of such periodic testing is to provide:

- assurances that the conditions that enable a food process to produce safe products are being maintained.
- a basis for analyzing performance trends so that corrective actions can be taken before loss of control.

- insights into the cause for loss of control (e.g., periodicity of contamination).
- a warning that conditions have changed sufficiently such that the original HACCP plan may need to be reviewed.

One of the most important features of control charts is that they allow visual tracking of results over multiple lots. As trends develop, not only does it become apparent when action is required, but it is also possible, by careful study of the data, to determine approximately when the trend started. Cumulative sum (CUSUM) charts (Duncan, 1986), which use the same data as in the control charts but plot them in a cumulative fashion, provide a clearer visual display for this purpose. Moving sum (MOSUM) charts are another means of presenting cumulative data. Knowledge of when a nonrandom influence entered the process can yield valuable clues to the identification and elimination of the problem.

It is not good practice to wait until a control chart indicates an out-of-control situation before taking action. Some control charts have warning limits inside the control limits to signal impending loss of control. CUSUM charts are especially useful for early detection of trends. Many processes can gradually go out of control, providing plenty of warning as this occurs. Control devices may drift, mechanical components will wear (causing increasing variability as they do so), cleaning and disinfection procedures may be less stringently performed, or the microbiological quality of an ingredient may gradually worsen. Sometimes even an abrupt change in some procedure may result in a gradual change elsewhere.

In many situations, the need for both upper and lower control limits may not be readily apparent. One-sided control charts could be constructed for situations where the upper limit is of concern. Examples include the pH of an acidified food that is to be distributed at ambient temperature or the retort temperature for a canned food that must be kept above a certain minimum requirement, but too high a temperature is wasteful of resources. Frequently at CCPs in a food process, one limit controls product safety while the other limit represents economy of manufacture or some aesthetic feature of the food.

In addition to monitoring existing, well-established processes, control charts are useful in developing new processes. By their nature, control charts draw attention to unusual results that have occurred due to the influence of some nonrandom and controllable factor. Hence, in process design, control charts can be used to help isolate sources of process variation so that these sources can be traced and, if possible, eliminated.

Acquiring data in this manner can provide important insights concerning the performance of a food safety system and the types of problems being encountered. As a means of demonstrating this point, Figures 13–1 through 13–5 depict graphic examples of hypothetical microbiological data acquired to verify performance of a food safety system in relation to an established acceptance criterion.

Figure 13–1 depicts a system that is under control. Even with a well-controlled system, deviations characteristic of the system occasionally occur. Setting the criterion for what is or is not a failure is dependent on the level of performance that can be expected (i.e., the variability of a process) and achieved (i.e., technological capability), and the consequences of not meeting the acceptance criterion. Second, in a well-controlled processing system, the majority of data tend to cluster around a central value.

Figure 13–2 depicts the same system with greater variability. This is reflected in an increased number of samples above the acceptance criterion and an increase in the scatter of values that meet the acceptance criterion. Such a scenario could indicate one or more factors not being controlled.

Figure 13-3 depicts a situation where a component of the process is losing its effectiveness over time. It is apparent, through trend analysis of the data, that it might be possible to detect and correct this deficiency before the acceptance criterion is exceeded. This trend indicates a gradual loss of control at an important processing step. An example of this type of failure might be the buildup of cooked egg on the holding tubes of a liquid egg pasteurizer, leading to decreasing effectiveness of the heat treatment. By comparison, Figure 13-4 depicts an example of a loss of control such as when a key piece of equipment failed abruptly.

Finally, Figure 13-5 provides an example of a system with distinct periodicity. Such data indicate the existence of a recurring, intermittent problem. This pattern can occur with a seasonal effect, if the data were to represent a summary chart spanning years, or lots produced on Mondays due to microbial growth on inadequately cleaned equipment over the weekend.

While many food producers use microbiological testing to establish microbiological profiles for their products, these data are often not analyzed and evaluated in a rigorous manner. The knowledge that can be gained from these data can be greatly enhanced through use of the relatively simple statistical tools that are described in this chapter. Considering the cost incurred in conducting microbiological testing, the use of these statistical tools can be viewed as increasing the manufacturer's return on investment.

13.5.1 Variables Charts of Quantitative Data

As mentioned above, control charts are the primary means for arraying process control data for visualization and analysis. However, the type of control chart used will depend on the type of data being evaluated. As discussed in Chapter 7, microbiological analyses generate two types of data, *attribute* data (nonquantified, presence/absence assays) and *variable* data (quantitative, population density determinations), and there are different classes of control charts for evaluating those data. Attribute charts will be discussed in section 13.5.7.

13.5.2 General Principles of Variable Control Chart Construction

Control charts are plots of data over time (Ryan, 1989) and may represent data from one lot or from across multiple lots. The x-axis on the control chart is usually the time at which the sample or samples were collected during the process, and the y-axis is the value obtained for the measurement. The control chart consists of 3 parallel lines: a lower control limit (LCL), a center (or central) line, and an upper control limit (UCL) (Figure 13-6). In some instances where the LCL is below the lower limit of detection (i.e., absence), the LCL is assumed to be zero or some predesignated value below the lower limit of detection.

Each step in a process has a degree of inherent variability. When combined, the variability of each step contributes to the overall variability of the system. In a well-controlled system, data points tend to cluster around a central value. Traditional statistical measures of central tendency include the mean, median, and mode, and the measures of variability, such as the range, standard deviation, and standard error of the mean. The centerline in a control chart typically is a measure of central tendency, whereas the flanking lines are limits based on the expected degree of variability.

To appreciate how the LCL and UCL are often established, it is necessary to understand the standard deviation, or *sigma* (σ), concept as it applies to control charts. The σ

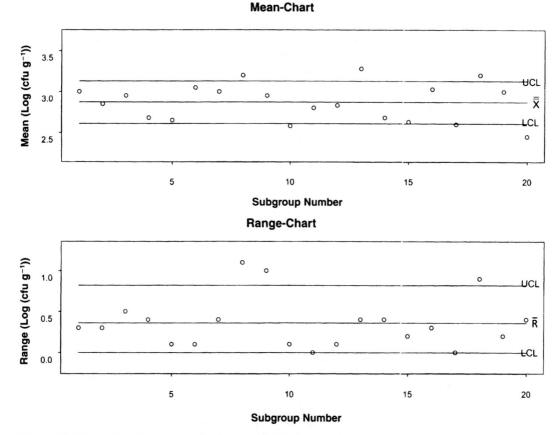

Figure 13-6 Hypothetical example of $\bar{X}$ and R control charts developed using a process capability study that measured the "Total Aerobic Plate Count" for 20 data subgroups each with 4 replicates (Table 13-1). The middle solid line represents the "target value" ($\bar{\bar{X}}$ and $\bar{R}$) and the flanking dotted lines represent ± 3σ.

concept is similar to that used in describing the variability of batches or lots. Control charting generally assumes that the distribution of data collected during the process is normal or approximately normal. Based on the normal distribution, approximately 68% of values will fall within plus or minus 1 standard deviation of the mean, approximately 95% within 2 standard deviations of the mean, and approximately 99.7% within 3 standard deviations of the mean (Figure 13-7). "One sigma" refers to one standard deviation from the mean; "2σ" to two standard deviations from the mean; and "3σ," to three standard deviations from the mean. Control limits are most commonly set at plus or minus 3σ from the mean. When 3σ control limits are used, the probability of any particular datum point being outside the control limits by chance alone is 0.3% when the process is actually *in control*. Thus, if the frequency at which values fall above or below 3σ is greater than 0.3%, then the process is deemed to be out of control.

In addition to determining when a process is out of control, control charts and their measures of central tendency and variation can be used to predict the frequency with which failures will occur despite a process being in control (i.e., frequency of Type I errors). For example, Peleg and co-workers (Nussinovitch *et al.*, 2000; Peleg *et al.*, 2000)

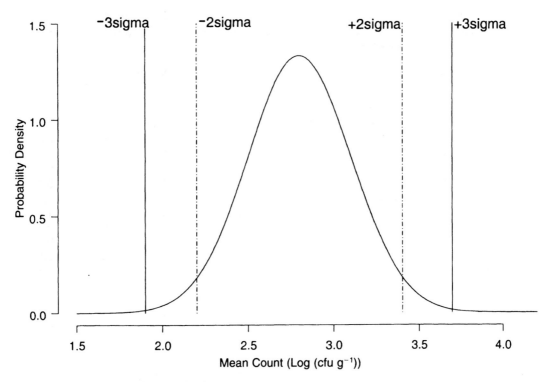

Figure 13-7 Density of a normal distribution.

used control charting in combination with a probabilistic model to predict the frequency of unusually elevated bacterial counts in foods.

As a means of introducing general concepts related to process control charting, the $\bar{X}R$ control chart will be used as an example. $\bar{X}R$ denotes the use of the mean ($\bar{X}$) of a subgroup of samples and the range (R) between the smallest and largest values in the subgroup of samples. It is commonly used in variables charting. The $\bar{X}R$ chart is actually two charts, one a measure of central tendency (i.e., the mean), and the other a measure of variability. It is based on the comparison of small subgroups, thus fulfilling the original intent or precision of the normal distribution, i.e., comparison of means. $\bar{X}$ and R are usually included on the same control chart to facilitate their use in tandem to establish whether the process is in control. The $\bar{X}R$ chart is one of the most useful from a practical standpoint because it is so simple to construct.

The first step in developing an $\bar{X}R$ chart is to define a subgroup, frequently a set of results from verification samples or monitoring data that are collected during a short time when the process is known to be under control. The variable of interest is measured for each item in the subgroup. Before the control chart is constructed, the general recommendation is to collect data on 20–30 subgroups, with each subgroup consisting of samples (n) of 4–5 units per subgroup. As an illustrative example, Table 13–1 shows a set of hypothetical total aerobic plate count data taken to verify the microbial quality of a ready-to-eat food when all manufacturing processes are in control. In this example, the control limits are based on 3σ and an equal number of items in each subgroup.

Table 13–1 Hypothetical Results of a Process Capability Study for a Ready-to-Eat Food Wherein Total Aerobic Plate Count Data ($\log_{10}$ cfu g^{-1}) Are Used To Verify the Acceptability of a Product

Subgroup	Sample 1	Sample 2	Sample 3	Sample 4	Mean	Range
1	3.1	3.0	2.8	3.1	3.00	0.3
2	2.7	2.9	3.0	2.8	2.85	0.3
3	2.8	3.3	2.8	2.9	2.95	0.5
4	2.9	2.5	2.6	2.7	2.68	0.4
5	2.6	2.7	2.7	2.6	2.65	0.1
6	3.1	3.0	3.0	3.1	3.05	0.1
7	3.2	2.9	2.8	3.1	3.00	0.4
8	3.8	2.8	2.7	3.5	3.20	1.1
9	3.5	3.0	2.5	2.8	2.95	1.0
10	2.6	2.5	2.6	2.6	2.58	0.1
11	2.8	2.8	2.8	2.8	2.80	0.0
12	2.9	2.9	2.7	2.8	2.83	0.1
13	3.0	3.3	3.4	3.4	3.28	0.4
14	2.9	2.5	2.7	2.6	2.68	0.4
15	2.5	2.6	2.7	2.7	2.63	0.2
16	3.1	2.8	3.1	3.1	3.03	0.3
17	2.6	2.6	2.6	2.6	2.60	0.0
18	2.8	3.4	3.7	2.9	3.20	0.9
19	2.9	3.0	3.1	3.0	3.00	0.2
20	2.4	2.3	2.7	2.4	2.45	0.4
				$\bar{\bar{X}}$	2.87	
				$\bar{R}$		0.36

1. Calculate the mean ($\bar{X} = [X_{1+} \ldots +X_n]/n$) for each subgroup (Table 13–1). If evaluating microbiological population density data, the $\log_{10}$ of the individual values would be used, thus converting the log-normal distribution associated with microbial "concentrations" to a normal distribution of log values.
2. Calculate the range (R) for each subgroup (Table 13–1). The range is the difference between the maximum and minimum values of items in the subgroup.
3. Compute the mean range ($\bar{R}$) for all subgroups. The mean range is the sum of the ranges for all of the subgroups divided by the number of subgroups (n_{sg}). In the current example, the $\bar{R}$ value is 0.36.
4. Compute the grand mean for all subgroups. The grand mean ($\bar{\bar{X}}$) is the sum of the means of all the subgroups divided by the total number of subgroups ($\bar{\bar{X}} = [\Sigma \bar{X}]/n_{sg}$). In the current example, the $\bar{\bar{X}}$ value is 2.87.
5. Set the centerline of the $\bar{X}$ chart at $\bar{\bar{X}}$, and set the centerline of the R chart at $\bar{R}$ (Figure 13–6).
6. Calculate the 3-σ control limits around the target value for the $\bar{X}$ chart. The LCL is equal to $[\bar{\bar{X}} - (A_2 \cdot \bar{R})]$. The UCL is equal to $[\bar{\bar{X}} + (A_2 \cdot \bar{R})]$. The value of A_2 can be obtained from Table 13–2 for subgroups with sample size (n) of 2–10. In the current example, the LCL = 2.61 and the UCL = 3.13.
7. Calculate the control limits for the R chart. The LCL is equal to $D_3 \cdot \bar{R}$. The UCL is $D_4 \cdot \bar{R}$. The values of D_3 and D_4 can be obtained from Table 13–2 for subgroups with sample size (n) of 2–10. In the current example, the LCL = 0.00 and the UCL = 0.82.

Table 13-2 Parameters for Determining "3-σ" Control Limits for $\bar{X}$ and R Charts

Sample Size (n) per Subgroup	A_2	D_3	D_4
2	1.880	0	3.268
3	1.023	0	2.574
4	0.729	0	2.282
5	0.577	0	2.114
6	0.483	0	2.004
7	0.419	0.076	1.924
8	0.373	0.136	1.864
9	0.337	0.184	1.816
10	0.308	0.223	1.777

8. Plot the $\bar{X}$ values on the $\bar{X}$ chart and the R values on the R chart (Figure 13–6). Plotting the current example makes it apparent that there is a fair degree of variability associated with both the central tendency and dispersion in this hypothetical process.

The $\bar{X}$R chart is one of the simplest variable control charts. Detailed descriptions of the procedures and uses of the wide variety of other variable control charts are available from standard references such as ASTM (1990) and Duncan (1986).

13.5.3 Corrections When Data Are Autocorrelated

An underlying assumption in the development of most control charts is that as the individual data points are plotted, they are randomly distributed according to a normal ($\log_{10}$ values) or log-normal (arithmetic) distribution over time. However, in actual production settings, successive measurements of product parameters collected sequentially over time are often correlated with one another (i.e., the measurements are autocorrelated). Autocorrelation is especially likely when values are collected close to one another in time. Autocorrelation affects the pattern of data on control charts, with implications for setting control limits. Wheeler (1995) and other standard references provide in-depth coverage of autocorrelated data and control charts. Ryan (1989) notes that autocorrelation is primarily a problem for CUSUM procedures and acceptance, difference, median, and midrange charts.

13.5.4 Special Considerations for Charting Individual Counts of Microbial Population Density

Data on concentrations of pathogens or indicator organisms in individual sample units can be presented in simple variables charts such as the hypothetical example in Figure 13–8. This fictional example shows numerical estimates of concentrations of indicator organisms in samples of a finished product. This chart shows fairly normal fluctuations in viable counts up to week 40. After week 40, the same sort of fluctuations occur, but about a much lower midpoint. This could reflect a change in raw materials or equipment, a seasonal effect, a change in the analytical method, etc. Even though the change appears welcome, investigation is warranted since it could represent either an analytical problem or a means for identifying a factor that could consistently enhance the performance of the system.

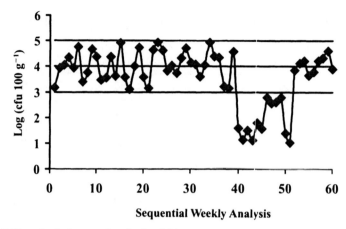

Figure 13-8 Hypothetical example of a "variables control chart" for a microbial indicator assay conducted weekly. The center horizontal line represents the mean value and the two flanking lines represent the upper and lower "warning" limits. An apparent period of significant improvement is depicted between weeks 40 and 51.

One of the most straightforward means of considering a set of quantitative microbiological data is to chart the individual test results. Such a control chart, commonly called an X chart, is simple and rarely misleading. Unusually high counts, or patterns of high counts, are typically the criteria used to designate loss of control and raise safety concerns. Conversely, unusually low counts, or patterns of low counts, may reflect a potential process improvement or a need to review analytical protocols. Obvious trends of either sort would merit attention. Subtle trends in the values for an indicator should not necessarily be viewed as trends at all, since some drifting up and down of indicator levels over weeks or months is to be expected. However, such trends may warrant a review of the process since they may give early warning of a gradual loss of process control (Figures 13–3 and 13–4), and thus allow corrective action before loss of control.

13.5.5 Selecting Limits for Variable Control Charts of Microbial Counts

Although professional judgments of food scientists, food microbiologists, and process managers are the most important factors in setting the upper and lower limits used for this type of variables control chart, there are SPC procedures that can help to select the initial choices of limits. The guidelines below will provide a reasonable starting point.

One of the key factors in selecting the limits is the test results for which no microorganism was detected, i.e., test value = 0. However, such a value is not directly useful for two reasons. First, the need to convert microbial population density data to $\log_{10}$ values results in the 0 values being undefined (and zero values are ignored by some software). Second, the inability to detect a microorganism in any specific sample reflects two possibilities: that the microorganism truly is not there, or that the microorganism is present, but at levels that the sampling protocol or analytical methods are incapable of detecting.

Consider first the situation in which at least 90% of the test results are expected to be quantifiable above the lower limit of detection for the method employed. Any results below the limit of detection (i.e., no microorganisms are detected) are assigned a value of one-half of the limit of detection and transformed into a $\log_{10}$ basis. (This might slightly underestimate the standard error. That would make the resulting "warning rules" and "stopping rules" slightly overcautious, but this can be adjusted when a large number of test results is available.) The limit-setting study would begin with testing at a high frequency to acquire enough results to study their average and their variability. Typically at least 10 quantifiable results, and preferably 30 or more, would be acquired and charted. The mean ($\bar{X}$) and the standard error (σ) are computed, "warning limits" are set at $\bar{X} \pm 2\sigma$, and upper and lower "stopping limits" are set at $\bar{X} \pm 3\sigma$.

When more than 10% of the limit-setting data are zero values (i.e., too low to be quantified), it is better initially to derive the warning and stopping limits by comparing the gap between percentiles of the test data with that which would be predicted by the normal distribution.

If less than half of the results are below the lower limit of detection (i.e., zero), a simple approach is to find the difference between the $\log_{10}$ of the 50th percentile (i.e., 50% of the samples have a lower value) and the $\log_{10}$ of the 75th percentile. With a normal distribution, the difference between the $\log_{10}$ value at the 75th percentile minus the $\log_{10}$ at the 50th percentile (i.e., mean) should cover 67% of the σ. Thus, dividing the difference of these percentile $\log_{10}$ values by 0.67 provides an estimate of σ. For example, if the 50% value is 3.2 and the 75% value is 4.6, then $\sigma = (4.6 - 3.2)/0.67 = 2.1$. Thus, dividing the difference of these percentile $\log_{10}$ values by 0.67 provides an estimate of σ. The 50th percentile value is used in place of $\bar{X}$ drawing the limit lines. Using these estimates of $\bar{X}$ and σ, the limits for the control chart are drawn as described above.

If more than half of the results are below detection, the analyst should consider whether to proceed with a variables chart. It might be more appropriate to use an attributes (presence/absence) chart. If that is not the preferred option, however, any table of cumulative probabilities of the normal distribution will provide the 50th percentile and σ from the gaps in the percentiles of the data. For example, the gap between the 70th and 90th percentiles should be 76% of σ. As a rough rule of thumb, percentiles greater than the 90th should not be used for this purpose.

Once the initial limits have been established, the first test results acquired subsequently should be scrutinized carefully as a means of ensuring that the data used to set the initial limits were acquired when the process was in control. Ordinarily, one would hope not to find any values outside the upper and lower control limits at this stage. If deviations are observed, then the data should be studied for any trend, and any trend identified should be investigated. If any result is below $\bar{X} - 3\sigma$ or above $\bar{X} + 3\sigma$, or if three of any five successive results are below $\bar{X} - 2\sigma$ or above $\bar{X} + 2\sigma$, possible causes should be investigated, such as a change in the process or ingredients or a change in the sampling or analytical protocols. The process capability study will have to be rerun with the new data until the warning and stopping limit values stabilize. When the process capability study has been completed without any deviations or apparent trends, then these limits at $\bar{X} \pm 2\sigma$ and $\bar{X} \pm 3\sigma$ should be selected for initial control limits for verification tests.

As verification testing proceeds, if it appears that the limits selected are giving an unreasonable number of false alarms, then the limits should be compared against the likelihood of Type II errors (i.e., not catching a safety defect) and expanded if appropriate. Such an expansion of the limits should be done cautiously with continued or increased scrutiny until sufficient data are acquired such that the process can be adequately judged in relation to:

- how truly rare is deviation when the process is in control, and
- how frequently is control lost.

Two examples of reasons follow for why the initial limit-setting exercise might have proved too restrictive.

1. Seasonality might cause $\bar{X}$ or σ to shift in ways that make the limits too tight. If seasonality has no effect on $\bar{X}$, then it probably has no effect upon σ. If it does affect $\bar{X}$, then (a) the charts may need to be adjusted by drawing seasonal $\bar{X}$ lines on them, and (b) the data must be studied to see if the σ has also changed. The seasonal changes in $\bar{X}$ and in the control lines can define a series of zones with different straight lines or $\bar{X}$ can be fit by a curving line with σ-determined lines curving at fixed distances above and below $\bar{X}$. However, if seasonality produces an unacceptable risk to safety, the appropriate response would be to eliminate the variability due to seasonality and not to modify the control chart limits.
2. If the limit-setting study is done with a single source of raw materials and the process then proceeds to use a variety of sources, an additional source of variation has been added, and the initial limits will be too narrow. When more data, covering more sources of raw materials, are in hand, the $\bar{X}$ and σ should be recalculated. Again, if the additional sources of raw material represent an unacceptable risk to safety, the appropriate response would be to eliminate the unacceptable variability associated with the new source of raw material and not to modify the control chart limits.

13.5.6 Caution in Interpreting Certain Types of Variables Charts

Moving Range (R) charts are useful for process control monitoring of parameters such as pH and temperature. These charts track the absolute value of the difference between each result and the previous one. Declines and increases are treated the same. When dealing with levels of undesirable microorganisms, a significant decline in microbial concentration may not have the same significance as an increase. This is because downward trends are generally not a cause for alarm, and a low value preceding a high value looks significant as a range but doesn't make the high value any more important.

Standard procedures are described in general quality control texts for converting the average of the moving ranges into an estimate of σ. However, this is also not advisable for pathogens and indicator organisms. Weather, seasonal changes, shifting harvest zones, early or late crops, spot markets for ingredients, and other factors will force $\bar{X}$ up or down over periods of multiple tests in ways that cannot be predicted very precisely. This will induce a degree of autocorrelation. As described above, autocorrelation is the tendency for results to be more like those near them in test sequence and less like those farther away in sequence. For this reason, the distribution of differences between successive test results will usually underestimate the true variability of microbial levels in foods.

For similar reasons, CUSUMs of differences in the observed $\log_{10}$ microbial concentrations and some target log numbers are not appropriate for these variable charts. The mean, or target log number, is not truly a target value, since one does not intend to achieve a target value of pathogens. In this instance, the uncontrolled shifting of $\bar{X}$ results in an inability to set standard limits for a chart of the mean of the previous n observations. However, such charts, for small n, would still be useful as a means of illustrating shifting patterns, such as those that might occur as a result of seasonal variation. (Note that CUSUMs are useful for attribute charting.)

13.5.7 Attribute Charts of Presence/Absence Determinations

One of the simplest, most widely used microbiological tests for specific pathogens or indicator organisms is presence/absence testing. An analytical unit of known weight (mass) is tested by a standard method to determine whether the microorganism of interest is, or is not, detected in that food sample. Over time, this type of microbiological assay can be used to assess the maintenance of food safety control systems. However, interpretation of such tests is highly dependent on the method used to determine the presence of the microorganism, particularly its lower limit of detection. Consequently, it is generally not possible to compare or combine the results from one data set with those from another unless both data sets were acquired using the same standard method that has consistent analytical performance characteristics.

Consider a hypothetical example of a single microbiological presence/absence assay for *Salmonella* spp. performed once per day for a total of 50 successive days as a means of process verification. The microbiological test requires 48 hours to complete so there is an effective 2.5-day delay between the sample being taken and the results becoming available. Figure 13-9 depicts these data as a simple control chart. Let us assume that when the process being evaluated is under control, about 1 of every 4 assays is positive for *Salmonella*. Further, the stringency of the system has already been established based on the criterion that if *Salmonella* is detected on four successive days, this is considered a loss of control due to the introduction of an attributable source of variability. In this example, we have included an abrupt failure of a critical part of the process on day 29 that is not detectable by normal process control monitoring procedures but is detected by the supplemental verification testing. While evaluation of the control chart ultimately allows the user to estimate when the failure was likely to have taken place, it shows that there is a substantial delay before an action can be taken. Based on the established criterion of four consecutive positive assays, if the results were available instantaneously, the

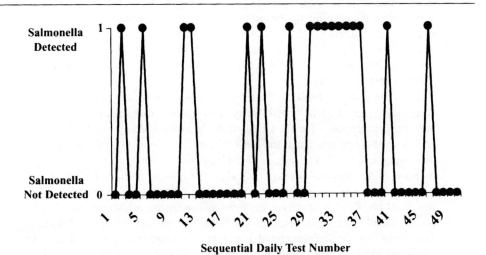

Figure 13-9 Hypothetical example of an "attributes control chart" for a single daily detection/nondetection assay for *Salmonella* spp. microbe where in-control detection rate is 0.25 (25%). In this example, control is lost on Day 29 and restored on Day 37.

chart would indicate a loss of control at day 33. However, because of the 2.5-day delay in acquiring the data, the actual day when the loss of control would be acted upon would be day 37. This is reflected in the control chart (Figure 13-9) when the values return to the normal pattern on day 37. It is apparent from this simple example that the response time associated with a control chart is based on the frequency of testing, the decision criterion established, and the time required to analyze the samples. For example, if the decision criterion had been three consecutive positive samples, the failure would have occurred a day earlier. However, this more restrictive criterion would increase the risk of Type I errors.

Presence/absence charts illustrate a sequential pattern of *yes/no* results. The usual assumption behind these charts is that when a process or system is operating under control there is some fraction p of sample units that will yield a *yes* result when analyzed. It is further assumed that successive test results are independent of each other. Analysis of charts shows whether these assumptions hold, and might lead to process improvements. For example, if the day shift has a higher p than the night shift at the same plant, this would suggest that there might be an additional source of variability during the day shift that may be reducible. Conversely, if one observes a longer than expected series of *no* results, the assumption of the independence of observations may not be true. This potentially could lead to identification of means for further process improvements.

If there is a change in a process that causes p to increase markedly, the frequency of detecting the microorganism of interest would be expected to increase (i.e., an increased frequency of *yes* results). Verifying that p is not changing is accomplished by tracking the sums of detection results over some interval of results. Starting from some beginning point in collection of data, one can determine for a number of future observations, n, an upper limit, denoted $U(n)$, that should not be exceeded by the cumulative sum of the samples in which the microorganism is detected. Standard statistical software included in most spreadsheet programs will compute the probability of getting j positive isolations out of n tests when the probability of the presence of the microorganism is p. One needs only to decide the degree of rarity that will be tolerated before intervening in the process. Then that criterion will determine $U(n)$ for every n. The process capability study that estimated p is followed by creation of a table of values of $U(n)$. Then the process manager notes after each test the CUSUM and compares it with $U(n)$. As long as CUSUM does not exceed $U(n)$, the statistic has not shown a loss of control.

There are practical limits to the length of sequential results that are closely related to each other. The CUSUM is also fairly complex. For these reasons, another statistic is sometimes preferred. After the CUSUM has been followed up to a fairly large n, there is little loss of discrimination if the most recent (fixed) n results are summarized, without increasing n further. The number of times the presence of the microorganism was detected in the previous n results is called the moving sum, or MOSUM, because the count is derived from a zone that moves along the string of observations. The limit for this fixed MOSUM is of course the same as that for the nth step in the CUSUM. If one picks a small n for evaluating the MOSUM, one can achieve a rapid detection of any major shift in p. On the other hand, if p shifts slowly upward, the sum over a small n will not be sensitive enough to detect it. A larger n is more discriminating, and therefore is more sensitive to even gradual shifts in p, but a large n can delay the discovery of the shift. One of the MOSUM techniques that has been used extensively with microbial food safety verification testing is the Moving Window assay (see section 13.6).

This dilemma between quick response and high sensitivity can be overcome by using multiple MOSUMs. These are a compromise between a single MOSUM and the CUSUM. For many processes, a choice of two MOSUM charts, a short interval for responsiveness and a long interval for sensitivity, can be very useful. MOSUMS, by their nature, are highly correlated with their predecessors, so apparent long strings of MOSUMS moderately above the mean value have no special importance. Increased frequency of testing can be especially valuable when MOSUMS or CUSUMS get close to their action limits. When these limits are exceeded, the situation changes to one of investigational sampling (Chapter 9) and problem solving to determine the new source of attributable variability and return the process to an in-control state.

13.5.8 Advantages of Multiple Charting of Data Subsets

Verification sampling of the end product is an essential data set for effective control charting because it reflects the integration of all the process control steps and reveals failures due to problems that affect the entire production. However, additional verification sampling at key intermediate steps within the process can provide valuable information for isolating special causes of loss of control. In addition, plotting subsets of a single verification data set may more rapidly identify an emerging problem. Returning to an example discussed above, suppose a plant operated three shifts per day, and workers analyzed one verification sample per shift for *E. coli*. In addition to the chart of all of the test results, it would be beneficial to generate individual charts of the results for each shift. In this way, new sources of variability that were specifically associated with a single shift would be identified and isolated substantially faster.

13.5.9 Caution Regarding Control Chart Software

Control charting software is increasingly available and greatly simplifies the process of setting up a variety of control charts. Caution must be observed, however, in using any of the more sophisticated options that are included in most software packages for advanced applications. Food safety processing rarely satisfies all the assumptions that underlie many industrial quality control computations. Care must be taken to determine whether options selected are pertinent to the development of either attribute or variable control charts related to microorganisms or their metabolic products in foods. This is particularly important when designing variables control charts.

Caution should also be exercised in keeping control records only through software (and not on hardcopy). Some software products automatically readjust $\bar{X}$ and σ as the data are inputted. This will hide slow trends in $\bar{X}$ or σ by leaving no traces of the adjustments (Wise & Fair, 1997).

13.6 USE OF PROCESS CONTROL TESTING AS A REGULATORY TOOL

Food control agencies have traditionally relied on batch-by-batch testing for lot acceptance of foods, particularly in relation to international trade. However, with the increased emphasis on the adoption of food safety systems such as HACCP, there is likely to be increased focus on the validation, monitoring, and verification of process control measures as the means that food control agencies use to assess safety. An example of how these techniques can be used in a regulatory framework is the US Department of Agriculture Food Safety and Inspection Service's Pathogen Reduction/HACCP regulation (McNamara, 1995; FSIS, 1996). This regulation included two forms of microbiological testing as a means of verifying the effectiveness of HACCP programs required for the production of meat and poultry at slaughter. The first was testing of carcasses by industry for the presence of biotype I *Escherichia coli* as an indicator of fecal contamination. The second was testing for *Salmonella* spp. performed by USDA. Both were based on the application of a MOSUM control charting technique, the Moving Window.

In the case of *E. coli* testing, the technique was adapted for variables testing (i.e., consideration of quantitative data) using a limit that could not be exceeded (M-value) and a warning value (m-value) that could not be exceeded more than three times (c-value) in a moving window of 13 tests ($n = 13$). The m-value and M-value were based on national baseline surveys of the various types of meat and poultry and are specific for that commodity (FSIS, 1996). The rate of sampling (one sample taken for a set number of animals slaughtered) is also commodity-specific.

The method for evaluating the carcasses for *Salmonella* was limited to presence/absence testing. Accordingly, the moving window technique was adapted for attribute testing. Again, the limits that indicated that the HACCP system was no longer under control were based on national baseline survey results for each commodity. The number of samples within the window and the number of positive isolations of *Salmonella* wherein the system is still considered in control varies among the different types of meat and poultry (Table 13–3).

13.7 INVESTIGATING AND LEARNING FROM PREVIOUSLY UNRECOGNIZED FACTORS OR UNFORESEEN EVENTS

In part, HACCP verification can be viewed as additional testing conducted for the purposes of ensuring that the conditions and requirements identified in the hazard analysis upon which a HACCP program was developed are still valid. As an example, consider pasteurization of liquid egg products. An obvious CCP for production of pasteurized liquid egg products is the time/temperature conditions of the heating step. This CCP is effectively monitored on a real-time basis by recording temperature and duration of heating. Parameters for this processing step were developed using data acquired from inoculated pack studies with *Salmonella* and have been highly successful for controlling this

Table 13–3 *Salmonella* Performance Standards Associated with the USDA Pathogen Reduction/HACCP Regulation

Class of Products	Performance Standard (% positive for Salmonella)	Number of Samples Tested (n)	Maximum Number of Positives to Achieve Standard (c)
Steers/Heifers	1.0	82	1
Cows/Bulls	2.7	58	2
Ground Beef	7.5	53	5
Broilers	20.0	51	12
Hogs	8.7	55	6
Ground Turkey	49.9	53	29
Ground Chicken	44.6	53	26

pathogen in these products. However, these studies were originally conducted at a time when most eggs used to make liquid egg products were eggs left over from the marketing of shell eggs. These eggs were typically several days to weeks old. When initially laid, the white of an egg has a pH of between 7–8, but within several days pH rises to between 10–11. This elevated pH substantially decreases the heat resistance of *Salmonella*. Since development of the original heat resistance data for *Salmonella* in egg products, there has been a substantial change in the demand for liquid egg products and an increasing portion of the eggs used come directly from the egg producer. The time between laying and pasteurization has decreased to the point where the pH of egg white is still in the range of pH 7–8. At this pH, *Salmonella* is substantially more heat resistant. Monitoring the CCP for temperature and duration of heating alone would not detect this gradual shift in a key characteristic of the egg. Instead, the egg industry was alerted by an increased incidence of salmonellae observed in periodic samples taken as part of effective verification programs.

13.8 REFERENCES

ASTM (American Society for Testing and Materials) (1951). *Manual on Quality Control of Materials*. Philadelphia, PA: American Society for Testing and Materials.

ASTM (American Society for Testing and Materials) (1990). *Manual on Presentation of Data and Control Chart Analysis*, 6th edn. Committee E-11 on Quality and Statistics. Philadelphia, PA: American Society for Testing and Materials.

Bothe, D. R. (1997). *Measuring Process Capability. Techniques and Calculations for Quality and Manufacturing Engineers*. New York: McGraw-Hill.

BSI (British Standards Institution). *Guide to process control using quality control chart methods and cusum techniques* (BS5700:1984); *Guide to number-defective charts for quality control* (BS5701:1980); *Guide to data analysis and quality control using cusum techniques* (BS5703, Parts 1, 2, 3, and 4); *Drafting specifications based on limiting the number of defectives permitted in small samples* (BS2635:1955). London: BSI.

Buchanan, R. L. (2000). Acquisition of microbiological data to enhance food safety. *J Food Prot* 63, 832–838.

CAC (Codex Alimentarius Commission) (1997). *Joint FAO/WHO Food Standards Programme, Codex Committee on Food Hygiene. Food Hygiene, Supplement to Volume 1B-1997. Hazard Analysis and Critical Control Point (HACCP) System and Guidelines for Its Application*. CAC/RCP 1–1969, Rev. 3.

DeVor, R. E., Chang, T-H. & Sutherland, J. W. (1992). *Statistical Quality Control and Design. Contemporary Concepts and Methods*. New York: Macmillan Publishing Company.

Duncan, A. J. (1986). *Quality Control and Industrial Statistics*, 5th edn. Boston, MA: Irwin McGraw-Hill.

FSIS (US Department of Agriculture-Food Safety and Inspection Service) (1996). Pathogen reduction; hazard analysis and critical control point (HACCP) systems; final rule. *Fed Reg* 61, 38806–38989.

Grant, E. L. & Leavenworth, R. S. (1972). *Statistical Quality Control*, 4th edn. New York: McGraw-Hill.

Hubbard, M. R. (1990). *Statistical Quality Control for the Food Industry*. New York: Van Nostrand Reinhold.

ICMSF (International Commission on Microbiological Specifications for Foods) (1988). *Microorganisms in Foods 4. Application of the Hazard Analysis Critical Control Point (HACCP) System to Ensure Microbiological Safety and Quality*. Oxford: Blackwell Scientific Publications Ltd.

ISO/TC 69. (International Organization for Standardization/Technical Committee 69). *Introduction to control charts* (ISO 7880:1993); *Shewhart control charts* (ISO 8258:1991); *Control charts for arithmetic average with warning limits* (ISO 7873:1993); *Acceptance control charts* (ISO 7966:1993); *Cumulative sum control charts* (ISO/TR 7871:1997). Geneva, Switzerland: International Organization for Standardization, Secretariat.

Juran, J. M., Godfrey, A. B., Hoogstoel, R. E. & Schilling, E. G. (1999). *Juran's Quality Control Handbook*, 5th edn. New York: McGraw-Hill.

Kramer, A. & Twigg, B. A. (1982). *Quality Control for the Food Industry*, 3rd edn. Westport, CT: AVI Publishing.

Ledolter, J. & Burrill, C. W. (1999). *Statistical Quality Control. Strategies and Tools for Continual Improvement.* New York: John Wiley & Sons.

Massart, D. L., Dijkstra, A. & Kaufman, L. (1978). *Evaluation and Optimization of Laboratory Methods and Analytical Procedures*, Vol. 1. Amsterdam: Elsevier.

McNamara, A. M. (1995). Establishment of baseline data on the microbiota of meats. *J Food Safety* 15, 113–119.

NACMCF (National Advisory Committee on Microbiological Criteria for Foods) (1997). Hazard analysis and critical control points and application guidelines. *J Food Prot* 61, 762–775.

NRC (National Research Council) (1985). *An Evaluation of the Role of Microbiological Criteria for Foods and Food Ingredients.* Subcommittee on Microbiological Criteria. Washington, DC: National Academy Press.

Nussinovitch, A., Currasso, Y. & Peleg, M. (2000). Analysis of the fluctuating microbial counts in commercial raw milk—case study. *J Food Prot* 63, 1240–1247.

Peleg, M., Nussinovitch, A. & Horowitz, J. (2000). Interpretation of and extraction of useful information from irregular fluctuating industrial microbial counts. *J Food Sci* 65, 740–747.

Ryan, T. P. (1989). *Statistical Methods for Quality Improvement.* New York: John Wiley & Sons.

Smith, G. M. (1998). *Statistical Process Control and Quality Improvement*, 3rd edn. Columbus, Ohio: Prentice Hall.

Steiner, E. H. (1984). Statistical methods of quality control. In *Quality Control in the Food Industry*, Vol. I, 2nd edn., pp. 169–298. Edited by S. M. Herschdoerfer. London: Academic Press.

Wheeler, D. J. (1995). *Advanced Topics in Statistical Process Control.* Knoxville, TN: SPC Press.

Wise, S. A. & Fair, D. C. (1997). *Innovative Control Charting.* Milwaukee, WI: American Society for Quality.

Chapter 14

Aflatoxins in Peanuts

14.1 Introduction
14.2 Risk Assessment
14.3 Risk Management
14.4 Acceptance Criteria for Final Product
14.5 References

14.1 INTRODUCTION

Aflatoxins are the most potent liver carcinogens known, capable of causing cancer in all animal species studied, including birds, fish, and humans. Aflatoxins are also acute, chronic, genotoxic, and immunosuppressive poisons, but these effects are usually considered to be less important.

The primary sources of aflatoxins are the common fungi *Aspergillus flavus* and the closely related species *A. parasiticus*. *A. flavus* is very common in tropical and subtropical regions of the world and is particularly associated with peanuts and other nuts, and with maize and other oilseeds. *A. parasiticus* is primarily associated with peanuts and has a more restricted distribution.

The four major naturally produced aflatoxins are known as B_1, B_2, G_1, and G_2. "B" and "G" refer to the blue and green fluorescent colors produced under UV light on thin layer chromatography plates, while the subscript numbers 1 and 2 indicate major and minor compounds, respectively. *A. flavus* produces only B aflatoxins and only about 40% of isolates from nature are toxin producers. *A. parasiticus* produces both B and G aflatoxins, and virtually all natural isolates are toxin producers (Klich & Pitt, 1988).

Because aflatoxins are extremely toxic, regulatory levels in many countries are very low, ranging from 1 to 25 $\mu g\ kg^{-1}$ (van Egmond, 1989). The Codex Committee on Food Additives and Contaminants (CCFAC) of the Codex Alimentarius Commission (Codex) has recommended to Codex that the limit for foods in international trade be set at 15 $\mu g\ kg^{-1}$ total aflatoxins.

Aflatoxins are readily detected by a very strong fluorescence in ultraviolet light. Analyses may be performed using thin layer chromatography, high performance liquid chromatography, or immunoassays (AOAC, 1984; Nesheim & Trucksess, 1986).

In commodities, especially peanuts, aflatoxins are very irregularly distributed; consequently, aflatoxin determinations must be based on large and representative samples obtained using carefully constructed sampling plans. Samples must be ground or finely divided, then subsampled for assay. Assays based on small samples are unreliable. In more homogeneous products, e.g., peanut butter, this is less of a problem.

14.2 RISK ASSESSMENT

Aflatoxins are among the most potent mutagenic and carcinogenic substances known. Extensive experimental evidence has shown that aflatoxins are capable of inducing liver cancer in most animal species. However, translating that information to human risks has proven to be extremely difficult. Some evidence suggests that humans are at substantially lower risk of liver cancer from aflatoxins than other species. Some epidemiologic studies have suggested that intake of aflatoxin alone poses a detectable risk, but others show that the presence of other factors, such as hepatitis B virus (HBV), are necessary to induce liver cancer.

The identification of hepatitis C virus (HCV) is an important recent advance in understanding the etiology of liver cancer. Epidemiologic studies are largely consistent in showing a strong association between antigens to HCV and the occurrence of liver cancer. The risk linked to HCV is independent of HBV and other risk factors. HCV is likely to be the major cause of liver cancer in countries at low to intermediate risk of liver cancer, such as the US, Europe, and Australia. The epidemiologic evidence of the carcinogenicity of HCV has been reviewed and endorsed by an international group under the leadership of the International Agency for Research on Cancer (IARC, 1994).

Viral hepatitis is a major worldwide public health problem. It is estimated that over 300 million individuals are chronically infected with HBV and perhaps 100 million with HCV. Although the evidence remains inconclusive, it is estimated that 50–100% of liver cancer cases worldwide are associated with persistent infection with HBV and/or HCV. HBV is prevalent in the developing parts of the world, and HCV is emerging as a major cause of hepatocellular cancer in Japan and western societies (Bosch, 1995).

The Joint FAO/WHO Expert Committee on Food Additives (JECFA; [Food and Agricultural Organization/World Health Organization]), which is responsible for assessing the toxicology of chemicals in foods, has summarized this position.

> Risks from specific exposures to aflatoxins are difficult to estimate and predict, despite extensive information available from epidemiological studies, mutagenicity tests, animal bioassays, in vitro and in vivo metabolic studies, and p53 mutation studies. Many questions remain regarding the independence of aflatoxin as a human carcinogen, the extent to which hepatitis B, hepatitis C, and other factors modify the effect of aflatoxin, how findings from countries with high liver cancer rates and high prevalence of hepatitis B may be compared to those from countries with low rates, how to deal with the wide range of susceptibility to aflatoxin carcinogenesis among experimental animals, and how to describe the dose–response curve over the wide range of aflatoxin exposure found worldwide (JECFA, 1997).

JECFA has also reviewed dose–response analyses performed on aflatoxins. All of these analyses suffer limitations, of which three are especially important. First, all of the epidemiologic data from which a dose response relationship can be developed are confounded by concurrent HBV infection. The epidemiologic data are from geographic areas where both the prevalence of HBV positive individuals and aflatoxins is high; the relationship between these risk factors in areas of low aflatoxin contamination and low HBV prevalence is unknown. Second, the reliability and precision of the estimates of aflatoxin exposure in the relevant study populations are unknown. In particular, the biological markers currently used to indicate aflatoxin intake by humans do not reflect long-term aflatoxin intake, and in most cases, analyses of crops for aflatoxins do not take account of reduction in levels of aflatoxins consumed in foods after processing. Third, the shape of the dose–response relationship is unknown, which introduces an additional element of uncertainty when choosing mathematical models for interpolation.

Observations concerning the interaction of HBV and aflatoxins suggest that two separate aflatoxin potencies exist, one in populations in which chronic hepatitis infections are common and the second where such infections are rare. In consequence, JECFA divided potency estimates for analyses based on toxicologic and epidemiologic data into two basic groups, applicable to individuals with and without HBV infection. Despite differences in mathematical models, JECFA found these estimates useful, as a broad range of possible values is covered. Epidemiologic data for which HBV infection status was unknown and for which potencies were calculated were also reviewed and found to be in the range of potencies for HBV infected or uninfected individuals. The recent JECFA review also looked at the extrapolation of animal data to estimate potency in humans; these also generally fell within the range of the potency estimates derived from the epidemiologic data.

To permit discussion of population risks from the intake of aflatoxins, specific potency estimates must be developed and used. JECFA has reviewed the potencies of the aflatoxins estimated from the positive epidemiologic studies and chosen separate central tendency estimated potencies and ranges for HBV positive and HBV negative individuals. Potency values were chosen, in positive individuals, of 0.3 cancers per year per 100,000 population per ng aflatoxin ingested per kg body weight per day with an uncertainty range of 0.05 to 0.5. In negative individuals, potency values were 0.01 cancers per year per 100,000 population per ng aflatoxin per kg body weight per day with an uncertainty range of 0.002 to 0.03.

The fraction of the incidence of liver cancer in a population attributable to intake of aflatoxins is derived by combining aflatoxin potency estimates (risk per unit dose) and estimates of aflatoxin intake (dose per person). In one calculation, JECFA assumed a population with a European diet, from which all samples containing over 20 $\mu g\ kg^{-1}$ aflatoxin had been removed. The mean aflatoxin intake for this population was 19 ng per person per day. Assuming a 60 kg person, the mean cancer risk for that population was 0.004 cancers per 100,000 population per annum.

At the other end of the scale, these figures should be contrasted with those of Lubulwa & Davis (1994), who estimated deaths from aflatoxins in Indonesia, a country of high risk. They used data of Pitt & Hocking (1996) and Pitt *et al.* (1998) on the incidence of aflatoxin in Southeast Asian commodities, and estimated that the liver cancer rate from aflatoxins in Indonesia was 10 per 100,000 population per annum, a rate 1000 times higher than the JECFA estimate for European populations. Peanuts accounted for most of the ingested aflatoxins. Given that the Indonesian population approaches 200 million people, those figures indicate 20,000 deaths per annum from liver cancer due to aflatoxins in Indonesia.

Because some data on contamination levels used in these examples were derived from nonrandom samples (which might be biased upward because studies focused on contaminated lots of commodity) and some data are not likely to be based on current Codex sampling recommendations for aflatoxins, these figures must be used with caution.

14.3 RISK MANAGEMENT

Because aflatoxin is a well known chemical hazard (albeit from a microbial source), risk management has taken a different path from that expected for bacteria or bacterial toxins. In the years following the discovery of aflatoxins, the limits set for aflatoxins in foods at first amounted to the limit of detection by chemical assay. In importing countries, this was at first 5 µg kg^{-1}, then in some cases reduced to as low as 1 µg kg^{-1} (van Egmond, 1989). However, it soon became clear that producing countries could not meet such limits: the US set 25 µg kg^{-1} and Australia 15 µg kg^{-1} as practical limits that would reduce aflatoxin ingestion as far as possible without destroying the peanut industries in those countries.

Epidemiologic and animal studies that followed established that aflatoxins were genotoxic carcinogens, lacking an apparent no-effect level. Equations for cancer incidence in relation to aflatoxin consumption were developed, based of necessity on epidemiologic evidence, but were not universally accepted because of the probable interaction with HBV and HCV, as outlined above. In consequence, limits continued to be set more on the basis of perceived risk in importing countries or on attainable levels in developed exporting countries. Limits established in other producing countries were seldom met in practice.

14.3.1 Tolerable Level of Risk

As noted above, the development of a tolerable level of risk (TLR) for aflatoxins has proven very difficult, both because a no-effect level has not been established, and because of the interaction, or even synergy, of aflatoxins with HBV and HCV. On the one hand, it could be argued that a TLR logically should be at the limit of detection of cancer in man, say the amount of aflatoxin in the total diet that would induce one case of liver cancer per 10^6 population per annum. On the other, it can be argued that any level of cancer from aflatoxins is too high, so that the TLR might just as logically be lower than that by a factor of 1000. The calculations above made by JECFA, based on European intakes of aflatoxins, would suggest that in practice the actual risk, 0.04 cancers per 10^6 population per annum, lies near the mean of those two hypothetical figures. In contrast, the figures calculated for Indonesian people, 100 cancers per 10^6 population per annum, clearly imply an unacceptable level of risk from aflatoxins.

14.3.2 Food Safety Objective

In the case of a chemical toxin such as aflatoxin, the limits set by a country for aflatoxins in foods can be logically considered also to have the status of a food safety objective (FSO). If it is accepted that a maximum permitted level established within a country is equivalent to an FSO, then by 1990 each major country importing or exporting peanuts had established a *de facto* FSO set, not on the basis of risk analysis, but on more pragmatic approaches.

During the mid-1990s, JECFA and CCFAC carried out a thorough reexamination of the toxicity of aflatoxins, especially in the light of newer evidence on the influence of hepatitis viruses on carcinogenicity of aflatoxins. After protracted discussion, CCFAC recommended to Codex that the maximum permitted level for total aflatoxins in foods in international trade should be 15 µg kg^{-1}. If Codex adopts this recommendation, and it is accepted that the FSO in this case is equal to that limit, then an FSO of 15 µg kg^{-1} has been established for peanuts in international trade.

This FSO is based on a number of factors. First, statistical analyses of detailed surveys of the levels of aflatoxins occurring in European foods have shown that reduction of the permitted limit to 10 or even 5 µg kg^{-1} has only a marginal effect on the risk associated with aflatoxin consumption in importing countries. Second, it is difficult for producing countries to reliably supply nuts below 15 µg kg^{-1} to consuming countries. Third, JECFA has stated that the evidence that aflatoxin is a confirmed Class I carcinogen in humans *in the absence of hepatitis B virus* remains inconclusive. The FSO may have to be adjusted downward if conclusive evidence becomes available. The FSO is considered to be technologically achievable by major exporting countries, including the US, China, and Australia, but is currently out of reach of a number of producing countries in the tropics.

14.3.3 Control Measures

Controlling aflatoxins in peanuts is not easy. Under the drought stress conditions that often prevail when peanuts are grown as a dry culture crop, aflatoxins may be produced before the nuts are pulled from the ground. Under these conditions, it is clear that control of aflatoxin formation by measures taken after that point cannot be totally effective. Aflatoxins are also quite resistant to normal food processing, including heating (ICMSF, 1996), so processes used to reduce bacterial or fungal contamination cannot be relied on to remove aflatoxins.

14.3.3.1 Good Hygienic Practices (GHP)/Hazard Analysis Critical Control Point (HACCP) System

Quantitative aflatoxin assays. The primary control measure taken to limit aflatoxins in peanuts in developed producing countries, i.e., the US and Australia, is quantitative assay of representative samples. Although not considered a conventional control point as defined in bacteriological terms, this step is indeed the critical control point (CCP) in controlling aflatoxins in peanuts.

Sampling peanuts for the presence of aflatoxins is very difficult because usually only a small proportion of nuts is infected with *A. flavus*. However, such nuts may contain high concentrations of toxin, so it is essential to use well devised sampling plans to ensure that assayed nuts are representative. In consequence, the primary concern in the implementation of this HACCP strategy is the adequacy of the sampling plan used. Sampling accounts for more than 90% of the variability associated with aflatoxin assays (Whitaker *et al.*, 1994a).

A variety of mathematical expressions have been put forward for suitability for accurate assessment of the distribution of aflatoxins in peanuts (e.g., Jewers *et al.*, 1989; Whitaker & Dickens, 1989; Giesbrecht & Whitaker, 1998), and a variety of sampling plans developed (e.g., Brown, 1982, 1984; Coker, 1989; Read, 1989; Whitaker *et al.*, 1994b). Sampling plans in current use include 3 × 8 kg samples, 3 × 21.8 kg, 4 × 7.5 kg, and 1 × 10 kg, usually per 10 ton lot (Read, 1989; Whitaker *et al.*, 1995).

The probability of acceptance of lots failing to meet specifications is quite high and inherent in sampling plans (Brown, 1982, 1984; Whitaker *et al.*, 1994a), so it is strongly recommended that all lots of peanuts accepted by manufacturing companies or importers be again assayed on receipt, using appropriate sampling plans. This provides a much higher measure of assurance of acceptability.

Producing countries that export peanuts either carry out their own assays or rely on importing authorities for testing. However, in many producing countries, nuts destined for domestic consumption are sold and consumed at the village level, so testing or regulation of aflatoxin levels is often poor to nonexistent.

Color sorting. The second major control measure used for aflatoxin reduction in peanuts in developed countries is color sorting, carried out before the assay step. In this procedure, nuts are inspected individually by electronic or laser sorting systems, and discolored nuts removed. The rationale for aflatoxin reduction by color sorting is that the growth of a fungus in a peanut results in discoloration, so removal of discolored nuts sorts out those containing aflatoxins as well. In the United States and Australia, it is standard commercial practice that every individual shelled peanut entering commercial streams has been color sorted at least once and in some cases up to five times. Color sorting is not a CCP, but rather a major feature of Good Hygienic Practices (GHP).

The color sorting process is sometimes ineffective. This occurs when severe drought stress causes peanuts to commence drying in the soil before harvest, under which conditions growth of *A. flavus* may occur between shrinking cotyledons. Discoloration (and aflatoxin production) is often invisible to color sorters inspecting intact shelled nuts. When this problem occurs, it is common practice to blanch to remove skins, then roast and color sort again. This accentuates the darkening process and facilitates color sorting.

Total control of aflatoxins in peanuts is possible by these combined processes: color sort, assay, blanch and roast, then assay again. However, this procedure is expensive and results in nuts with a limited storage or shelf-life, requiring packaging under inert atmospheres. Much effort continues to be expended on techniques for limiting aflatoxin by other means.

Basic approaches to aflatoxin control include biocontrol by competitive exclusion and genetic engineering, such as incorporation of the gene that codes for chitinase and other enzymes in peanut plants to inhibit fungal infection.

14.3.3.2 Performance Criteria

The performance criterion is to use color sorting and other procedures as necessary to reduce the levels of aflatoxins in peanuts so that assays on representative samples indicate that an acceptable level of $< 15\ \mu g\ kg^{-1}$ has been achieved consistently.

14.3.3.3 Process Criteria

Farm management practice. Farm management practice has an important role in limiting aflatoxin in peanuts. Management of drought stress by irrigation is the best preventive measure, but most of the world's peanuts are grown under dry culture conditions where irrigation is expensive or impractical. Factors such as weed control, increased plant and row spacing, and any other technique increasing water holding capacity in soils is important. Cultivars of reduced susceptibility to *A. flavus* infection have long been sought, but with little success.

Improved drying practice, especially the use of improved pulling techniques, rapid threshing, and mechanical drying are all valuable measures, but applicability varies widely with farm type. Adequate storage on-farm is essential, a minor problem in developed economies such as the US and Australia, but very difficult to implement in subsistence agriculture in the humid tropics.

Total control of aflatoxin formation in peanuts is impossible with current knowledge, primarily because in bad seasons, i.e., seasons with severe drought stress in the 2–3 weeks before harvest, aflatoxin forms in nuts before pulling from the soil. In regions where dry land farming is practiced and irrigation impossible, good farm management cannot overcome this problem. Hence GHP practiced on-farm can assist in aflatoxin reduction, but not in complete prevention.

The most important good farm practices are:

- maintaining soil moisture by weed control and other appropriate measures;
- harvesting as early as possible to reduce the time and severity of drought stress;
- drying to safe moisture contents (0.70 a_w, equivalent to 8% moisture) as rapidly as possible, either in the field or by mechanical means;
- storing at constant temperatures, in well designed bins, in shade, and with good moisture control, preferably with monitoring of humidity;
- cleaning before transport to shellers to remove shriveled and damaged nuts more likely to contain high levels of aflatoxins.

Shelling plants, which generally control commercial peanut handling after harvest, have good storage practice for shelled peanuts in developed countries, but practice in developing countries often leaves much to be desired. Transport of peanuts also may cause problems due to moisture migration.

Major GHP measures in the shelling plant include:

- random sampling at intake to assess moisture and aflatoxin content. Loads with excess moisture should be rejected and returned to the farm for drying; loads with excess aflatoxin should be segregated.
- careful storage on-farm, with control of insects, temperature gradients, and moisture;
- color sorting after shelling, preferably using instruments that can distinguish more than one color and that can be set to segregate larger or smaller proportions of the nuts, depending on the aflatoxin status of the raw material;
- aflatoxin assays carried out with an adequate sampling plan, preferably on a continual basis;
- as necessary, roasting and blanching before color sorting and reassay;
- storage of acceptable product under carefully controlled conditions until shipped.

14.3.3.4 Monitoring

Monitoring consists of aflatoxin assays to ensure that the processing system in the shelling plant can consistently deliver sorted nuts with $<15~\mu g~kg^{-1}$ total aflatoxins. This system, which provides validation for the GHP/HACCP processes, would be carried out on color-sorted nuts, either continuously from the process stream or using a recognized sampling plan.

14.4 ACCEPTANCE CRITERIA FOR FINAL PRODUCT

14.4.1 Organoleptic

Peanuts are graded on size and color for distribution in international trade. No other organoleptic criteria are currently used for peanut acceptance.

14.4.2 Chemical and Physical

The primary criterion for acceptance of peanuts in international trade is certification of total aflatoxins as $<15~\mu g~kg^{-1}$. No specific assay methods have been internationally approved, but the methods of the Association of Official Analytical Chemists (AOAC, 1984) are a de facto standard. No sampling plans have been internationally recognized to date.

14.4.3 Microbiological

Although well standardized procedures exist for the examination of peanuts for the presence of *A. flavus* (e.g., Pitt & Hocking, 1997), these are unlikely to find a place in international trade.

14.5 REFERENCES

AOAC (Association of Official Analytical Chemists) (1984). Natural Poisons. In *Official Methods of Analysis*, 14th edn., pp. 477–500. Washington, DC: Association of Official Analytical Chemists.

Bosch, F. X. (1995). HCV and liver cancer: the epidemiological evidence. In *Hepatitis C and Its Involvement in the Development of Hepatocellular Carcinoma*, pp. 15–25. Edited by K. Kobayashi *et al.* Princeton, NJ: Princeton Science Publishers.

Brown, G. H. (1982). Sampling for "needles in haystacks". *Food Technol Austral* 34, 224–227.

Brown, G. H. (1984). The distribution of total aflatoxin levels in composited samples of peanuts. *Food Technol Austral* 36, 128–130.

Coker, R. D. (1989). Control of aflatoxin in groundnut products with emphasis on sampling, analysis, and detoxification. In *Aflatoxin Contamination of Groundnut: Proceedings of the International Workshop, 6–9 October, 1987, ICRISAT Centre, India*, pp. 123–132. Patancheru, India: International Centre for Research in the Semi-Arid Tropics.

Giesbrecht, F. G. & Whitaker, T. B. (1998). Investigations of the problems of assessing aflatoxin levels in peanuts. *Biometrics* 54, 739–753.

IARC (International Agency for Research on Cancer) (1994). *Some Naturally Occurring Substances: Food Items and Constituents, Heterocyclic Aromatic Amines and Mycotoxins*. Monograph 56. Lyon, France: International Agency for Research on Cancer.

ICMSF (International Commission on Microbiological Specifications for Foods) (1996). Toxigenic fungi: Aspergillus. In *Microorganisms in Foods 5. Characteristics of Food Pathogens*, pp. 347–381. Gaithersburg, MD: Aspen Publishers, Inc.

JECFA (Joint FAO/WHO Expert Committee on Food Additives) (1997). *Aflatoxins B, G and M*. 49th Joint FAO/WHO Expert Committee on Food Additives. WHO Document PCS/FA/97.17. Geneva, Switzerland: World Health Organization.

Jewers, K., Coker, R. D., Jones, B. D. *et al.* (1989). Methodological developments in the sampling of foods and feeds for mycotoxin analysis. *Soc Appl Bacteriol Symp Ser* 18, 105S–116S.

Klich, M. A. & Pitt, J. I. (1988). Differentiation of *Aspergillus flavus* from *A. parasiticus* and other closely related species. *Trans Br Mycol Soc* 91, 99–08.

Lubulwa, A. S. G. & Davis, J. S. (1994). Estimating the social cost of the impacts of fungi and aflatoxins. In *Stored Product Protection. Proceedings of the 6th International Working Conference on Stored-Product Protection*, pp. 1017–1042. Edited by E. Highley *et al.* Wallingford, UK: CAB International.

Nesheim, S. & Trucksess, M. W. (1986). Thin-layer chromatography/high performance thin-layer chromatography as a tool for mycotoxin determination. In *Modern Methods in the Analysis and Structural Elucidation of Mycotoxins*, pp. 239–264. Edited by R. J. Cole. Orlando, FL: Academic Press.

Pitt, J. I. & Hocking, A. D. (1996). Current knowledge of fungi and mycotoxins associated with food commodities in Southeast Asia. In *Mycotoxin Contamination in Grains*, pp 5–10. Edited by E. Highley & G. I. Johnson. ACIAR Technical Reports 37. Canberra: Australian Centre for International Agricultural Research.

Pitt, J. I. & Hocking, A. D. (1997). *Fungi and Food Spoilage*, 2nd edn. London: Blackie Academic and Professional.

Pitt, J. I., Hocking, A. D., Miscamble, B. F. *et al.* (1998). The mycoflora of food commodities from Indonesia. *J Food Mycol* 1, 41–60.

Read, M. (1989). Removal of aflatoxin contamination from the Australian groundnut crop. In *Aflatoxin Contamination of Groundnut: Proceedings of the International Workshop, 69 October, 1987, ICRISAT Centre, India*, pp. 133–140. Patancheru, India: International Centre for Research in the Semi-Arid Tropics.

Van Egmond, H. P. (1989). Current situation on regulations for mycotoxins. Overview of tolerances and status of standard methods of sampling and analysis. *Food Addit Contam* 6, 139–188.

Whitaker, T. B. & Dickens, J. W. (1989). Simulation of aflatoxin testing plans for shelled peanuts in the United States and in the export market. *J Assoc Offic Analyt Chem* 72, 644–648.

Whitaker, T. B., Dowell, F. E., Hagler, W. M. *et al.* (1994a). Variability associated with sampling, sample preparation, and chemical testing for aflatoxin in farmers' stock peanuts. *J AOAC Int* 77, 107–116.

Whitaker, T. B., Wu, J., Dowell, F. E. *et al.* (1994b). Effects of sample size and sample acceptance level on the number of aflatoxin-contaminated farmers stock lots accepted and rejected at the buying point. *J AOAC Int* 77, 1672–1680.

Whitaker, T. B., Springer, J., Defize, P. R. *et al.* (1995). Evaluation of sampling plans used in the United States, United Kingdom, and the Netherlands to test raw shelled peanuts for aflatoxin. *J AOAC Int* 78, 1010–1018.

Chapter 15

Salmonella in Dried Milk

15.1 Introduction
15.2 Risk Evaluation
15.3 Risk Management
15.4 Product and Process Criteria
15.5 GHP and HACCP
15.6 Acceptance Criteria for Final Product
15.7 References

15.1 INTRODUCTION

Milk is widely used worldwide as an important part of the diet, especially for younger children. Raw milk is a very perishable product and efforts to preserve it have led to the development of a range of dairy products (Anon., 1995). One of the earliest means of preserving milk was natural fermentation (ICMSF, 1998), but modern technologies have improved the old preservation methods and added new ones (Robinson, 1986a, b; Varnam & Sutherland, 1994; ICMSF, 1998 at Chap. 16). In particular, drying has developed into one of the most used technologies. In addition to preserving the most nutritious elements, drying offers the advantage of weight reduction, making transportation more convenient. Dried milk has been shipped in large quantities around the world for decades. In many tropical countries, producing and/or importing dried milk provides children with this nutritious food.

Milk is pasteurized, or even sterilized, before drying to render the product safe because drying cannot be relied upon to kill pathogens in sufficient numbers (Daemen & van der Stege, 1982). Unfortunately, very occasionally recontamination has occurred and, especially when dried milk was used in infant formula, this led to cases or outbreaks of salmonellosis (Anon., 1997; Becker & Terplan, 1986; Gelosa, 1994; Louie, 1993; Usera *et al.*, 1996; Rowe *et al.*, 1987). Vast quantities of dried milk are consumed daily worldwide without evidence of causing salmonellosis. This may indicate that recontamination seldom occurs, or that low numbers of salmonellae do not cause an infection and that multiplication of *Salmonella* in the reconstituted milk must occur to cause illness. Data on the infective dose of salmonellae in children are not well documented (FAO/WHO, 2000) because infections due to dried milk are rare, and when they occur (Collins *et al.*, 1968; Furlong *et al.*, 1979), the number of salmonellae in the reconstituted product is rarely determined. Several aspects of the risk evaluation of *Salmonella* in dried milk will be discussed and serve as an example of setting a food safety objective (FSO).

15.2 RISK EVALUATION

15.2.1 Hazard Identification

All *Salmonella* serovars, with the exception of *Salmonella typhi* and *S. paratyphi*, which behave very differently, are considered in this risk evaluation. Salmonellae are

Gram-negative rods that can grow in milk with $a_w > 0.95$ and pH > 4.9 at temperatures between 8–45 °C (ICMSF, 1998). Salmonellae are not particularly heat resistant and when present in milk are readily killed by heat treatments such as the various time-temperature combinations used in commercial pasteurization (Thomas et al., 1966; Read et al., 1968). Salmonellae are widespread in most regions of the world and animals can be infected by a wide range of serovars (ICMSF, 1996; Ziprin, 1994). During infection, animals may excrete salmonellae in high numbers. Consequently, water and dust may become contaminated and, particularly when cows are shedding salmonellae, milk may be contaminated during milking. When contaminated water or dust enters a dairy plant, salmonellae may gain entry to production lines and the product may become contaminated (Van Schothorst & Kleiss, 1994).

15.2.2 Exposure Assessment

Pasteurization of milk was one of the first examples of a technology applied to render safe a potentially hazardous food. During pasteurization, at least a 10-log reduction of salmonellae is normally obtained by minimal processing conditions (72 °C for 15 s) (ICMSF, 1996). Often an even greater killing effect is achieved in the production of whole milk powder, because temperatures of 80–85 °C are applied for at least 5 s to inactivate most lipolytic enzymes.

Milk pasteurization is a process that can be closely controlled. Handling the dried milk is, however, more critical because many potential routes for recontamination exist (IDF, 1991, 1994). The frequency of recontamination is unpredictable, but when good hygienic practices are maintained, it is rare and the numbers of salmonellae involved are very low. Normal end-product analysis of dried milk will not reveal such low levels of recontamination; consequently, accurate estimates of the frequency of contamination and the true concentration are difficult to establish. In addition, milk powder must be reconstituted before consumption. Although the procedures recommended for preparing, storing, and using reconstituted milk are well documented, the potential opportunities for multiplication are not known. However, based on commercial experience and epidemiologic evidence, it is reasonable to assume that the presence of *Salmonella* in the product is infrequent and the concentration at the moment of consumption is low.

15.2.3 Hazard Characterization

This part of the risk evaluation describes the effect of the identified hazard on an individual or particular groups within a population. Salmonellosis is normally caused by the ingestion of salmonellae in food or water (Silliker & Gabis, 1986). The incubation period may vary from 5 h to 5 days (ICMSF, 1996). Symptoms include nausea, stomach cramps, vomiting, diarrhea, and fever (D'Aoust, 1991). Especially the very young, very old, and immunocompromised persons may become infected with low doses (< 100 cfu g^{-1}) of salmonellae, and their symptoms may be more severe (Blaser & Newman, 1982; Glynn & Palmer, 1992). Salmonellae may cause death among the aged and other vulnerable persons. Moreover, salmonellosis may leave sequelae, such as rheumatoid arthritis (Archer, 1985; Keat, 1983; Smith, 1994). When estimating risk, it is thus very important to specify which type of illness is considered and the target population.

The type of food also may influence the dose-response relationship. Experience has shown that in foods with a high fat content (e.g., chocolate, cheese, or fermented meats), low numbers of salmonellae (less than 10 cfu g^{-1}) may provoke illness (D'Aoust et al., 1975; D'Aoust, 1985; Graven et al., 1975; Greenwood & Hooper, 1983; Hedberg et al., 1992; Lehmacher et al., 1995). On the other hand, classical volunteer studies conducted with salmonellae in fluids indicated that up to 10^6 salmonellae were necessary (McCullough & Eisele, 1951a, b, c,). While data are not available for the number of cases of salmonellosis caused by reconstituted milk powder, it is reasonable to state that the number is very low in terms of number of servings per annum. Since reconstituted dry milk is often consumed by children, a vulnerable population group, a worst case scenario approach is used in this risk evaluation, i.e., it is assumed that one *Salmonella* per serving of milk powder may lead to gastroenteritis in children.

15.2.4 Risk Characterization

The world production of whole milk powder is estimated to be more than one million tons (10^{12} g) per annum. Assuming a serving size of 10 g, 10^{11} servings are consumed annually. Outbreaks and cases of salmonellosis due to the direct consumption of dried milk are rare (Collins et al., 1968; Furlong et al., 1979), but in this example it will be assumed that fewer than 1000 cases of salmonellosis result from contaminated milk powder annually. This gives a probability of illness of < 1 in 10^8 servings. In the absence of accurate data, it is not possible to validate this estimate, but this is most likely an overestimation of the actual risk.

15.3 RISK MANAGEMENT

15.3.1 Acceptable Level of Consumer Protection

In most industrialized countries, the prevalence of salmonellosis from all sources averages about 50 cases per 100,000 population per annum (Gomez et al., 1997). For this example, a public health goal could be stated as less than 1 case per 100,000 population per annum associated with consumption of dried milk. This means the product will not significantly add to the current burden of disease.

15.3.2 Establishing a Food Safety Objective

If the number of servings consumed annually is assumed to be one glass of milk per day (i.e., 10 g powder) by one-third of the world's population, this would mean that about 10 million servings per 100,000 persons would be consumed per annum or 10^8 g. Considering the estimated probability of infection to be < 1 in 10^8 servings, this would mean that the consumption of dried milk would be associated with < 1 case per 100,000 population per annum.

Using a worst-case scenario, i.e., that 1 *Salmonella* cell per serving could provoke illness, the following FSO could be proposed:

The concentration of *Salmonella* in dried milk powder must be < 1 cuf 100 t^{-1} at the time it is reconstituted for serving and consumed.

This FSO reflects a public health goal of < 1 case per 100,000 population per annum. The same FSO also could have been established by reviewing annual production records and the number of illnesses reported or estimated, as was calculated in the risk characterization section.

For this particular food-pathogen combination, no change in risk is expected until the milk powder is reconstituted, after which, the time and temperature of holding will determine whether salmonellae can grow and the rate of growth. The potential for growth and increase in risk, such as during subsequent storage, must be addressed through education, such as labeling and other means. In this example, the FSO is intended for milk powder that is consumed immediately following reconstitution.

The safety of reconstituted milk also can be affected by contamination introduced from other sources, such as the water used to reconstitute the milk, utensils, container, and the person involved in preparing the milk. Other pathogens may be of significance (e.g., viruses, shigellae) in addition to salmonellae. These concerns are not considered in this example.

15.3.3 Control Measures

In principle, several control measures are available to help prevent *Salmonella* contamination in powdered milk and thus to meet the FSO. First, elimination of salmonellae at the dairy farm would prevent contamination during milking and further handling. However, this is very difficult to achieve, because salmonellae are widely distributed in nature, including farm environments (Van Schothorst, 1986). A second option is to eliminate salmonellae from the whole factory environment, but this also is difficult to achieve. Consequently, efforts to eliminate salmonellae should be concentrated on the equipment and environment following pasteurization (Stadhouders, 1975; Jarl & Arnold, 1982; Terplan & Becker, 1989). In current industrial practice, a series of control measures are applied, from collection of the milk until the powdered milk is packaged, to avoid contamination to the extent possible. Collectively, the control measures ensure the FSO has been met. In this example, the control measures are described only briefly. More detailed information can be found in the literature (IDF, 1991, 1994). Control measures include:

- *Controlling initial levels in raw materials*—Effective hygienic measures are applied at the farm level and the collected milk is refrigerated to below 7 °C to prevent multiplication.
- *Reducing levels during pasteurization of raw milk*—Raw milk is pasteurized to eliminate any salmonellae present (initial concentration).
- *Preventing an increase in levels by avoiding recontamination*—Recontamination during drying, handling, storing, and packaging is prevented by adopting a number of good hygienic practices (GHPs) that minimize the chance of contamination, e.g., use of filtered air; good separation of the wet and dry processing steps.
- *Providing consumer information*—A risk management option is to provide consumers with information on the product's label. This information provides instructions on reconstituting the milk powder, including precautions for what should not be done during and after reconstitution. Governments may want to include this information in general food hygiene information programs.

In this example, testing the final product cannot be used as a control measure to achieve the required FSO. Even with the most stringent sampling plan recommended (i.e., two-class sampling plan, $n = 60$, analytical unit $= 25$ g), the mean concentration of *Salmonella* would have to be at 1 cfu in 526 g or greater before a positive lot could be detected with a 95% probability, assuming a log-normal distribution and a standard deviation of 0.8 (Legan et al., 2001).

15.3.4 Performance Criterion

A performance criterion is the required outcome of a step, or combination of steps, that contributes to ensuring that an FSO is met. Performance criteria are usually applied at steps where hazards can either be reduced or increased if appropriate control measures are not applied. Performance criteria may be more stringent than the FSO to ensure that the FSO is met. For milk powder that must be reconstituted before consumption, a more stringent performance criterion may be adopted to account for the slight abuse during preparation and use.

To arrive at a performance criterion for the control measures during manufacturing, the equation from Chapter 3 can be used:

$$H_0 - \Sigma R + \Sigma I \leq \text{Performance criterion}$$

$$H_0 - \Sigma R + \Sigma I \leq -8$$

where:
ΣR = total (cumulative) decrease of the hazard
ΣI = total (cumulative) increase of the hazard
H_0 = initial level of the hazard
Performance criterion, H_0, R, and I are expressed in $\log_{10}$ units.

The basic control measures to achieve the FSO and its related performance criterion consist of killing the salmonellae that may occasionally be found in fresh milk and preventing recontamination of the heat-treated raw material.

15.3.4.1 Controlling Initial Levels in Raw Materials

Since the effectiveness of the heat treatment depends on the number of salmonellae, hygienic measures must be taken at the farm level to keep contamination of salmonellae as low as reasonably possible. The collected milk must then be cooled rapidly and maintained below 7 °C to prevent multiplication. Tankers must be refrigerated and/or insulated. Upon receipt at the dairy plant, the temperature should be checked as one of the several physico-chemical tests. Total viable counts and coliforms may be determined as a check on the effectiveness of the chilling and the general hygienic measures taken. In many countries, maximum total counts have been established as microbiological standards.

The first stages of milk powder production have little or no influence on the number of salmonellae in the milk. During clarification, filth, clumps of milk particles, etc., as well as an excess of animal cells (for instance leukocytes), are removed. A separator is used to remove a portion of the cream. When this equipment is adequately and frequently cleaned, separation has no influence on the number of salmonellae. Depending on the type of milk powder produced, some of the cream may be added back to the fluid milk to obtain the required fat content. Storing and handling the fat, holding and standardizing the milk, etc. should be performed in a manner that prevents contamination with salmonellae and does not permit their number to increase.

Given the above hygienic precautions, the initial number of salmonellae in raw milk is usually very low, and no more than 1 cfu ml^{-1} (i.e., $H_0 = 0$).

15.3.4.2 Reducing Levels during Pasteurization of Raw Milk

Pasteurization is of great importance and is normally identified as a critical control point (CCP) in a Hazard Analysis Critical Control Point (HACCP) system. The temperature is monitored with a thermograph. Most equipment has a flow diversion valve and when the temperature drops below the target value, the milk is diverted to a holding vat for repasteurization; thereby, the flow rate is also monitored and maintained at the required rate and survival of salmonellae is prevented. During cooling, an overpressure should be kept on the pasteurized milk side of the heat exchanger to prevent recontamination. If the pasteurizer does not function correctly or if salmonellae are introduced during cooling, the required low level of salmonellae may not be achieved. Fouling of equipment and buildup of biofilms must be prevented. Inspections and/or tests to detect micro-leaks are used in good hygiene and maintenance practices.

If the initial concentration of salmonellae is assumed to be no more than 1 cfu g^{-1}, then to achieve a performance criterion of 1 cfu per 10^8 g, an 8 $\log_{10}$ reduction needs to be achieved through pasteurization, i.e., $\Sigma R = 8$ (see Figure 15–1).

$$H_0 - \Sigma R + \Sigma I \leq \text{Performance criterion}$$

$$0 - 8 + 0 = \text{Performance criterion}$$

$$\Sigma R = 8$$

15.3.4.3 Preventing an Increase in Levels by Avoiding Recontamination in the Factory

Following pasteurization, the milk moves through an evaporator to concentrate it before spray drying. The temperatures used for condensing may further reduce the number of salmonellae, but evaporation is normally not important for the safety of the milk powder. Balance tanks are often used between condensing and spray drying to compensate for differences in flow rates. The balance tanks should be located in clean, *Salmonella*-free environments to prevent recontamination. The air entering the balance tank could be filtered as an additional preventive measure.

At this stage, it is critical to maintain a barrier between the wet operations of the factory and the dry operations where the spray dryer and the rest of the production line are

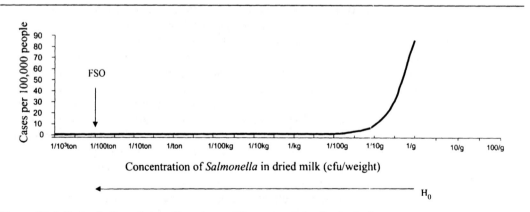

Figure 15–1 Pasteurization of raw milk and preventing recontamination in the factory.

located. People, forklift trucks, etc. should not enter the dry zone without taking precautions to prevent introducing contamination. Air entering this dry sensitive area should be of controlled hygienic quality, entry of pests should be prevented, humidity should be low, and dry cleaning should be the routine procedure used.

The condensed milk is transported via a high-pressure pump to a spray nozzle or another spraying device installed at the top of the spray dryer. Hot air is blown into the aerosol of fluid milk that has been generated by the spray nozzle. The water is evaporated and milk powder falls to the bottom of the dryer. The hot air, which may still contain very fine particles of milk powder, is passed through cyclones to remove the milk particles from the air. These particles are returned to the milk powder at the bottom of the spray dryer.

The powder needs further cooling, and additional air is blown over the film of milk powder leaving the spray dryer. This air must be of high hygienic quality, because if it were to contain salmonellae, the cells would contaminate the final product. After cooling, the powder is passed over a sieve to remove clumps. It is then transported, either mechanically or pneumatically, to a hopper, which feeds the packaging machine or intermediate storage bins. Maintaining a clean, dry environment that consistently tests negative for salmonellae is critical for all the steps that occur within the dry areas of the plant, including the packaging operation.

All the lines where the milk is still in a fluid phase can be cleaned with Cleaning In Place installations or manually using water, cleaning agents, brushes, etc. The environment in this area of the plant is also wet cleaned. In contrast, the spray tower and its environment, as well as the sections of the line and surrounding environment after the spray dryer, are all dry cleaned whenever possible. Occasionally wet cleaning is necessary, but in these instances the equipment or surrounding environment are dried as soon as possible afterward. Drains and pipes containing water should be properly sealed. Condensation on cold pipes should be prevented; when insulation is used, care should be taken to prevent buildup of moisture and microbial growth, a potentially significant source of contamination. The principle applied here is that salmonellae cannot multiply when water is not available.

If the initial concentration of salmonellae is assumed to be no more than 1 cfu g^{-1} and an 8 $\log_{10}$ reduction is achieved through pasteurization (i.e., $\Sigma R = 8$), then, to achieve a performance criterion of 1 cfu per 10^8 g, there must be no contamination (i.e., $\Sigma I = 0$) (see Figure 15–1).

$H_0 - \Sigma R + \Sigma I \leq$ Performance criterion

$0 - 8 + \Sigma I \leq -8$

$\Sigma I = 0$

15.4 PRODUCT AND PROCESS CRITERIA

The process criterion for the pasteurization would consist of heating the milk for at least 15 s at 72 °C, or for 5 s at 80 °C.

The most serious problem in controlling the hazard from salmonellae in dried milk is preventing recontamination. To achieve that, many GHP-related measures are necessary. In practice, this means eliminating salmonellae from the equipment and its immediate environment and preventing salmonellae from entering this environment with air, dust, pests, packaging materials, personnel, etc. Although it is difficult to quantify such low contamination rates, a recontamination rate of < 1 *Salmonella* per 10^8 g would be required. To put this into perspective, 10^8 gram is 100 tons of product, and that implies preventing recontamination during 30 h of production using a medium-sized spray-drying tower.

15.5 GHP AND HACCP

The basis for the production of safe milk powder is the application of GHPs that have been developed through many decades of industrial production of this product. All milk powders receive a heat treatment during one or more stages of production, and consequently preventing salmonellae in the end product depends mainly on preventing recontamination after the final heat treatment.

During development of the HACCP plan, the sources of *Salmonella* contamination and possible multiplication have to be identified, control measures put into place, and monitoring procedures described and implemented. A well-known source of salmonellae is milk residues, which may be found on the roof of spray tower buildings. Birds may contaminate this powder, and when the secondary air intake is close to the roof, this air may transfer salmonellae to the powder if the air filtration is not controlled. Salmonellae may also occur in the dust or mud outside the factory and may even gain access in noncritical areas, such as warehouses for raw materials or finished goods. Operators, maintenance personnel, visitors, etc. should, therefore, change their shoes before entering the dry areas of the milk powder production line. Movement of air throughout the various rooms should be well controlled. A slight overpressure should be maintained in the spray tower building, as well as in the packaging room. Again, the quality of this air should be carefully controlled. The packaging material could be a source of salmonellae, so control measures to ensure its high hygienic quality should be in place. Many other good hygienic practices are put in place to prevent recontamination, as described elsewhere, for instance in the IDF bulletin, *Recommendations for the Hygienic Manufacture of Milk and Milk Based Products* (IDF, 1994).

15.5.1 Monitoring and Verification from the Raw Milk to the Consumer

The quality of the incoming milk, and thus the potential level of contamination with *Salmonella*, is not under direct control of the dairy plant. However, the plant can influence the hygienic conditions at the farm and during transportation of the milk. The hygienic condition of incoming milk can be monitored by a number of tests, such as freezing point determination, methylene blue tests, filth tests, tests for aerobic mesophilic microorganisms, cell counts, acidity, etc. If necessary, the incoming milk could be tested microbiologically to check, for example, that a level of 1 *Salmonella* ml^{-1} is not normally exceeded, as a verification procedure.

The time and temperature specified as process criteria are monitored in most dairy plants with continuous temperature and flow rate recorders of different designs and levels of sophistication.

Elimination of *Salmonella* from the equipment and its environment and preventing their introduction requires a number of hygienic measures that need careful surveillance (see Chapter 10). Testing for indicator microorganisms, as well as for salmonellae, can check the effectiveness of these hygienic measures. If no salmonellae are found in the most likely locations and if the numbers of indicators, such as Enterobacteriaceae, are in the normal order of magnitude, it can be assumed that the performance criteria are achieved.

Testing end products for salmonellae is, in principle, not necessary as a routine monitoring procedure. Greater reliance must be placed on monitoring the HACCP system and assessing control of the environment through visual inspections and microbiological testing. When the process and environment records indicate that the operation is under control, microbiological testing would not be sufficiently sensitive to determine whether the performance criterion has been met. Nevertheless, many in industry often test a few samples for the presence of salmonellae as part of their verification program. While a positive result would indicate that the process was not under control, a negative result cannot be relied upon to ensure that the performance criterion and FSO have been met (see sections 15.3.3 and 15.6.3).

15.6 ACCEPTANCE CRITERIA FOR FINAL PRODUCT

15.6.1 Organoleptic

Milk powder has certain organoleptic characteristics that can easily be checked at any stage in the food chain. Milk powder should be dry, free flowing, and easily dissolve in cold or lukewarm water. It has a distinct white or creamy white color, without brown (Maillard reaction) or black (burned) discoloration or particles. It should not have a rancid or other abnormal aroma. Unopened packages should not be damaged or show any sign of insect infestation. Although these organoleptic acceptance criteria are more related to quality than to safety, they indicate that GHP were adhered to during production, storage, and distribution. As mentioned above, deviations from GHP could lead to safety problems. Atypical product should, therefore, not be accepted without further inquiries or investigational sampling.

15.6.2 Chemical and Physical

A number of quality-related tests, such as moisture level, fat and protein content, solubility, etc., are used in commerce (CAC, 1999). Most of them are unrelated to the safety of the product. Sometimes tests for aflatoxin M, antibiotics, heavy metals, pesticides, etc. are performed to check compliance with legislation. Moisture or water activity tests are normally not necessary to determine whether growth of pathogens might have occurred. When moisture is raised to a_w levels allowing microbial growth, the powder becomes totally unacceptable and would be rejected through normal organoleptic examination.

15.6.3 Microbiological

Microbiological examination often includes a test for aerobic plate counts (APC), coliforms, or Enterobacteriaceae, and sometimes *Salmonella* (see also ICMSF Book 2, (1986)). For APC, case 2 is selected with a three-class plan consisting of $n = 5$, $c = 2$, $m = 3 \times 10^4$, and $M = 3 \times 10^5$. Case 2 is appropriate to this situation in which these counts indicate the normal level of microorganisms found in these products, without direct link to GHP or HACCP.

Table 15–1 Performance of Sampling Plans for *Salmonella* Expressed as the Mean Concentration Detected with 95% Probability Using a 25 g Sample Size and Assuming a Log-Normal Distribution and a Standard Deviation of 0.8

Case	10	11	12
No. of Samples	$n = 5$	$n = 10$	$n = 20$
Mean Concentration	32/1000g (1 cfu/32g)	12/1000g (1 cfu/83g)	5.4/1000g (1 cfu/185g)
Case	13	14	15
No. of Samples	$n = 15$	$n = 30$	$n = 60$
Mean Concentration	7.4/1000g (1 cfu/135g)	3.6/1000g (1 cfu/278g)	1.9/1000g (1 cfu/526g)

Case 5 can be used for coliforms with the following values: $n = 5$, $c = 1$, $m = 10$, and $M = 10^2$. Case 5 relates to indicators of adherence to GHP and HACCP. Lots of milk powder meeting this criterion should not be rejected, unless other findings raise concern.

The following guidance can be used for *Salmonella* testing. Cases 10, 11, and 12 normally apply to foods being tested for *Salmonella* and result in 5, 10, and 20 samples, respectively. The analytic unit from each sample is typically 25 g. For milk powder, case 11 is most appropriate because there may be no change in the level of *Salmonella* between when the powder is tested at the factory or port-of-entry and when the powder is reconstituted for serving. When the powder is intended for a high-risk population, the number of samples can be increased to 15, 30, and 60, with 30 samples being the most appropriate. The performance of these sampling plans can be related to a mean concentration of bacteria that could be detected with a probability of 95% using the approach of Foster (1971) and Legan *et al.* (2001) and assuming a log-normal distribution with a standard deviation of 0.8. The mean concentrations controlled appear in Table 15–1. Thus, for case 11, there would be 95% confidence of detecting *Salmonella* if present at a level of 1 cfu 83 g^{-1} or greater and if the level of contamination is log normally distributed throughout the lot with a standard deviation of 0.8.

These criteria are applicable when no information is available on the history of the lot or the supplier's control system. For a full discussion on the need for testing and choice of sampling plans, the reader is referred to the previous chapters.

15.7 REFERENCES

Anonymous (1995). *Dairy Processing Handbook*. Tetra Pak Processing Systems A. B. Lund, Sweden: LP Graviska.

Anonymous (1997). *Salmonella anatum* infection in infants linked to dried milk. *Communic Dis Rpt* 7, 33, 36.

Archer, D. L. (1985). Enteric microorganisms in rheumatoid diseases: causative agents and possible mechanisms. *J Food Prot* 48, 538–545.

Becker, H. & Terplan, G. (1986). Salmonellen in Milchtrockenprodukten. *Deutsche Molkerei Zeitung* 42, 1398–1403.

Blaser, M. J. & Newman, L. S. (1982). A review of human salmonellosis: 1. Infective dose. *Rev Infect Dis* 4, 1096–1106.

CAC (Codex Alimentarius Commission) (1999). *Codex Standard for Milk Powders and Cream Powder*. Joint FAO/WHO Food Standards Programme. Codex Stan 207–1999. Rome: Food and Agriculture Organization of the United Nations.

Collins, R. N., Treger, M. D., Goldsby, J. B. et al. (1968). Interstate outbreak of *Salmonella newbrunswick* infection traced to powdered milk. *J Am Med Assoc* 203, 838–844.

D'Aoust, J. Y., Aris, B. J., Thisdele, P. et al. (1975). *Salmonella eastbourne* outbreak associated with chocolate. *J Inst Can Food Sci Technol Aliment* 8, 181–187.

D'Aoust, J. Y. (1985). Infective dose of *Salmonella typhimurium* in cheddar cheese. *Am J Epidemiol* 122, 717–720.

D'Aoust, J. Y. (1991). Pathogenicity of foodborne *Salmonella*. *Int J Food Microbiol* 12, 17–40.

Daemen, A. L. M. & van der Stege, H. J. (1982). The destruction of enzymes and bacteria during the spray drying of milk and whey. 2. The effect of the drying conditions. *Netherlands Milk & Dairy J* 36, 211–229.

FAO/WHO (Food and Agriculture Organization/World Health Organization) (2000). *Joint FAO/WHO Expert Consultation on Risk Assessment of Microbiological Hazards in Foods*. FAO Food and Nutrition Paper 71. Rome: FAO.

Foster, E. M. (1971). The control of salmonellae in processed foods: A classification system and sampling plan. *J AOAC* 54, 259–266.

Furlong, J. D., Lee, W., Foster, L. R. & Williams, L. P. (1979). Salmonellosis associated with consumption of nonfat powdered milk—Oregon. *Morbidity & Mortality Wkly Rpts* 28, 129–130.

Gelosa, L. (1994). Latte in polvere per la prima infanzia contaminato da *Salmonella bovis morbificans*. *Industrie Alimentari* 33, 20–24.

Glynn, J. R. & Palmer, S. R. (1992). Incubation period, severity of disease, and infecting dose: evidence from a *Salmonella* outbreak. *Am J Epidemiol* 136, 1369–1377.

Gomez, T. M., Motarjemi, Y., Miyagawa, S. et al. (1997). Foodborne salmonellosis. *World Health Stats Quarterly* 50, 81–89.

Graven, P. C., Mackel, D. C., Baine, W. B. et al. (1975). International outbreak of *Salmonella eastbourne* infection traced to contaminated chocolate. *Lancet* i, 788–793.

Greenwood, M. H. & Hooper, W. L. (1983). Chocolate bars contaminated with *S. napoli*: an infectivity study. *Brit Med J* 286, 1394.

Hedberg, C. W., Korlath, J. A., D'Aoust, J. Y. et al. (1992). A multistate outbreak of *Salmonella javiana* and *Salmonella oranienburg* infections due to consumption of contaminated cheese. *J Am Med Assoc* 268, 3203–3207.

ICMSF (International Commission on Microbiological Specification for Foods) (1986). *Microorganisms in Foods 2. Sampling for Microbiological Analysis: Principles and Specific Applications*, 2nd edn. Toronto: University of Toronto Press.

ICMSF (International Commission on Microbiological Specification for Foods) (1996). *Microorganisms in Foods 5. Characteristics of Microbial Pathogens*. Gaithersburg, MD: Aspen Publishers, Inc.

ICMSF (International Commission on Microbiological Specification for Foods) (1998). *Microorganisms in Foods 6. Microbial Ecology of Food Commodities*. Gaithersburg, MD: Aspen Publishers, Inc.

IDF (International Dairy Federation) (1991). *IDF Recommendations for the Hygienic Manufacture of Spray Dried Milkpowders*. Bulletin of the International Dairy Federation No. 267. Brussels, Belgium: International Dairy Federation.

IDF (International Dairy Federation) (1994). *IDF Recommendations for the Hygienic Manufacture of Milk and Milk Based Products*. Bulletin of the International Dairy Federation No. 292. Brussels, Belgium: International Dairy Federation.

Jarl, D. H. & Arnold, E. A. (1982). Influence of drying plant environment on *Salmonella* contamination of dry milk products. *J Food Prot* 45, 16–18, 22.

Keat, A. (1983). Reiter's syndrome and reactive arthritis in perspective. *New Engl J Med* 309, 1606–1615.

Legan, J. D., Vandeven, M. H., Dahms, S. & Cole, M. B. (2001). Determining the concentration of microorganisms controlled by attributes sampling plans. *Food Control* 12, 137–147.

Lehmacher, A., Bockmhl, J. & Aleksic, S. (1995). A nationwide outbreak of human salmonellosis in Germany due to contaminated paprika and paprika-powdered potato chips. *J Infect Diseases* 115, 501–511.

Louie, K. K. (1993). *Salmonella* serotype *tennessee* in powdered milk products and infant formula—Canada and the United States. *J Am Med Assoc* 270, 432.

McCullough, N. B. & Eisele, C. W. (1951a). Experimental human salmonellosis. I. Pathogenicity of strains of *Salmonella meleagridis*, and *Salmonella anatum* obtained from spray-dried whole egg. *J Infect Diseases* 88, 278–289.

McCullough, N. B. & Eisele, C. W. (1951b). Experimental human salmonellosis. III. Pathogenicity of strains of *Salmonella newport, Salmonella derby* and *Salmonella bareilly* obtained from spray-dried whole egg. *J Infect Diseases* 89, 209–213.

McCullough, N. B. & Eisele, C. W. (1951c). Experimental human salmonellosis. IV. Pathogenicity of strains of *Salmonella pullorum* obtained from spray-dried whole egg. *J Infect Diseases* 89, 259–266.

Read, R. B., Bradshaw, R. W., Dickerson, R. W. & Peeler, J. T. (1968). Thermal resistance of salmonellae isolated form dry milk. *Appl Microbiol* 16, 998–1001.

Robinson, R. K., ed. (1986a). *Modern Dairy Technology, Vol. 1. Advances in Milk Processing.* New York: Elsevier Applied Science Publishers.

Robinson, R. K., ed. (1986b). *Modern Dairy Technology, Vol. 2. Advances in Milk Processing.* New York: Elsevier Applied Science Publishers.

Rowe, B., Hutchinson, D. N., Gilbert, R. J. *et al.* (1987). *Salmonella ealing* infections associated with consumption of infant dried milk. *Lancet* ii, 900–903.

Silliker, J. H. & Gabis, D. A. (1986). *Salmonella.* In *Advances in Meat Research, Vol. 2: Meat and Poultry Microbiology*, pp. 209–229. Edited by A. M. Pearson & T. R. Dutson. Westport, CT: AVI Publishing Co.

Smith, J. L. (1994). Arthritis and foodborne bacteria. *J Food Prot* 57, 935–941.

Stadhouders, J. (1975). Microbes in milk and milk products, an ecological approach. *Netherlands Milk & Dairy J* 29, 104–126.

Terplan, G. & Becker, H. (1989). The occurrence of *Salmonella* in dairy plants and dairy products. Consequences for operational quality control. *Deutsche Milchwirtschaft* 40, 368–371.

Thomas, C. T., White, J. C. & Longree, K. (1966). Thermal resistance of salmonellae and staphylococci in foods. *Appl Microbiol* 14, 815–820.

Usera, M. A., Echeita, A., Aladuena, A. *et al.* (1996). Interregional foodborne salmonellosis outbreak due to powdered infant formula contaminated with lactose-fermenting *Salmonella virchow. Euro J Epidemiol* 12, 377–381.

Van Schothorst, M. (1986). Do we have to live with Salmonella? *Food Technol Austral* 38, 64–67.

Van Schothorst, M. & Kleiss, T. (1994). HACCP in the dairy industry. *Food Control* 5, 162–166.

Varnam, A. H. & Sutherland, J. P. (1994). *Milk and Milk Products. Technology, Chemistry, and Microbiology.* London: Chapman & Hall.

Ziprin, R. L. (1994). Salmonella. In *Foodborne Disease Handbook. Diseases Caused by Bacteria*, Vol. 1, pp. 253–318. Edited by Y. H. Huy, J. R. Gorham, K. D. Murrell & D. O. Cliver. New York: Marcel Dekker, Inc.

Chapter 16

Listeria monocytogenes in Cooked Sausage (Frankfurters)

16.1 Introduction
16.2 Risk Evaluation
16.3 Risk Management
16.4 Process and Product Criteria
16.5 GHP and HACCP
16.6 Acceptance Criteria for Final Product
16.7 References

16.1 INTRODUCTION

This chapter considers *Listeria monocytogenes* in cooked sausage (i.e., frankfurters) as an example of a microbial hazard that is capable of growth in a wide variety of perishable ready-to-eat (RTE) foods. The use of performance criteria, process criteria, and validation in relation to the Hazard Analysis Critical Control Point (HACCP) plan is described. In addition, the importance of Good Hygienic Practices (GHP) and an example of a sampling program to assess control of a pathogen in the environment is discussed. In this example, the pathogen of concern is psychrotrophic and can establish itself as a resident and multiply in the refrigerated areas of the food operation, as well as in the refrigerated food. This chapter describes the application of principles introduced in previous chapters. Hypothetical values are used throughout the chapter wherever assumptions were necessary to illustrate a concept or procedure. No attempt has been made to validate their accuracy.

Sausage is one of the oldest forms of processed food and may actually have been the first processed meat product. Thousands of years ago, primitive civilizations began processing meat when they discovered that procedures such as salting, sun drying, and cooking could prolong the edible qualities of meat products, in addition to improving flavor. Almost all races and societies have used this type of food processing method at some point during their development. The term "sausage" is derived from the Latin word *salus*, meaning chopped or minced meat preserved by salting. In 1987, Frankfurt-am-Main, over the protests of Vienna (Wien), which considers itself the originator of the "frankfurter," celebrated the 500th birthday of its claim to being the birthplace of the frankfurter (hot dog). Hundreds of different sausage varieties are available, including cooked, uncooked, smoked, fermented, cured, etc., as well as varieties using combinations of these processes.

Cooked sausages, which include the ever-popular frankfurter as well as bologna and various luncheon meats, are prepared from mixtures of comminuted beef, pork, chicken, and/or turkey to which salt, sugar, sodium nitrite, and spices are normally added. When making frankfurters, the meat/ingredient mixture, commonly referred to as the sausage emulsion, is stuffed into natural or artificial casings, which are then twisted to form sausage links. Frankfurters are typically cooked to an internal temperature of ≥ 70 °C to (a) coagulate protein, (b) fix the cured color, and (c) destroy certain pathogens and bacteria that would cause spoilage. Although not an absolute requirement, frankfurters and other similar sausages are commonly subjected to natural smoke. Alternatively, a commercially available liquid smoke solution can be added to the sausage emulsion or atomized onto the surface of frankfurters before or during heating. After cooking, frankfurters are cooled, packaged, and refrigerated for shipment to wholesale and retail markets (Figure 16–1). Skinless frankfurters, which are very popular, are produced in a similar manner, except that the artificial casing is mechanically peeled from the sausage after cooking.

Cooking renders the sausages free of *L. monocytogenes*. However, post-processing contamination can occur and in rare instances has led to illness, such as the 1998–1999 multistate outbreak with 101 confirmed cases, including 15 adult deaths and six miscarriages/stillbirths (CDC, 1999). It should be noted that multiplication of *L. monocytogenes* on frankfurters is probably necessary for disease to occur, but data on the dose-response relationship in humans are not available. This and other aspects of the risk analysis of *L. monocytogenes* in frankfurters are discussed below. Several reports dealing with risk assessment and *L. monocytogenes* have been published and contain additional background information (Farber *et al.*, 1996; Hitchins, 1996; Buchanan *et al.*, 1997; Bemrah *et al.*, 1998; FAO/WHO, 2000; Buchanan & Lindqvist, 2000; Ross *et al.*, 2000; FDA, 2001).

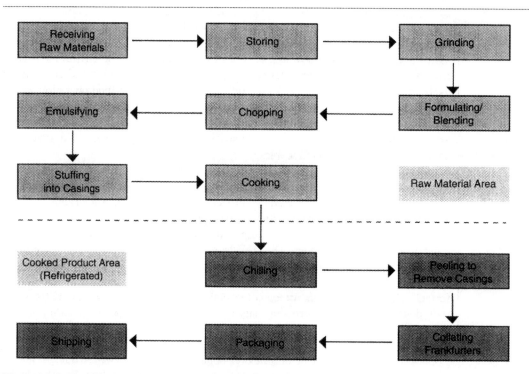

Figure 16–1 Flow diagram for typical frankfurter operation.

16.2 RISK EVALUATION

16.2.1 Hazard Identification

Epidemiologic data suggest that most exposures to *L. monocytogenes* are foodborne (Ciesielski *et al.*, 1988; Broome *et al.*, 1990; Farber & Peterkin, 1991; McLauchlin, 1993; Mead *et al.*, 1999). Although listeriosis occurs infrequently, at somewhere between 2–6 cases annually per million of population, between 20–30% of the cases are fatal (McLauchlin, 1993; Rocourt, 1996; Mead *et al.*, 1999). Thirteen serotypes of *L. monocytogenes* have been identified, but most illnesses have been associated with only three, namely serovars 4b, 1/2a, and 1/2b. Serovar 4b accounts for close to half of all the cases in Europe, while serovars 4b, 1/2a, and 1/2b are more equally represented in North America (Farber & Peterkin, 1991; McLauchlin, 1993; Rocourt, 1996; Rocourt & Bille, 1997). However, the potential for all serovars to cause human illness cannot be discounted.

All *L. monocytogenes* serotypes are considered in this risk evaluation. *L. monocytogenes* is a Gram-positive, non-spore-forming rod. It is facultatively anaerobic, growing at temperatures between -0.4 and 45 °C, pH values 4.39–9.4, a water activity of 0.92 or higher, and a salt concentration of 10% or less (ICMSF, 1996). It is catalase positive, oxidase negative and produces β-hemolytic colonies on blood agar. The organism is widely distributed in nature and can be found on fruits and vegetables, in soil, water, sewage, silage, slaughterhouse waste, milk from healthy and mastitic cows, as well as human and animal feces (Farber & Peterkin, 1991, 1999). *L. monocytogenes* can be found as part of the normal flora of many animal species and humans, and it has been estimated that between 2 and 6% of humans and animals are transient carriers of the organism. Fecal excretion of the organism is not necessarily an indication of disease, and the role of healthy carriers in the epidemiology of listeriosis is still unclear (Rocourt, 1996).

L. monocytogenes is inactivated by heat treatments of 70 °C or above, used in cooking or pasteurization. However, products, such as frankfurters, that are purchased as a precooked sausage are sometimes eaten without further cooking or may be reheated by a variety of methods (frying pan, grilling, simmering water, microwave). The practice of reheating in a microwave oven was associated with one sporadic case of listeriosis in an elderly woman who had cancer (CDC, 1989). Meat products, such as cooked sausages, can be recontaminated between cooking and packaging in the processing facility. During extended refrigerated storage (e.g., $\geq$ 35 days), the organism could multiply to potentially hazardous levels on the surface of the frankfurters and in the exudate, especially if the product is temperature-abused. The free exudate in the package has a higher concentration of cells than the product itself and may be a source of cross-contamination in the kitchen.

16.2.2 Hazard Characterization

Human listeriosis is now recognized as a disease primarily caused by ingestion of *L. monocytogenes* in food. This also includes secondary transmission between a mother and her fetus or neonate. Despite its widespread presence in the environment, illness due to this organism occurs infrequently. The outcome of listeriosis can be severe, with an estimated case fatality rate of between 20–30% in the segments of the population most at risk, i.e., immunocompromised individuals, pregnant women, the very young (i.e., fetuses and neonates), and the elderly. This spectrum of individuals is estimated to comprise 15–20% of the population (Buchanan et al., 1997) and is expected to increase, reflecting the trend in longer life expectancy and an overall aging population.

Unlike the symptoms caused by many foodborne pathogens, diarrhea and other gastrointestinal symptoms are not common symptoms of listeriosis, although the patient may experience malaise and a mild fever. In several foodborne listeriosis outbreaks, patients have exhibited only these mild symptoms (Salamina et al., 1996; Dalton et al., 1997). However, the majority of reported cases have been the invasive type of listeriosis, symptoms of which include meningitis, encephalitis, and septicemia. In pregnant mothers, unrecognized and untreated listeriosis can lead to abortion of the fetus, stillbirth, or premature delivery of a sick child. Serious sequelae, such as mental retardation and hydrocephalus, also have been reported following cases of foodborne listeriosis (Büla et al., 1995).

There are no experimental dose-response data available for *L. monocytogenes* in humans (i.e., the minimum infectious dose (MID) is uncertain). As with many other foodborne pathogens, the MID will depend on factors such as the virulence of the strain, the amount of food consumed, the levels of the organism in the food, and the immune state of the host. Some animal data are available, mainly using the mouse model (Audurier et al., 1980; Golnazarian et al., 1989; FAO/WHO, 2000), but extrapolation of mouse data to the human situation is tenuous at best. While dose-response studies using human volunteers for certain pathogens have sometimes been conducted, this is not feasible with *L. monocytogenes*, since the risk of severe morbidity or mortality is too great. An alternate approach used with RTE foods is to combine epidemiologic data with estimates of the number of organisms present in the incriminated food. However, because outbreaks are an uncommon occurrence and it is difficult to obtain sufficient material in most cases, the amount of reliable data is scant and the information insufficient to draw reliable conclusions. The type of food consumed, perhaps, may influence the dose-response relationship.

When a risk estimate for *L. monocytogenes* is being made, it is important to determine which type of illness is being considered (i.e., mild versus invasive) and for which part of the population (i.e., "normal" versus "high-risk"). Since frankfurters are consumed by all segments of the population, different risk estimates ideally should be used for normal versus high-risk individuals.

Foodborne listeriosis appears generally to follow one of three scenarios. Scenario 1 consists of isolated cases for which information about the food is seldom available. Scenario 2 consists of an outbreak or cluster of cases involving a single lot of contaminated food. Both of these events typically involve errors in food handling that lead to a single lot of food becoming contaminated and an opportunity for multiplication before the food is consumed. Once the implicated food is eliminated, further cases cease to occur.

Scenario 3 consists of an outbreak involving anywhere from a few cases to several hundred cases scattered by time and location. This type of outbreak typically involves an unusually virulent strain that has become established in the food processing environment and contaminates multiple lots of food over days or months of production. Experience in cooked meat and poultry operations indicates that a niche or site within the cooked product environment wherein *L. monocytogenes* becomes established and multiplies is commonly involved. The sites may be impossible to reach and clean with normal cleaning and sanitizing procedures. In fact, the processing environment typically appears visually clean and acceptable. The sites serve as a reservoir from which the pathogen is dispersed during a food processing operation, contaminating food contact surfaces and, thereby, the food. The niche usually affects only the product being handled (e.g., filled, sliced, conveyed, assembled, packaged) along one packaging line and not the product on an adjacent line. Microbiological testing is necessary to detect the niche. Examples of niches include hollow rollers on conveyors, cracked tubular support rods on equipment, the space between close fitting layers of metal-to-metal or metal-to-plastic, worn or cracked rubber seals around doors, on-off valves and buttons for equipment, and saturated insulation.

In all three scenarios, there is an opportunity for *L. monocytogenes* to multiply in the food before it is consumed. Food processors should establish systems to prevent scenario 3 events and minimize the risk of scenarios 1 and 2.

16.2.3 Exposure Assessment

L. monocytogenes is found as a frequent contaminant of RTE foods (Farber & Peterkin, 1991, 1999). Many different foods have been involved in both sporadic cases and foodborne outbreaks, with the approximate levels detected ranging from 100 to 1×10^9 cfu g^{-1} (Table 16–1). The meat industry has a long history of providing safe food, including frankfurters, notwithstanding the recent large listeriosis outbreak (CDC, 1999). Qvist & Liberski (1991) found *L. monocytogenes* to be present in four of 67 (6%) frankfurter samples, while Wang & Muriana (1994) detected the organism in seven of 93 (7.5%) packages of retail frankfurters examined. The US Department of Agriculture (USDA) monitoring program for RTE meats after packaging at the manufacturing facility found the prevalence of *L. monocytogenes* in small-diameter cooked sausages to be 5.3, 4.8, 4.1, 3.7, 3.3, 3.5, and 1.8% annually from 1993 to 1999, respectively.

Frankfurter processing involves a step-heating schedule to allow for the formation of the "skin," for setting of the emulsion, and to allow the curing reaction to occur. This step also reduces *L. monocytogenes* numbers. Zaika et al. (1990) prepared frankfurters from a sausage emulsion inoculated to contain around 10^8 cfu g^{-1} of *L. monocytogenes*. After stuffing, all frankfurters were thermally processed (without smoke) according to a standard commercial heating schedule. Zaika and colleagues found that *L. monocytogenes* decreased approximately 1000-fold in frankfurters heated to an internal temperature of 71.1 °C. Hence, based on those data, cooking frankfurters to an internal temperature of 71.1 °C should eliminate higher levels of *L. monocytogenes* ($\sim 10^3$ cfu g^{-1}) than are likely to occur in raw frankfurter emulsions. In addition, post-process pasteurization can eliminate *L. monocytogenes* from the surface of frankfurters that have been contaminated between cooking and packaging. This procedure is currently practiced commercially by certain manufacturers throughout the world.

The cooking step for frankfurters can be controlled in facilities operating with effectively designed and implemented HACCP plans, but preventing contamination of the cooked product during cooling and packaging is much more difficult. Data gathered by the American Meat Institute in 1988 pointed to frankfurters as a likely carrier of *L. monocytogenes* and suggested that uncontrolled environmental conditions before packaging may play a major role in contaminating the finished product (Anon., 1989). Environmental sampling of a turkey frankfurter plant, whose product was linked to a case of listeriosis, found that contamination of the majority of frankfurters occurred at a single point during the peeling process, just prior to packaging (Wenger et al., 1990). In retrospect, a properly applied environmental sampling program (see Chapter 10) might have detected the site and, if corrected, prevented contamination of the final product.

Table 16-1 Foodborne Outbreaks due to *L. monocytogenes*

Location (year)	No. of Cases (deaths)	Perinatal/nonperinatal	No. cfu g^{-1}	Foods associated	Serotype	Reference
Boston, USA (1979)	20 (5)*	0/20	NK[†]	Raw celery, tomatoes, lettuce[‡]	4b	Ho et al., 1986
New Zealand (1980)	29 (9)	22/7	NK	Shellfish, raw fish	most cases 1/2b	Lennon et al., 1984
Maritime Prov., Canada (1981)	41 (17)	34/7	NK	Coleslaw[‡]	4b	Schlech et al., 1983
Massachusetts, USA (1983)	49 (14)	7/42	NK	Pasteurized milk[‡]	4b	Fleming et al., 1985
California, USA (1985)	142 (48)	93/49	$10^3 - 10^4$	Queso blanco fresco cheese	4b	Linnan et al., 1988
Switzerland (1983–1987)	122 (31)	65/57	$10^4 - 10^6$	Raw milk cheese	4b	Bille, 1990
Connecticut, USA (1989)	9 (1)	2/7	NK	Shrimp[‡]	4b	Riedo et al., 1994
United Kingdom (1987–1989)	355 (94)[§]	185/129[∥]	$< 10^2 - 10^6$	Pâté[‡]	4b; 4b(x)	McLauchlin et al., 1991
W. Australia (1990)	11** (6)	11/0	8.8×10^3	Pâté[‡]	1/2a	Kittson, 1992
Tasmania, Australia (1991)	4 (0)	0/4	1.6×10^7	Smoked mussels	1/2b	Mitchell et al., 1991
New Zealand (1992)	3 (2)	2/1	NK	Smoked mussels	1/2a	Baker & Wilson, 1993; Brett et al., 1998
France (1993)	39 (12)	31/8	$< 100 - 1.0 \times 10^4$	Pork rillettes (pâté)	4b	Goulet, 1995
France (1992)	279 (85)	92/187	NK	Pork tongue in jelly	4b	Goulet et al., 1993

Country (Year)	Cases (Deaths)	Concentration	Food	Serotype	Reference
Italy (1993)	23 (0)	0/23	NK	Rice salad[‡]	1/2b	Salamina et al., 1996
USA (1994)	45 (0)	1/44	2.9×10^{11}	Chocolate milk	1/2b	Dalton et al., 1997; Proctor et al., 1995
France (1995)	20 (4)	9/11	NK	Raw milk soft cheese	4b	Goulet et al., 1995
Sweden (1994–1995)	6 (1)[††]	2/4	$<100 - 2.5 \times 10^6$	Cold-smoked rainbow trout	4b	Ecklow et al., 1996
Canada (1996)	2 (0)	0/2	2×10^9	Imitation crab meat	1/2b	Farber et al., 2000
Sweden (NK)	5 (0)	0/5	1.9×10^5	Cold-smoked rainbow trout	1/2a	Miettinen et al., 1999
Finland (1999)	18 (4)	0/18	$<100 - 1.1 \times 10^4$	Butter	3a	Lyytikäinen et al., 1999
USA (1998–1999)	101 (20)	NK	NK	Frankfurters, sliced deli meats	4b	CDC, 1999

* For 2 of these 5 deaths, an underlying disease, not listeriosis was apparently the cause of death.
† Not known.
‡ Foods only epidemiologically linked.
§ Outcome only known for 252 cases.
‖ Only 314 patients categorized.
** Includes two sets of twins.
†† Not known if death due to listeriosis or leukemia.

Glass & Doyle (1989) reported that this pathogen can proliferate on vacuum-packaged, artificially contaminated (~0.01 L. monocytogenes cfu g^{-1}) retail frankfurters during storage at 4.4 °C; populations increased by 2–5 log$_{10}$ on samples that were judged organoleptically acceptable after 4 weeks. In another study, *L. monocytogenes* increased in number from 5×10^2 to 2.1×10^5 MPN g^{-1} on vacuum-packed frankfurters stored at 4 °C for 20 days. Interestingly, non-inoculated control products contained 1.2×10^2 MPN g^{-1} after the 20-day storage period, after initially testing negative for the organism (Bunčić, *et al.*, 1991). In a more detailed study of vacuum-packaged all-beef, poultry, or beef/pork frankfurters stored at 5 °C for up to 28 days, McKellar *et al.* (1994) found 40 of 61 (65.6%) frankfurters supported growth of *L. monocytogenes*, with an average increase of 1.26 log$_{10}$ within a 14-day period. Initial lactic acid bacteria concentrations, as well as phenol and nitrite levels, varied considerably during storage, unlike sodium chloride levels. In addition, average pH levels decreased significantly during storage by 0.19 pH units. Several statistical models were derived in an attempt to describe adequately the growth and death of the organism in all frankfurter samples. Although no one model was found to be sufficient, the best model implicated initial and final lactic acid bacteria counts and initial pH as factors that influence the growth of *L. monocytogenes*.

16.2.4 Risk Characterization

Cooked sausages, of which the frankfurter is but one example, are consumed in large quantities throughout the world. It is estimated that 20 billion frankfurters are consumed each year in the US, for an average of 60 frankfurters per person per year. A retrospective case-control study conducted in 1986–1987 involving 154 listeriosis patients found that sporadic cases were significantly more likely to have eaten either frankfurters that had not been fully reheated or chicken meat that was still pink (Schwartz *et al.*, 1988). It was estimated that 20% of sporadic cases were linked to the two food products. A subsequent case-control study identified as potential causes Mexican-style and feta cheeses, foods purchased from the deli section of grocery stores and, again, undercooked chicken, but not frankfurters (Schuchat *et al.*, 1992).

The population of the US in 1998 was 270 million. From 1996–1999, the estimated incidence of listeriosis from the FoodNet system was 5, 5, 6, and 5 cases per million per year, respectively (CDC, 2000) or about 1,350 cases per year. If it is assumed that 1% (13.5 cases) of the total listeriosis cases might be due to frankfurters and there are 20×10^9 servings of frankfurters per year, then that equals 13.5 cases for every 20×10^9 servings among the total population.

Approximately 20% of the US population (54 million) is immunocompromised. If it is assumed that this high-risk population consumes frankfurters at the same rate as the general population (i.e., about 60 frankfurters per year), then the total number of frankfurters consumed by this subpopulation would be $(54 \times 10^6) \times 60 = 3240 \times 10^6$ or 3.24×10^9. If it is assumed that listeriosis is restricted to this immunocompromised population and 1% (13.5) of the total listeriosis cases might be due to frankfurters, then there would be 13.5 cases for every 3.24×10^9 servings among the high risk population.

The risk associated with frankfurters, however, is likely to be related to growth of *L. monocytogenes* and handling/preparation practices. Between 1993–1999, the prevalence of *L. monocytogenes* among the lots sampled by USDA-Food Safety and Inspection Service declined from 5.3 to 1.8%. Using a single dose approach (Buchanan *et al.*, 1997), one could assume that all cases of listeriosis are associated with the proportion of frankfurters having a high level of *L. monocytogenes* (e.g., $\geq 1 \times 10^3$ cfu g^{-1}). Although there are no data for the proportion of frankfurters that may have this level, it will be assumed that 5% contain *L. monocytogenes* but only 10% of those have a high level (i.e., $\geq 1 \times 10^3$ cfu g^{-1}) when purchased. The total number of contaminated frankfurters having a high level of *L. monocytogenes* purchased by the high-risk population then would be $3.24 \times 10^9 \times 0.05 \times 0.10 = 16.2 \times 10^6$.

If it is further assumed that the method of handling and reheating is inadequate to eliminate the high level of *L. monocytogenes* in 10% of the servings, then the frequency of exposure to a high level of *L. monocytogenes* would be $16.2 \times 10^6 \times 0.10 = 1.62 \times 10^6$. If 1% of all listeriosis is associated with eating franks, then the 13.5 cases would more likely result from the estimated 1.62×10^6 franks having a high level of *L. monocytogenes* that are handled and prepared in such a manner that illness can result among susceptible individuals.

Thus, the probability (P) that the high-risk population would acquire listeriosis from frankfurters containing high levels of *L. monocytogenes* would be P = 13.5 cases per 1.62×10^6 servings (i.e., approximately one case per 120,000 servings) or P = 8.3×10^{-6}.

Since numerous assumptions have been made and reliable data are lacking, it is impossible to validate this estimate or to calculate attendant uncertainties. For example, the estimate assumes that all the strains of *L. monocytogenes* detected on frankfurters are equally virulent, which is unlikely. The estimate does suggest, however, that the risk of acquiring listeriosis from frankfurters on a per serving basis is exceptionally low, even among the high-risk population. Experience indicates that when frankfurters are contaminated with a highly virulent strain, cases may occur among the high-risk population (CDC, 1999).

The draft risk assessment conducted jointly by the US Food and Drug Administration and the USDA estimated the median number of cases of listeriosis per serving of frankfurters to be 3.0×10^{-6}, 5.0×10^{-8} and 5.9×10^{-9} for the perinatal, elderly, and intermediate age-based populations, respectively (FDA, 2001). The perinatal population consisted of fetuses and neonates (16 weeks after fertilization to 30 days after birth) with exposure occurring in utero from contaminated food eaten by the pregnant mother. The elderly group included individuals who were 60 or more years of age. The intermediate age group included the remaining population, including susceptible individuals not included as perinatal or elderly, such as cancer, AIDS, and transplant patients, for whom there was insufficient data for them to be considered as a separate population.

On a per serving basis, *reheated* frankfurters were of low risk relative to 19 other categories of foods studied, ranking 15th out of 20 for the three population groups. Frankfurters consumed *without reheating* were of high risk, ranking first or second among the 20 categories for the three population groups, assuming that from 1–14% of frankfurters are consumed without reheating. Overall, frankfurters ranked number 4 or 5 among the 20 food categories for the three population groups when the median estimates were used to predict the relative risk for listeriosis on a per annual consumption basis, again assuming that 1–14% of frankfurters are consumed without reheating (FDA, 2001).

16.3 RISK MANAGEMENT

16.3.1 Acceptable Level of Consumer Protection

In most industrialized countries, the annual incidence of listeriosis is between 2–6 cases per million of the population (Rocourt & Bille, 1997). For the risk characterization, an estimate of 1% of the total cases of listeriosis was used for the cases that might be associated with frankfurter consumption, i.e., 13.5 cases. An example of a public health goal could be to reduce the number of cases attributable to the consumption of frankfurters by some value (e.g., 50% reduction to no greater than 6.75 cases per million of the population).

16.3.2 Establishing a Food Safety Objective

Due to its widespread presence in the environment, eradication of *L. monocytogenes* from the food supply is impossible. There is general agreement that when the organism is ingested in low numbers, even susceptible individuals have little chance of developing listeriosis. A realistic food safety objective (FSO) must therefore be established that would, as far as possible, recognize that it is not possible with current technology to eliminate contamination of foods with *L. monocytogenes*. When total prevention is not possible, measures must be put into place to control recontamination to an acceptable level. To achieve these goals, it is essential that GHP and HACCP programs that are specific to the control of *L. monocytogenes* be applied at all stages of manufacture, storage, transport, and retail. In addition, research should continue to develop additional barriers to control the growth of *L. monocytogenes* in meat and poultry products. This might include the addition of food additives and/or the use of a competitive flora (McMullen & Stiles, 1996).

It is currently recognized that on a daily basis, humans are consuming *L. monocytogenes* in foods at levels of at least 100 cfu g^{-1} and not becoming ill. For example, data from Denmark show an overall contamination rate for cold-smoked salmon products of around 25–32%, with 4–5% of those samples containing greater than 100 cfu g^{-1} (Huss, 1997). Even with the large amounts of Danish cold-smoked salmon being consumed worldwide, there have been no recorded cases of listeriosis due to this product. In addition, countries such as Canada have established an action level for *L. monocytogenes* of 100 cfu g^{-1} for low risk products and have not seen an increase in the baseline levels of human listeriosis. This is also consistent with the estimated dose-response curve for *L. monocytogenes* from Buchanan et al. (1997), which includes an estimate for immunocompromised individuals, is reasonably conservative, and allows for the possibility of a single cell causing serious illness (Figure 16–2). Finally, epidemiologic data indicate that foods involved in listeriosis outbreaks are those in which the organism has multiplied and in general have contained levels well in excess of 100 cfu g^{-1} (see Table 16–1). Thus, on the basis of epidemiologic and prevalence data, the following FSO is proposed:

The concentration of *L. monocytogenes* in frankfurters should not exceed 100 cfu g^{-1} at the time of consumption.

This proposal is consistent with an earlier recommendation from the International Commission on Microbiological Specifications for Foods (ICMSF) (ICMSF, 1994). It also is compatible with a conclusion of the World Health Organization (WHO)/Food and Agriculture Organization (FAO) risk assessment for *L. monocytogenes* in RTE foods that a more strict tolerance of "not detected in 25 g" does not provide a higher level of protection (Ross et al., 2000). Indeed, the incidence rate for listeriosis in the US, which has maintained the "negative in 25 g" tolerance, is not lower than other industrialized countries that have applied a tolerance of 100 cfu g^{-1}.

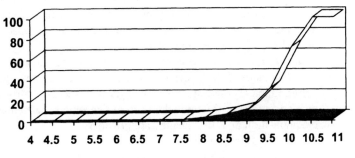

Figure 16–2 Estimated dose response curve for *Listeria monocytogenes*.

16.3.3 Control Measures

It is anticipated that industry can easily meet the proposed FSO for the product at the time of manufacture. However, if frankfurters are recontaminated after the commercial cooking, by the time they are consumed, there is potential for concentrations to exceed 100 g^{-1}, especially with a long refrigerated shelf-life, no additional barriers to growth, and/or temperature abuse. To reduce the incidence of listeriosis (e.g., to fewer than 5 cases per million population) by specifically reducing the number of cases due to frankfurters (e.g., from 13.5 to no greater than 6.75 cases), then other risk management options need be considered.

A flow diagram for a typical frankfurter operation is shown in Figure 16–1. HACCP systems and GHP procedures must be used to achieve the expected level of control and meet the FSO. It is important to consider available control measures in the context of the steps used in manufacture. In this example, controlling *L. monocytogenes* so the concentration will not exceed 100 cfu g^{-1} when frankfurters are consumed will involve combinations of the following control measures:

- *Controlling initial levels in raw materials*—Handling raw materials during storage and preparation to minimize an increase in numbers due to contamination or growth.
- *Reducing levels during cooking of frankfurters in the manufacturing process*—Cooking to eliminate *L. monocytogenes* in the raw meat emulsion. Establishing a performance criterion for cooking that results in process criteria (i.e., critical limits) that can be incorporated into the HACCP plan.
- *Preventing recontamination between cooking and packaging*—Minimizing recontamination between cooking and packaging by adopting GHP measures, e.g., separating raw from cooked product, and an environmental management and monitoring program.
- *Reducing levels in cooked product after packaging (in-pack pasteurization)*—Applying commercially feasible pasteurization processes for pasteurizing the product after packaging.
- *Preventing increase in levels between packaging and preparation for serving*—Controlling the increase of *L. monocytogenes* that subsequently may occur with

recontamination and growth in packaged frankfurters during storage and distribution. Examples of control measures might include: adding safe, acceptable additives to control growth of *L. monocytogenes* on frankfurters, using code-dating practices and improved chill chain management to preclude an unacceptable increase in *L. monocytogenes*, or freezing the product.
- *Reducing levels prior to consumption*—Improving storage and handling practices and inactivating *L. monocytogenes* through effective cooking and reheating prior to consumption. These outcomes could be achieved as control measures in food service operations or through education of consumers, particularly the more susceptible population and their health care providers.

16.3.4 Performance Criteria

A performance criterion is the required outcome of one or more control measures at a step or combination of steps applied to achieve an FSO (see Chapter 3). Performance criteria are usually applied at steps where hazards can either be reduced or where hazards may increase.

To arrive at a performance criterion for the control measures needed to meet an FSO for frankfurters, the equation from chapter 3 can be used:

$H_0 - \Sigma R + \Sigma I \leq FSO$

$H_0 - \Sigma R + \Sigma I \leq 2.0$

where:
FSO = Food safety objective
H_0 = Initial level of the hazard
ΣR = Total (cumulative) reduction of the hazard from processing, etc.
ΣI = Total (cumulative) increase of the hazard

FSO, H_0, R, and I are expressed in $\log_{10}$ units.

Manufacturers of frankfurters should assume that *L. monocytogenes* will be present in all species and types of raw meat and poultry. This is evident from reports in the literature and results from surveys such as the USDA baseline studies for ground beef, ground turkey, and ground chicken (Table 16-2) (USDA, 1995a; USDA, 1995b; USDA, 1996). Data of this nature can be used to establish the initial concentration (H_0). Another option would be to generate similar data from raw emulsions collected at the facility prior to cooking.

The USDA baseline studies reported the prevalence of *L. monocytogenes* in ground beef, chicken, and turkey to be in the range of 18–35% when analyzing 25 g samples. Further analysis of the positive samples yielded low numbers of *L. monocytogenes*. It can be concluded from the data that concentrations of 100 g^{-1} or fewer would be expected under normal conditions. Further reducing the likelihood of higher concentrations is the effect of blending and emulsification that occurs in the manufacturing process. These steps disperse the cells more uniformly throughout the meat emulsion and result in an overall lower number g^{-1} with little likelihood of localized pockets of higher concentrations.

16.3.4.1 Controlling Initial Levels in Raw Materials

Consideration should be given to multiplication of *L. monocytogenes* that might occur as raw meat and poultry is stored and prepared for cooking. It is important to note that

Table 16–2 USDA Baseline Survey Results for Ground Beef, Ground Turkey, and Ground Chicken

	Ground beef (563 samples)	Ground turkey (165 samples)	Ground chicken (162 samples)
Prevalence based on 25 g samples	18%	23%	35%
Geometric mean for positive samples only	3.9/g	3.8/g	2.24/g
95% confidence level	2.2–7.2/g	0.4–2.9/g	1.06–4.7/g

Samples found positive by the qualitative method were further analyzed to determine the number of *L. monocytogenes* g^{-1} of ground meat. The results for the positive samples were as follows:

Number of *L. monocytogenes*/g	Ground beef (99 samples)	Ground turkey (52 samples)	Ground chicken (59 samples)
< 0.03	45.2	82.5	56.1
0.03–0.29	0.0	0.0	0.0
0.3–2.9	30.2	6.7	29.3
3.0–29.9	15.0	0.0	10.5
30–299.9	9.6	10.9	3.2
300 or higher	3 samples had > 110/g	0.0	0.9

Note: These data indicate, for example, that in the case of ground beef, 563 samples were analyzed using 25 g each. Only 99 samples were found to contain *L. monocytogenes*. Upon further analysis of the 99 samples, 90.4% had fewer than 30 cfu gram^{-1}. From an overall perspective, only 3 of the total 563 samples had greater than 110 *L. monocytogenes* gram^{-1}, the upper limit of detection used for testing.

In the case of ground turkey, 165 samples were analyzed using 25 g each. Only 52 were found to contain *L. monocytogenes*. Upon further analysis of the 52 samples, 89.2% had fewer than 30 cfu gram^{-1}. From an overall perspective, the highest number detected among the 165 samples was a sample containing 93 cells gram^{-1}.

---

the samples for the baseline studies were collected, packaged, and shipped in refrigerated containers to the laboratory. Samples were considered acceptable for analysis if they were received at the laboratory by the day after collection and with a temperature of 10 °C or less. Hence, the samples were at least one day older than when collected and some temperature abuse may have occurred.

To further place the possibility of multiplication during preparation into perspective, the data in Table 16–3, derived from the USDA Pathogen Modeling Program, indicate that from 4–9 days at 4 °C would be required for *L. monocytogenes* to increase 10-fold in meat or poultry with no added salt (0.5%) (Buchanan & Whiting, 1996). Food MicroModel (Ver 3.02) similarly predicts that it would take around 6 days for a 10-fold increase of *L. monocytogenes* (4 °C, 0.5% NaCl, pH 6.5). Longer times would be required in raw materials containing added salt. For example, the mechanically deboned poultry meat used for manufacturing frankfurters typically contains added salt and sodium nitrite and is chilled to well below 4 °C before shipping to the frankfurter manufacturing plant. The values in Table 16–3 include the lag phase estimated for each set of parameters. The length of the lag phase that actually occurs in raw meat under commer-

Table 16–3 Days for a 1 Log$_{10}$ (10-Fold) Increase in *L. monocytogenes* at 4 °C and pH 6.0

	Days for 10-fold increase			
	Aerobic conditions (e.g., meat at top of container)		Anaerobic conditions (e.g., meat below surface of container)	
Sodium chloride (%)	No NaNO$_2$ Added	150 µg g^{-1} NaNO$_2$ added	No NaNO$_2$ Added	150 µg g^{-1} NaNO$_2$ added
0.5	5.3	8.9	4.1	8.8
1.0	5.6	9.4	4.5	9.6
1.5	6.0	10.0	4.9	10.5
2.0	6.4	10.6	5.3	11.5
2.5	6.8	11.4	5.8	12.4
3.0	7.3	12.3	6.2	13.4

cial conditions has not been determined, but the raw meat and/or poultry used for the manufacture of frankfurters is normally less than 7 days from the date of slaughter.

From the data in Table 16–2 it is evident that the numbers of *L. monocytogenes* are very low in raw meat and poultry. Furthermore, the data in Table 16–3 indicate that significant multiplication during preparation is unlikely. For this example, it will be assumed that the initial concentration may under certain circumstances increase 10-fold prior to commercial cooking. Thus, in this example, the initial number in the raw materials prior to cooking is assumed to be no more than 1000 g^{-1} (i.e., H$_0$ = 3).

16.3.4.2 Reducing Levels during Cooking of Frankfurters in the Manufacturing Process

If it is assumed that initial numbers of *L. monocytogenes* in the raw materials could be as high as 1000 g^{-1} (i.e., H$_0$ = 3), then a performance criterion for the reduction step (i.e., cooking) to meet the FSO would, theoretically, only need to be a 1 log$_{10}$ reduction as given below and illustrated in Figure 16–3a.

$$H_0 - \Sigma R + \Sigma I \leq FSO$$
$$3 - \Sigma R + 0 \leq 2.0$$
$$\Sigma R = 1$$

However, if any viable cells remain after cooking, by the time the frankfurters are consumed, there is the potential for concentrations to exceed 100 g^{-1}, especially with a long refrigerated shelf-life, no additional barriers to growth, and/or if temperature abuse occurs. Therefore, to ensure that survivors do not remain and multiply during distribution and storage, it is necessary to apply a more stringent cooking process. For this example, it will be assumed that a 6 log$_{10}$ reduction (ΣR = 6) during cooking will achieve the desired result, given an initial concentration of no greater than 1000 g^{-1} (H$_0$ = 3). Thus, a 6 log$_{10}$ reduction would result in a final concentration of ≤ 1 cfu kg^{-1} after cooking. This could be expressed as a performance criterion:

The concentration of *L. monocytogenes* after cooking shall be ≤ 1 cfu kg^{-1}.

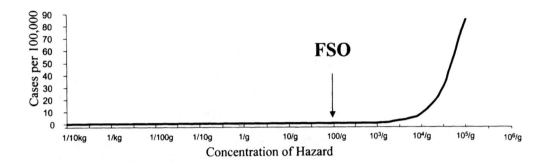

1) FSO could be met with only 1 $\log_{10}$ reduction:

$$\sum R = 1 \quad \longleftarrow \quad H_0$$

2) Cooking during manufacture (6 $\log_{10}$ reduction, $\sum R = 6$) is applied to reduce levels to meet a performance criterion of $\leq 1 \text{ kg}^{-1}$ ($\leq 10^{-3}$ per g)

$$\sum R = 6 \quad \longleftarrow \quad H_0$$

Figure 16–3a Control measure: cooking.

This is demonstrated in the following equation and illustrated in Figure 16–3a.

$$H_0 - \sum R + \sum I \leq 1/\text{kg} \ (\leq 10^{-3}/\text{g})$$

$$3 - \sum R + 0 \leq -3$$

$$\sum R = 6$$

If the initial number is less than 1000 g^{-1} through raw material selection and/or preventing any increase during handling and storage (e.g., ≤ 100 g^{-1}), then a 5 $\log_{10}$ reduction would be adequate to meet the performance criterion of ≤ 1 kg^{-1}.

16.3.4.3 Prevent Recontamination Following Cooking

L. monocytogenes is a very hardy organism and can be recovered from different locations in the plant environment, depending on the level of control. In the absence of an effective control program, the organism can persist for prolonged periods in frankfurter production environments. During a 1989 investigation, the same strain of *L. monocytogenes* was recovered from a patient, from leftover frankfurters in the patient's refrigerator, and 4 months later from the frankfurter peeler in the manufacturing facility (Wenger et al., 1990). In another outbreak in the US associated with frankfurters, it was reported that removing a refrigeration unit resulted in new or increased contamination of production equipment (CDC, 1999). This outbreak involved cases from August 2, 1998 through January 16, 1999, suggesting that multiple lots of product were implicated. Attempts to verify that the strain had become established in the environment, however, were unsuccessful. Thus, even if process criteria are met for cooking, recontamination of processed product is a possibility unless proper attention has been paid to the GHPs that are targeted toward control of *L. monocytogenes*.

Certain measures that fall within the realm of GHP have been found to be important for control of *L. monocytogenes* in cooked meat and poultry operations. The measures include plant layout, equipment design, maintaining equipment, cleaning and sanitizing procedures that are specific to listeriae control, maintaining a clean and dry environment, use of enclosed steam to sanitize equipment as a scheduled routine procedure, and low temperature storage (Tompkin *et al.*, 1999).

Two factors determine the effectiveness of a listeriae control program: routine environmental testing and the response to a positive finding. Without an environmental testing program, it is not possible to assess control. Furthermore, in the event a positive product contact sample is detected, corrective actions should be initiated to eliminate the source of contamination, thereby minimizing the risk of product contamination. To verify control, plants should implement an environmental monitoring program for *L. monocytogenes* or an indicator, such as *L. innocua* or *Listeria spp*. The program must be specific to each plant and should detail the areas to be sampled, the analytic method to be used, the frequency of sampling, and the action to be taken when *Listeria spp.* are detected (Tompkin *et al.*, 1992, 1999).

An effective monitoring program to assess control of the cooked product environment should consider the following strategies:

- preventing the establishment of *L. monocytogenes* in niches or other sites that can lead to contamination of RTE foods,
- implementing a sampling program that can assess in a timely manner whether the environment in which RTE foods are produced is under control,
- responding to each positive product contact surface sample as rapidly and as effectively as possible, and
- verifying that the problem has been corrected and providing data (e.g., tabulated, trends in graphical form) to facilitate short- and long-term assessment of control.

Experience in frankfurter operations indicates that sites within the cooked product environment wherein *L. monocytogenes* becomes established and multiplies (i.e., niches) are important contributing factors to product contamination. The sites serve as a reservoir from which the pathogen is dispersed and contaminates product contact surfaces. Furthermore, an environmental sampling program, such as described in Chapter 11, is necessary to assess whether the environment is under control and to detect the presence of a niche.

Response to a positive product contact sample (e.g., a sponge sample collected from a conveyor belt used in a frankfurter operation) is best when the following items are considered. Assuming an effective control program is in place, the primary source of *L. monocytogenes* in frankfurter operations is contamination from one or more niches. In general, contamination flows downstream with the process flow, much like a river. To regain control, it is necessary to locate the source where growth and contamination are occurring. This can best be accomplished by creating a map of the cooked product rooms and the equipment layout (see Chapter 11). The results of samples collected from each piece of equipment should be recorded on the map, including both positive and negative results. The layout map should then be reviewed for patterns. Which sites are more frequently positive? Where in the process flow do the first positives occur? It is important when seeking the source to analyze all the sponge samples separately and not as composites. In addition, sampling should be aggressive both in terms of location and frequency throughout the period of operation. Unfortunately, niches can rarely be detected unless the equipment is operating and product is being produced. While pursuing the source of contamination, consideration must be given to the possibility that a niche may not be involved (contamination could result, e.g., from someone touching the floor or other unclean surface and returning to handle exposed product).

When equipment has been identified as the source, the following steps can be effective. First, dismantle the equipment, collecting samples of suspicious material as this is being done. Second, replace obvious defective parts (e.g., hollow rollers) and clean and sanitize the equipment as it is being reassembled. If this procedure fails to eliminate the cause (or source) of contamination, it may be necessary to remove sensitive electronics, oil, and grease, and heat the equipment. This can be accomplished by placing the equipment in an oven and heating to an internal temperature of 71 °C with high humidity. Alternatively, the equipment can be shrouded with a tarp and steam can be injected into the space until the equipment reaches an internal temperature of 71 °C. Internal temperatures can be monitored with strategically placed thermocouples.

In frankfurter operations the following are examples of identified sources of contamination:

- rubber gaskets around the doors and other openings to the brine chill system
- saturated insulation on pipes carrying refrigerant to the brine chill system
- peelers that remove artificial casings before packaging
- casing removal systems
- hollow rollers on conveyors
- on/off valves and buttons for various equipment
- within complex collating equipment
- hollow support rods on the frames of equipment.

It is difficult to assign a performance criterion to recontamination, as any recontamination has the potential to reach high numbers through growth during the subsequent distribution and storage.

To ensure that the performance criterion of ≤ 1 kg^{-1} (i.e., $\leq 10^{-3}$ per g) is not exceeded due to recontamination, there must be no recontamination and "I" in the following equation must be zero. This is illustrated in Figure 16–3b.

$$H_0 - \Sigma R + \Sigma I \leq 1 \text{ kg}^{-1} (\leq 10^{-3} \text{ per g})$$
$$3 - 6 + \Sigma I \leq -3$$
$$\Sigma I = 0$$

16.3.4.4 Reducing Levels in Cooked Product after Packaging (In-Package Pasteurization)

Experience indicates that recontamination after cooking is the most common reason for the presence of *L. monocytogenes* in packaged, cooked sausages, such as frankfurters. If the increase in concentration due to recontamination is assumed to be as high as 10 g^{-1}, then an in-package pasteurization treatment could be applied as a means to achieve a 4 log reduction and still meet the performance criterion of ≤ 1 kg^{-1} (i.e., 10^{-3} per g). As demonstrated in the following equation, if cooking provides an initial 6 log reduction and recontamination results in 10 cfu g^{-1}, then an in-package pasteurization treatment could be used to provide a final 4 log reduction and meet the performance criterion of ≤ 1/kg.

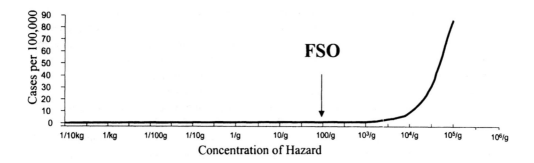

Preventing recontamination between cooking and packaging through effective GHPs and an environmental sampling program

$$H_0 - \Sigma R + \Sigma I \leq \text{Performance criterion } (1 \text{ kg}^{-1} \text{ or } \leq 10^{-3} \text{ per g})$$

$$3 - 6 + \Sigma I \leq -3$$

$$\Sigma I \leq 0$$

Figure 16–3b Control measure: preventing recontamination.

Hence, the combined reduction steps for cooking and in-package pasteurization result in a 10 log reduction (i.e., $\Sigma R = 10$). This is illustrated in Figure 16–3c.

$$H_0 - \Sigma R + \Sigma I \leq 1 \text{ kg}^{-1} \ (\leq 10^{-3} \text{ per g})$$

$$3 - \Sigma R + 4 \leq -3$$

$$\Sigma R = 10 \text{ (Cooking 6D, In-pack pasteurization 4D)}$$

16.3.4.5 Preventing Increase in Levels between Packaging and Preparation for Serving

In this and the next example, an in-package pasteurization treatment is not applied. If the increase in concentration due to recontamination is again assumed to be as high as 10 g^{-1} (i.e., a 4 $\log_{10}$ increase from 1 kg^{-1}) and if the FSO of 100 g^{-1} at time of consumption is to be met, there can be no greater than a 10-fold increase during storage and distribution before the food is consumed (see also Figure 16–3d).

$$H_0 - \Sigma R + \Sigma I \leq \text{FSO}$$

$$3 - 6 + 4 \leq 2$$

$$\Sigma I \leq 1$$

If recontamination occurs, then an increase of 10-fold or more must be controlled by adding inhibitory ingredients, freezing the product for distribution, and/or establishing use by date coding that are within the time for a 10-fold increase.

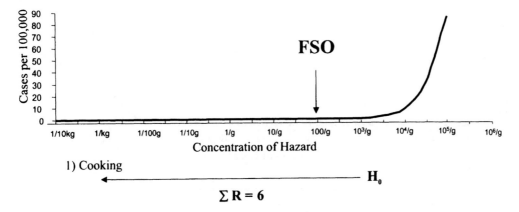

1) Cooking

$\Sigma R = 6$

2) Assume that if contamination occurs, it could be as as high as $10\ g^{-1}$ ($\Sigma I = 4$)

$\Sigma I = 4$

3) To eliminate the contamination that may occur between cooking and packaging and meet the performance criterion of $\leq 1\ kg^{-1}$, a 4-$\log_{10}$ reduction step is needed

$\Sigma R = 4$

Figure 16–3c Control measure: reducing levels using in-package pasteurization.

16.3.4.6 Reducing Levels Prior to Consumption

If the product is recontaminated with *L. monocytogenes* to a level of $10\ g^{-1}$ ($\Sigma R = 4$) and growth does occur during subsequent distribution and storage, then the handling and preparation procedures become very important for susceptible consumers. If it is assumed that there could be a 5 $\log_{10}$ increase due to growth ($\Sigma I = 5$) between when the frankfurters are cooked in the factory and when they are reheated for serving, then the combined increase after cooking would be $\Sigma I = 4 + 5$ for a total increase of 9 $\log_{10}$. At the time of preparation, the frankfurters would have a level of $10^6\ g^{-1}$. This could be represented as follows (see also Figure 16–3e).

$H_0 - \Sigma R + \Sigma I \leq FSO$

$3 - 6 + 9 =$ level at preparation

$6 =$ level at preparation

Therefore, to meet the required FSO of $\leq 100\ g^{-1}$, an additional reduction step during preparation for serving would be required prior to consumption, as follows.

$H_0 - \Sigma R + \Sigma I \leq FSO$

$3 - \Sigma R + 9 \leq 2$

$\Sigma R = 10$

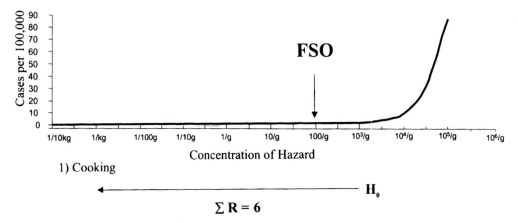

Figure 16–3d Control measure: use of inhibitor(s) to prevent growth after packaging.

After taking account of the 6 $\log_{10}$ reduction (i.e., $\Sigma R = 6$) that occurs when the frankfurters were cooked during manufacture, the reduction during preparation for serving must be at least as follows (Figure 16–3e).

$\Sigma R = 4$

The final reheating before serving then must reduce the level of *L. monocytogenes* by 4 $\log_{10}$ to ensure that the frankfurters will meet the FSO of 100 g^{-1} at the time they are consumed.

16.4 PROCESS AND PRODUCT CRITERIA

Each manufacturer must determine the parameters for cooking that will provide the desired product quality and cost, and, in this example, ensure that the performance criterion (i.e., < 1 kg^{-1}) is met for cooking. The parameters adopted by one manufacturer may differ from those adopted by others. This is due to differences in equipment, desired product quality, type of raw materials (e.g., beef, pork, chicken and/or turkey), type of casing (e.g., natural, artificial), method of applying smoke (e.g., natural, liquid), fat content, and other factors. To meet the performance criterion requires knowledge of the heat sensitivity of *L. monocytogenes*, particularly in frankfurter emulsions.

Historically, the minimum internal requirement set by USDA for cooked sausage was 64.4 °C. For frankfurters, industry typically applied processes to achieve at least 68.3 °C. Since about 1980–1985, many processors have increased the temperature to 71 °C, or

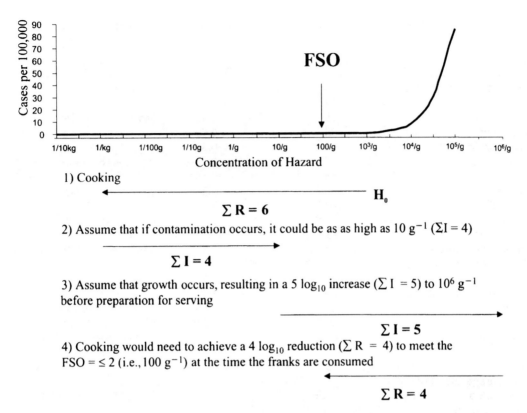

Figure 16–3e Control measure: reducing levels prior to consumption.

higher, to achieve greater consistency in product quality and refrigerated shelf-life. Even with the minimum requirement of 64.4 °C, however, there was no evidence that cooked sausages were implicated as a source of foodborne illness due to survival of enteric pathogens, probably due to the low concentration of enteric pathogens present initially and their sensitivity to heat. In the case of *L. monocytogenes*, the data in Table 16–4 indicate that survival also is unlikely to occur. Even with extensive testing for the pathogen since 1988, there has been no documented incident in which *L. monocytogenes* surviving the heat process was the underlying reason for a contaminated cooked sausage.

The thermal process can be validated by various means. One could be to conduct thermal destruction tests in the laboratory using frankfurter emulsion inoculated with a mixture of *L. monocytogenes* isolates from meat operations. Another may be to perform in-house challenge experiments with nonpathogenic surrogate organisms, such as *L. innocua*. Various items, such as provided in Chapter 3, should be considered when conducting such studies to validate a thermal process. A third approach to validation is to review the published literature for data on thermal resistance. For example, numerous studies have been published that report thermal destruction values for *L. monocytogenes* in various meat and poultry products (ICMSF, 1996; Farber & Peterkin, 1999; Zaika et al., 1990; Fain et al., 1991).

The data summarized in Table 16–4 provide lethality values for *L. monocytogenes* in meat, poultry, and seafood cooked to temperatures within the range of 60–65 °C. A number of factors influence the rate of death, but from the data, it can be concluded that even

Table 16–4 Reported Thermal Destruction Values for *L. monocytogenes* in Meat, Poultry, and Seafood Products

Product	Temp (°C)	D value* (min.)	Reference
Ground meat with cure mix	64	2.28	Farber (1989)
Ground meat	64	<1.01	
Beef	65	0.93	Mackey et al. (1990)
Chicken leg	65	0.53	
Chicken breast	65	0.52	
Buffer	65	0.29	Boyle et al. (1990)
Meat slurry	65	0.75	
Lean beef	62.8	0.6	Fain et al. (1991)
Fatty beef	62.8	1.2	
Beef steak	63.9	2.2	Gaze et al. (1989)
Chicken breast	63.9	1.6–1.8	
Ham—control culture	60	1.8	Carlier et al. (1996)
Ham—heat-shocked culture	60	3.5	
Ham—"resistant" culture	60	1.0	
Mussels	62.2	1.85	Bremer & Osborne (1995)
Crabmeat	60	1.3–2.6	Harrison & Huang (1990)
Lobster meat	62.8	1.1	Budu-Amoako et al. (1992)
Salmon	65	0.9–1.2	Embarek & Huss (1993)
Cod	65	0.27–0.28	
Cured sausage blend of 66% pork and 33% beef	63.9	3.3	Farber & Brown (1990)
Cured sausage blend of 66% pork and 33% beef	30 min. at 47.8 °C, then heat to 63.9 °C	4.2	
Cured sausage blend of 66% pork and 33% beef	60 min. at 47.8 °C, then heat to 63.9 °C	4.7	
Cured sausage blend of 66% pork and 33% beef	120 min. at 47.8 °C, then heat to 63.9 °C	8.0	

*D value = time in minutes at a specified temperature for a one $\log_{10}$ (i.e., ten-fold) reduction in the number of *L. monocytogenes*.

these lower temperatures are effective for killing *L. monocytogenes* in a wide variety of foods. Such data can be used to arrive at alternative thermal processes for sausage products, such as frankfurters, and to consider the disposition of products in the event of a deviation. The adequacy of a thermal process based on an internal temperature of 70 °C can be determined from the data in Table 16–5.

An example of a default process criterion to achieve a 6 $\log_{10}$ reduction for in-pack pasteurization can be found in the draft Code of Practice developed by a Belgium-Dutch Chilled Meals Working Group, which recommends an internal heat treatment of 70 °C for 2 minutes for elimination of *L. monocytogenes* in chilled, long-life pasteurized foods (Lund & Notermans, 1993). The advantage in using a performance criterion rather than a default process criterion is that it allows flexibility in how the performance criterion is achieved. This would facilitate the development of new technology that may offer novel ways to achieve the same end result.

Table 16–5 Reported D values for *L. monocytogenes* at 70 °C

Product	D value (min)*	Reference
Beef	0.14	Mackey et al. (1990)
Chicken leg	0.11	
Chicken breast	0.13	
Meats (predicted value)	0.13	
Meat slurry	0.23	Boyle et al. (1990)
Beef slurry (pH 5.9)	0.23	
Phosphate buffer (pH 7.2)	0.15	
Chicken	0.16, 0.20	Gaze et al. (1989)
Beef steak	0.20, 0.14	
Turkey meat emulsion	0.35	Fain et al. (1991)
Salmon, cod	0.13–0.17	Embarek et al. (1993)
Sucrose solution, a_w 0.98	0.15 @ 68.3C	Sumner et al. (1991)
Sucrose solution, a_w 0.96	0.30 @ 68.3C	
Raw milk	0.05 @ 68.9C	Bradshaw et al. (1985)
Raw milk	0.015@ 71. 7C	

*D value = time in minutes at a specified temperature for a one $\log_{10}$ (i.e., ten-fold) reduction in the number of *L. monocytogenes*.

Product criteria that are intended to prevent growth after packaging could be validated by performing challenge studies using surface-inoculated frankfurters to assess the likely increase (ΣI) of *L. monocytogenes* to ≥ 100 g^{-1} before the recommended use-by date. Challenge studies could involve, for example, using commercial product inoculated and packaged in a laboratory with a mixture of five isolates of *L. monocytogenes* and then stored at temperatures to which the product would be exposed during storage and distribution. Many factors should be considered when conducting such studies (see Chapter 3). Predictive modeling also can provide an estimate of the pathogen's behavior and possible growth or inactivation (McKellar et al., 1994; Armitage, 1997; McClure et al., 1997; McDonald & Sun, 1999; Buchanan & Whiting, 1996).

Examples of additives that have been found effective for inhibiting *L. monocytogenes* include sodium lactate, sodium acetate, and sodium diacetate (Schmidt, 1993; Schmidt & Leistner, 1993; Qvist et al., 1994; Wederquist et al., 1994, 1995; Blom et al., 1997; Tompkin, 1999; Juncher et al., 2000; Hoogenkamp & Wales, 2000).

A uniform procedure has not been developed to validate code-dating practices to ensure the safety of perishable foods with extended shelf-life (e.g., frankfurters and other cooked meat or poultry products). As discussed in Chapter 3, a number of factors must be considered when validating the effectiveness of one or more steps along the food chain. The following factors should be considered when performing a validation study for code-dating frankfurters:

- physiological state of the inoculum (e.g., 24 h broth culture)
- method of inoculation (surface)
- inoculum level (per package or per g)
- source and number of strains

- virulence of the strains (outbreak strains should be avoided in most laboratories)
- temperature(s) of storage
- natural competitive flora (commercial v. pilot plant product)
- product formulation (fermentable carbohydrate and rate of acid production).

16.5 GHP AND HACCP

Certain aspects of GHP must be controlled to minimize recontamination of cooked products, as has been discussed in this chapter and Chapter 11. In this example, a performance criterion of ≤ 1 kg^{-1} was established for the cooked product and for in-package pasteurization to ensure that the FSO of ≤ 100 g^{-1} is met. This led to the establishment of process criteria that could be adopted as critical limits in the HACCP plan. The critical limits typically consist of a holding time and internal temperature, their selection being influenced by the cooking system (e.g., continuous or batch oven). If, for example, 71 °C is adopted, the critical limit for cooking frankfurters during their manufacture could be stated as "an internal temperature of 71 °C", a hold time not being necessary.

With a properly designed and implemented HACCP plan, there would be no benefit to testing the cooked product for further validation of the process or to verify that each lot has been cooked correctly. If an event occurs requiring such testing, then a sampling plan of $n = 5$, $c = 0$, $m = 0/25$ could be used. This would provide 95% confidence of detecting a positive lot if the mean count was 0.03 cfu g^{-1} or greater.

16.6 ACCEPTANCE CRITERIA FOR FINAL PRODUCT

16.6.1 Organoleptic

No organoleptic criteria are applicable to assess the likely presence of *L. monocytogenes* in frankfurters.

16.6.2 Chemical and Physical

No chemical or physical criteria are applicable to assess the likely presence of *L. monocytogenes* in frankfurters.

16.6.3 Microbiological

When it is known that a product has been pasteurized in the pack, sampling final product would provide no added value and is not recommended. Likewise, for those plants that are using a validated kill step for cooking and where an effective environmental sampling program documents that the risk of recontamination is being controlled, there is little value in testing end product. The reason is that a comprehensive management system can maintain frequencies of contamination to less than 0.5%. Under these circumstances, the frequency of defective units is too low for detection with any practical sampling plan (see Chapters 6, 7, and 17).

Where product is crossing international borders and nothing is known about the product or the manufacturing process, end-product testing may be appropriate. The ICMSF has provided guidance on sampling plans for *L. monocytogenes* in a variety of foods and conditions (van Schothorst, 1996). For frankfurters that have not been in-pack pasteurized and, if contaminated, growth could occur (e.g., no added inhibitors), ICMSF sampling case 12 ($n = 20$, $c = 0$) would apply. When case 12 is applied and 25 g analytical units are used in a two-class sampling plan, which would be typical for *L. monocytogenes*, there is 95% probability that a positive lot would be detected when the mean concentration is 1 cfu in 186 g or greater. This assumes a log-normal distribution and standard deviation of 0.8. Thus, the sampling plan performance would be adequate to detect lots that exceed an FSO of ≥ 100 g^{-1}. If the frankfurters are intended specifically for highly susceptible individuals (e.g., hospitals), then case 15 would apply ($n = 60$, $c = 0$). When case 15 is applied and 25 g analytical units are used, there is a 95% probability that a positive lot would be detected when the mean concentration is 1 cfu in 526 g or greater.

16.7 REFERENCES

Anonymous (1989). 600, 000 pounds of turkey hot dogs recalled by Plantation. *Food Chem News* 32, 52–53.

Armitage, N. H. (1997). Use of predictive microbiology in meat hygiene regulatory activity. *Int J Food Microbiol* 36, 103–109.

Audurier, A., Pardon, P. & Martin, C. (1980). Experimental infection of mice with *Listeria monocytogenes* and *L. innocua*. *Annals Microbiol* (Paris) 131B, 47–57.

Baker, M. & Wilson, N. (1993). Microbial hazards of seafood: a New Zealand perspective. *Communic Diseases New Zealand* 93, 12–15.

Bemrah, N., Sanaa, M., Cassin, M. H. *et al.* (1998). Quantitative risk assessment of human listeriosis from consumption of soft cheese made from raw milk. *Prevent Vet Med* 37, 129–145.

Ben Embarek, P. K. & Huss, B. H. H. (1993). Heat resistance of *Listeria monocytogenes* in vacuum packaged pasteurized fish fillets. *Int J Food Microbiol* 20, 85–95.

Bille, J. (1990). Epidemiology of human listeriosis in Europe, with special reference to the Swiss outbreak. In *Foodborne Listeriosis*, pp. 71–74. Edited by A. J. Miller, J. L. Smith & G. A. Somkuti. Society for Industrial Microbiology. New York: Elsevier.

Blom, H., Nerbrink, E., Dainty, R. *et al.* (1997). Addition of 2.5% lactate and 0.25% acetate controls growth of *Listeria monocytogenes* in vacuum-packed, sensory-acceptable servelat sausage and cooked ham stored at 4C. *Int J Food Microbiol* 38, 71–76.

Boyle, D. L., Sofos, J. N. & Schmidt, G. R. (1990). Thermal destruction of *Listeria monocytogenes* in a meat slurry and in ground beef. *J Food Sci* 55, 327–329.

Bradshaw, J. G., Peeler, J. T., Corwin, J. J. *et al.* (1985). Thermal resistance of *Listeria monocytogenes* in milk. *J Food Prot* 48, 743–745.

Bremer, P. J & Osborne, C. M. (1995). Thermal death times of *Listeria monocytogenes* in green shell mussels (*Perna canaliculus*) prepared for hot smoking. *J Food Prot* 58, 604–608.

Brett, M. S. Y., Short, P. & McLauchlin, J. (1998). A small outbreak of listeriosis associated with smoked mussels. *Int J Food Microbiol* 43, 223–229.

Broome, C. V., Gellin, B. & Schwartz, B. (1990). Epidemiology of listeriosis in the United States. In *Foodborne Listeriosis*, pp. 61–65. Edited by A. J. Miller, J. L. Smith & G. A. Somkuti. Society for Industrial Microbiology, New York: Elsevier.

Buchanan, R. L. & Whiting, R. C. (1996). *USDA Pathogen Modeling Program*, version 5.1. Philadelphia, PA: Agricultural Research Service, U. S. Department of Agriculture.

Buchanan, R. L., Damert, W. G., Whiting, R. C. & van Schothurst, M. (1997). The use of epidemiologic and food survey data to estimate a conservative dose-response relationship for *Listeria monocytogenes* levels and incidence of listeriosis. *J Food Prot* 60, 918–922.

Buchanan, R. & Lindqvist, R. (2000). *Topic 1: Hazard Characterization of* Listeria monocytogenes *in Ready-to-Eat Foods*. Report of the Joint FAO/WHO Expert Consultation on Risk Assessment of Microbiological Hazards in Foods. Rome: FAO Headquarters.

Budo-Amoako, E. S., Toora, C., Walton, R. F. et al. (1992). Thermal death times for *Listeria monocytogenes* in lobster meat. *J Food Prot* 55, 211–213.

Büla, C., Bille, J. & Glauser, M. P. (1995). An epidemic of food-borne listeriosis in western Switzerland: description of 57 cases involving adults. *Clin Infect Diseases* 20, 66–72.

Bunčić, S., Paunovic, L. & Radisic, D. (1991). The fate of *Listeria monocytogenes* in fermented sausages and in vacuum-packaged frankfurters. *J Food Prot* 54, 413–417.

Carlier, V., Augustin, J. C. & Rozier, J. (1996). Heat resistance of *Listeria monocytogenes* (phagovar 2389/2425/3274/2671/47/108/340): D- and z-values in ham. *J Food Prot* 59, 588–591.

CDC (Centers for Disease Control and Prevention) (1989). Listeriosis associated with consumption of turkey franks. *MMWR Morb Mortal Wkly Rep* 38, 267–268.

CDC (Centers for Disease Control and Prevention) (1999). Update: Multistate outbreak of Listeriosis—United States, 1998–1999. *MMWR Morb Mortal Wkly Rep* 47, 1117–1118.

CDC (Centers for Disease Control and Prevention) (2000). Preliminary FoodNet data on the incidence of food-borne illnesses—selected sites, United States, 1999. *MMWR* 49, 201–205.

Ciesielski, C. A., Hightower, A. W., Parsons, S. K. et al. (1988). Listeriosis in the United States: 1980–1982. *Arch Intern Med* 148, 1416–1419.

Dalton, C. B., Austin, C. C., Sobel, J. et al. (1997). An outbreak of gastroenteritis and fever due to *Listeria monocytogenes* in milk. *New Engl J Med* 336, 100–105.

Ecklow, A., Danielsson-Tham, M.-L., Ericsson, H. et al. (1996). Listeriosis in the provinces of Värmland: the first reported outbreak of food-borne listeriosis in Sweden. In *Proceedings of the Symposium "Food Associated Pathogens,"* pp. 222–223. Uppsala, Sweden.

Fain Jr., A. R., Line, J. A., Moran, A. B. et al. (1991). Lethality of heat to *Listeria monocytogenes* Scott A: D-value and z-value determinations in ground beef and turkey. *J Food Prot* 54, 756–761.

FAO/WHO (Food and Agriculture Organization/World Health Organization) (2000). *Joint FAO/WHO Expert Consultation on Risk Assessment of Microbiological Hazards in Foods*. FAO Food and Nutrition Paper 71. Rome: FAO.

Farber, J. M. (1989). Thermal resistance of *Listeria monocytogenes* in foods. *Int J Food Microbiol* 8, 285–291.

Farber, J. M. & Brown, B. E. (1990) Effect of prior heat shock on heat resistance of *Listeria monocytogenes* in meat. *Appl Environ Microbiol* 56, 1584–1587.

Farber, J. M., Cai, Y. & Ross, W. H. (1996). Predictive modeling of the growth of *Listeria monocytogenes* in CO_2 environments. *Int J Food Microbiol* 32, 133–144.

Farber, J. M., Daley, E., Mackie, M. T. et al. (2000). A small outbreak of listeriosis potentially linked to the consumption of imitation crab meat. *Lett Appl Microbiol* 31, 100–104.

Farber, J. M. & Peterkin, P. I. (2000). *Listeria monocytogenes*. In *The Microbiological Safety and Quality of Food*, pp. 1178–1232. Edited by B. M. Lund, A. C. Baird-Parker & G.W. Gould. Gaithersburg, MD: Aspen Publishers, Inc.

Farber, J. M. & Peterkin, P. I. (1991). *Listeria monocytogenes*, a food-borne pathogen. *Microbiol Rev* 55, 476–511.

FDA (US Food and Drug Administration) (2001). *Draft Assessment of the Relative Risk to Public Health from Foodborne Listeria monocytogenes among Selected Categories of Ready-to-Eat Foods*. Center for Food Safety and Applied Nutrition. Washington, DC: US Department of Health and Human Services.

Fleming, D. W., Cochi, S. L., MacDonald, K. L. et al. (1985). Pasteurized milk as a vehicle of infection in an outbreak of listeriosis. *New Engl J Med* 312, 404–407.

Gaze, J. E., Brown, G. D., Gaskell, D. E. et al. (1989). Heat resistance of *Listeria monocytogenes* in homogenates of chicken, beef steak and carrot. *Food Microbiol* 6, 251–259.

Glass, K. A. & Doyle, M. P. (1989). Fate of *Listeria monocytogenes* in processed meat products during refrigerated storage. *Appl Environ Microbiol* 55, 1565–1569.

Golnazarian, C. A., Donnelly, C. W., Pintauro, S. J. et al. (1989). Comparison of infectious dose of *Listeria monocytogenes* F5817 as determined for normal versus compromised C57B1/6J mice. *J Food Prot* 52, 696–701.

Goulet, V. (1995). Investigation en cas d'épidémie de listériose. *Méd et Maladie Infect* 25, Spécial, 184–190.

Goulet, V., LePoutre, A., Rocourt, J. et al. (1993). Épidémie de listériose en France; bilan final et résultats de l'enquête épidémiologique. *Bull Épidémiol Hebdomaine* 4, 13–14.

Goulet, V., Jacquet, C., Vaillant, V. *et al.* (1995). Listeriosis from consumption of raw-milk cheese. *Lancet* 345, 1581–1582.

Harrison, M. A. & Huang, Y. W. (1990). Thermal death times for *Listeria monocytogenes* (Scott A) in crabmeat. *J Food Prot* 53, 878–880.

Hitchins, A. D. (1996). Assessment of alimentary exposure to *Listeria monocytogenes*. *Int J Food Microbiol* 30, 71–85.

Ho, J. L., Shands, K. N., Friedland, G. *et al.* (1986). An outbreak of type 4b *Listeria monocytogenes* infection involving patients from eight Boston hospitals. *Arch Intern Med* 146, 520–524.

Hoogenkamp, H. & Wales, J. (2000). Acetate and lactate blends fight bacteria with extra vigour. *Meat Int* 10, 24–25.

Huss, H. H. (1997). Control of indigenous pathogenic bacteria in seafood. *Food Control* 8, 91–98.

ICMSF (International Commission on Microbiological Specifications for Foods) (1994). Choice of sampling plan and criteria for *Listeria monocytogenes*. *Int J Food Microbiol* 22, 89–96.

ICMSF (International Commission on Microbiological Specifications for Foods) (1996). *Microorganisms in Foods 5. Characteristics of Microbial Pathogens*. Gaithersburg, MD: Aspen Publishers, Inc.

Juncher, D., Vestergaard, C. S., Søltfoft-Jensen, J. *et al.* (2000). Effects of chemical hurdles on microbiological and oxidative stability of a cooked cured emulsion type meat product. *Meat Sci* 55, 483–491.

Kittson, E. (1992). A case cluster of listeriosis in Western Australia with links to pâté consumption. In *Proceedings of the 11th International Symposium "Problems of Listeriosis,"* pp. 39–40. Copenhagen, Denmark.

Lennon, D., Lewis, B., Mantell, C. *et al.* (1984). Epidemic perinatal listeriosis. *Pediat Infect Diseases* 3, 30–34.

Linnan, M. J., Mascola, L., Lou, X. D. *et al.* (1988). Epidemic listeriosis associated with Mexican-style cheese. *New Engl J Med* 319, 823–828.

Lund, B. M. & Notermans, S. H. W. (1993). Potential hazards associated with REPFEDS. In *Clostridium botulinum: Ecology and Control in Foods*, pp. 279–303. Edited by A. H. W. Hauschild & K. L. Dodds. New York: Marcel Dekker, Inc.

Lyytikäinen, O., Ruutu, P., Mikkola, J. *et al.* (1999). An outbreak of listeriosis due to *Listeria monocytogenes* serotype 3a from butter in Finland. *Eurosurveillance Wkly* 11 March.

Mackey, B. M., Pritchet, C., Norris, A. *et al.* (1990). Heat resistance of *Listeria*: strain differences and effects of meat type and curing salts. *Lett Appl Microbiol* 10, 251–255.

McClure, P. J., Beaumont, A. L., Sutherland, J. P. & Roberts, T. A. (1997). Predictive modeling of growth of *Listeria monocytogenes*: The effects on growth of NaCl, pH, storage temperature and $NaNO_2$. *Int J Food Microbiol* 34, 221–232.

McDonald, K. & Sun, D-W. (1999). Predictive food microbiology for the meat industry: a review. *Int J Food Microbiol* 52, 127.

McKellar, R. C., Moir, R. & Kalab, M. (1994). Factors influencing the survival and growth of *Listeria monocytogenes* on the surface of Canadian retail frankfurters. *J Food Prot* 57, 387–392.

McLauchlin, J. (1993). Listeriosis and *Listeria monocytogenes*. *Environ Policy & Practice* 3, 201–214.

McLauchlin, J., Hall, S. M., Velani, S. K. *et al.* (1991). Human listeriosis and pâté: a possible association. *Brit Med J* 303, 773–775.

McMullen, L. M. & Stiles, M. E. (1996). Potential for use of bacteriocin-producing lactic acid bacteria in the preservation of meats. *J Food Prot Suppl*, 64–71.

Mead, P. S., Slutsker, L., Dietz, V. *et al.* (1999). Food-related illness and death in the United States. *Emerging Infect Diseases* 5, 607–625.

Miettinen, M. K., Siitonen, A., Heiskanen, P. *et al.* (1999). Molecular epidemiology of an outbreak of febrile gastroenteritis caused by *Listeria monocytogenes* in cold-smoked rainbow trout. *J Clin Microbiol* 37, 2358–2360.

Mitchell, D. L., Misrachi, A., Watson, A. J. *et al.* (1991). A case cluster of listeriosis in Tasmania. *Listeria* in smoked mussels in Tasmania. *Communic Disease Intell* 15, 427.

Proctor, M. E., Brosch, R., Mellen, J. W. *et al.* (1995). Use of pulsed-field gel electrophoresis to link sporadic cases of invasive listeriosis with recalled chocolate. *Appl Environ Microbiol* 61, 3177–3179.

Qvist, S. & Liberski, D. (1991). *Listeria monocytogenes* in frankfurters and ready-to-eat sliced meat products. *Danish Veterinaertidsskr* 74, 773–774, 776–778.

Qvist, S., Sehested, K. & Zeuthen, P. (1994). Growth suppression of *Listeria monocytogenes* in a meat product. *Int J Food Microbiol* 24, 283–293.

Riedo, F. X., Pinner, R. W., Tosca, M. de L. et al. (1994). A point-source foodborne listeriosis outbreak: Documented incubation period and possible mild illness. *J Infect Disease* 170, 693–696.

Rocourt, J. (1996). Risk factors for listeriosis. *Food Control* 7, 195–202.

Rocourt, J. & Bille, J. (1997). Foodborne listeriosis. *World Health Stats Quarterly* 50, 67–73.

Ross, T., Todd, E. & Smith M. (2000). *Exposure Assessment of Listeria monocytogenes in Ready-to-Eat Foods.* Preliminary Report for Joint FAO/WHO Expert Consultation on Risk Assessment of Microbiological Hazards in Foods. Rome: FAO.

Salamina, G., Dalle Donne, E., Niccolini, A. et al. (1996). A foodborne outbreak of gastroenteritis involving *Listeria monocytogenes*. *Epidemiol & Infection* 117, 429–436.

Schlech, W. F., III, Lavigne, P. M., Bortolussi, R. A. et al. (1983). Epidemic listeriosis—evidence for transmission by food. *New Engl J Med* 308, 203–206.

Schmidt, U. (1993). Behaviour of *Listeria monocytogenes* on sliced bruhwurst sausage. *Mitteilungsplatt der Bundesanstalt fur Fleischforschung* 119, 16–20.

Schmidt, U. & Leistner, L. (1993). Behavior of *Listeria monocytogenes* in unpackaged sliced bruhwurst sausage. *Fleischwirt* 73, 733–740.

Schuchat, A., Deaver, K. A., Wenger, J. D. et al. (1992). Role of foods in sporadic listeriosis. I. Case-control study of dietary risk factors. *J Am Med Assoc* 267, 2041–2045.

Schwartz, B., Ciesielski, C. A., Broome, C. V. et al. (1988). Association of sporadic listeriosis with consumption of uncooked hot dogs and undercooked chicken. *Lancet* 2 (8614), 779–782.

Sumner, S. S., Sandros, T. M., Harmon, M. C. et al. (1991). Heat resistance of *Salmonella typhimurium* and *Listeria monocytogenes* in sucrose solutions of various water activities. *J Food Sci* 56, 1741–1743.

Tompkin, R. B. (1999). *Petition to USDA-FSIS Seeking Approval for Use of Increased Concentrations of Sodium Diacetate and Sodium Acetate in Meat and Poultry Products.* Letter dated March 24. USDA-FSIS, Hearing Clerk's Office, Washington, DC.

Tompkin, R. B., Christiansen, L. N., Shaparis, A. B. et al. (1992). Control of *Listeria monocytogenes* in processed meats. *Food Austral* 44, 370–376.

Tompkin, R. B., Scott, V. N., Bernard, D. T. et al. (1999). Guidelines to prevent post-processing contamination from *Listeria monocytogenes*. *Dairy, Food & Environ San* 19, 551–562.

USDA (US Department of Agriculture) (1995a). *Nationwide Raw Ground Chicken Microbiological Survey. March 1995–May 1995.* Washington, DC: Microbiology Division, Food Safety and Inspection Service, US Department of Agriculture.

USDA (US Department of Agriculture) (1995b). *Nationwide Raw Ground Turkey Microbiological Survey. January 1995–March 1995.* Washington, DC: Microbiology Division, Food Safety and Inspection Service, US Department of Agriculture.

USDA (US Department of Agriculture) (1996). *Nationwide Federal Ground Beef Microbiological Survey. August 1993–March 1994.* Washington, DC: Microbiology Division, Food Safety and Inspection Service, US Department of Agriculture.

van Schothorst, M. (1996). Sampling plans for *Listeria monocytogenes*. *Food Control* 7, 203–208.

Wang, C. & Muriana, P. M. (1994). Incidence of *Listeria monocytogenes* in packages of retail franks. *J Food Prot* 57, 382–386.

Wederquist, H. J., Sofos, J. N. & Schmidt, G. R. (1994). *Listeria monocytogenes* inhibition in refrigerated vacuum packaged turkey bologna by chemical additives. *J Food Sci* 59, 498–500, 516.

Wederquist, H. J., Sofos, J. N. & Schmidt, G. R. (1995). Culture media comparison for the enumeration of *Listeria monocytogenes* in refrigerated vacuum packaged turkey bologna made with chemical additives. *Lebens Wissensch Technol* 28, 455–461.

Wenger, J. D., Swaminathan, B., Hayes, P. S. et al. (1990). *Listeria monocytogenes* contamination of turkey franks: Evaluation of a production facility. *J Food Prot* 53, 1015–1019.

Zaika, L. L., Palumbo, S. A., Smith, J. L. et al. (1990). Destruction of *Listeria monocytogenes* during frankfurter processing. *J Food Prot* 53, 18–21.

Chapter 17

E. coli O157:H7 in Frozen Raw Ground Beef Patties

17.1 Introduction
17.2 Risk Evaluation
17.3 Risk Management
17.4 Control Measures
17.5 Acceptance Criteria
17.6 Statistical Implications of the Proposed Sampling Plan
17.7 References

17.1 INTRODUCTION

This chapter is concerned with *Escherichia coli* O157:H7 in frozen raw ground beef patties. The information follows the principles outlined in earlier chapters. The unique severity of the pathogen and its impact among children under five years of age warrant special consideration. In this example, the merits of microbiological testing will be explored as a potential control measure to enhance the safety of ground beef. A significant portion of this material was published as a discussion paper (Tompkin & Bodnaruk, 1999). This example draws primarily on data from the US. The information may be applicable in other countries with similar prevalence rates in ground beef and where undercooking is a risk factor.

Ground beef is an important raw agricultural commodity. In 1992 in the US, the types of cattle from which ground beef was derived were steers, 39.1%; heifers, 19.2%; cows, 34%; and bulls, 7.7%. Imported boneless manufacturing beef composed 15% of the total US ground beef supply. Imports into the US account for almost half of the world trade in fresh, chilled, and frozen beef. Ninety percent of the beef imported into the US is from Australia, New Zealand, and Canada. Other sources include Costa Rica, Honduras, and the Dominican Republic. The annual average quantity of beef imported into the US has been over 600,000 tons (USDA, 1994). Annual per capita consumption was estimated to be about 11.8–12.7 kg from 1982–1998 (AMI, 2000). Ground beef is consumed in the form of hamburger and as a major ingredient in a variety of foods.

E. coli O157:H7 is among the more newly recognized foodborne pathogens, being first identified during an outbreak in 1982 in which ground beef was implicated. It has since been a steadily increasing cause of foodborne illness worldwide. It is recognized that there are numerous serotypes of enterohemorrhagic *E. coli* (EHEC), some of which may be more important in certain regions of the world than *E. coli* O157:H7 (WHO, 1997; Acheson, 2000).

Since 1982, ground beef has become recognized as a major vehicle for transmission of *E. coli* O157:H7 in the US. In general, the pathogen is deposited on the surface of beef carcasses during the slaughtering process. After chilling, the carcasses are cut into large portions for sale as bone-in or boneless beef cuts (e.g., round, loin, rib, chuck). During this process, the larger portions are trimmed of excess fat and other tissue. The trimmings commonly are used in the manufacture of raw ground beef and a wide variety of ready-to-eat products (e.g., sausages). The same process of trimming meat and fat occurs at other steps along the food chain, with much of it being used for ground beef. Wherever it may occur along the food chain, the process of trimming and subsequent grinding distributes the pathogen throughout the ground meat. The most common scenario leading to illness has involved undercooking, survival of the pathogen, and subsequent infection, particularly among the more susceptible consumers. Cross-contamination in kitchens and food service establishments from ground beef to other ready-to-eat foods also has occurred.

It is extremely difficult to compare data from different countries on incidence and prevalence due to the use of different methods of sampling/reporting. However, data for different countries indicate that *E. coli* O157:H7 is an international concern, with the prevalence in meat ranging from around 0.1%–5% and prevalence in cattle ranging from around 1.5%–28% (Table 17–1).

17.2 RISK EVALUATION

17.2.1 Hazard Identification

E. coli is a species of Gram-negative, facultatively anaerobic, rod-shaped bacteria commonly found in the lower part of the intestine of warm-blooded animals. *E. coli* O157:H7 is a particular serotype of the group referred to as enterohemorrhagic *E. coli* (EHEC). This is a subgroup of the verocytotoxigenic *E. coli* (VTEC) that has been shown to cause human illness. VTECs produce verotoxins, or shiga-like toxins, that are closely related to the toxin produced by *Shigella dysenteriae* (Cassin et al., 1998). Most EHEC isolates are acid-tolerant, capable of surviving in acid foods and during passage through the stomach (Benjamin & Datta, 1995; Arnold & Kaspar, 1995; Leyer et al., 1995; Cheville et al., 1996).

Table 17–1 Incidence of *E. coli* O157:H7 Infection and Prevalence in Meat and Cattle in Various Countries

Country	Incidence HUS/VTEC*†	Prevalence (%) in Meat	Prevalence (%) in Cattle
Argentina	HUS 7.8 per 100,000	0.4 bovine meat	
Australia	HUS 2.79 per 100,000		
Canada	VTEC 3.0 per 100,000		
Denmark	VTEC 0.1 per 100,000	0.13 minced beef	1.5
Germany		0.7 beef	
		5.0 minced beef	
Japan		0.3 carcass	1.4
Netherlands	30 per year	0.08 beef/pork	11.1
England and Wales	VTEC 1.29 per 100,000	1.5 beef	15.6
US	0.74 per 100,000	1.0	5
			28‡

*HUS = hemolytic uremic syndrome
†VTEC = verocytotoxigenic *E. coli*
‡using improved isolation methods

Cattle appear to be the main reservoir of *E. coli* O157:H7. Contamination of carcasses during slaughter is the primary route that ultimately leads to contamination of ground beef. Other foods (e.g., lettuce, sprouts, fruit juices, vegetables, raw milk) and water also have been implicated as vehicles of transmission. Person-to-person is an important mode of transmission, particularly in day care centers. Direct contact with animals carrying the organism is also a recognized source of infection (WHO, 1997).

17.2.2 Hazard Characterization

E. coli O157:H7 infection can result in moderate to severe disease or death, with most deaths occurring in children under five years of age and in the elderly (AGA, 1994; Tarr, 1994). Studies conducted in FoodNet sites in the US suggest that from 13–27 cases of infection occur in the community for each confirmed case that is reported (Mead *et al.*, 1999). Within the FoodNet sites, the annual laboratory-confirmed case rate per 100,000 population has been from 2.1–2.8 in the four years of 1996–1999 (CDC, 2000).

The time to onset of symptoms following exposure ranges from three to nine days, with an average of four. The duration of illness ranges from two to nine days; however, with complications, the illness may last many months and lead to permanent damage or even death. Symptoms may include hemorrhagic colitis: grossly bloody diarrhea, severe abdominal pain, vomiting, but no fever; hemolytic uremic syndrome (HUS): prodrome of bloody diarrhea, acute nephropathy, seizures, coma, and death; and thrombotic thrombocytopenic purpura: similar to HUS but also fever and central nervous system disorder (ICMSF, 1996).

The potent toxins of *E. coli* O157:H7 can initially cause a particularly severe form of human disease, hemorrhagic colitis. About 10% of these patients can go on to develop HUS, characterized by acute renal failure, hemolytic anemia, and thrombocytopenia, which is particularly serious in young children and elderly people. On average, 2–7% of patients with HUS die, but in some outbreaks among the elderly, the mortality rate has been as high as 50% (WHO, 1997). Long-term renal dysfunction occurs in about 10–30% of survivors of HUS. The estimate for the prevalence of HUS in North America is about 3/100,000 children under five years of age per year (Mahon *et al.*, 1997).

From outbreak data, it appears that fewer than 100 cells can cause disease among consumers (AGA, 1994), particularly those at greater risk (e.g., less than five years of age) (Mahon *et al.*, 1997). In the 1993 northwestern US outbreak involving undercooked hamburgers, it was estimated that as few as a dozen cells may have caused illness among children (Tarr, 1994).

There are estimated to be about 73,480 cases of illness attributed to *E. coli* O157:H7 per year in the US, of which 85% are estimated to be foodborne. The number of hospitalizations and deaths due to foodborne transmission of this pathogen was estimated to be 1,843 and 52 per year, respectively (Mead *et al.*, 1999). These values differ significantly from the earlier estimates of Mahon *et al.* (1997). The more recent estimate revealed more cases, but fewer deaths. The estimated costs of medical treatment and lost productivity range from $216–580 million/year (AGA, 1994).

17.2.3 Exposure Assessment

The source of *E. coli* O157:H7 on beef carcasses is the hide and intestines of the cattle being slaughtered. Prevalence among cattle being held for slaughter is similar to or slightly higher than the prevalence on the external surface of hides of recently slaughtered animals (Hancock, 1998).

The percent of cattle with *E. coli* O157:H7 in their feces was initially reported to be typically less than 5% (Hancock, 1998). A later study involving a more sensitive analytical method found 28% of the cattle entering slaughtering plants to be actively shedding *E. coli* O157:H7 or nonmotile *E. coli* O157 in their feces during July and August, the months of highest prevalence. Eleven percent of the hide surfaces also were positive (Elder et al., 2000).

Studies have shown that colonization of cattle is of short duration (1–2 months); long-term carriers have not been found. The typical pattern of shedding in a herd followed over time is one of epidemics of shedding interspersed with longer periods with rare or non-shedding animals. These epidemics occur mainly during warm weather, suggesting that environmental proliferation may play an important role in the epidemiology of *E. coli* O157:H7 (Hancock, 1998). It is important to note that *E. coli* O157:H7 does not cause any adverse effects in cattle. Its presence in a herd or individual animal can be detected only by microbiological testing.

While some cross-contamination may occur from carcass-to-carcass through contact with common equipment and workers' hands, there is no published evidence that *E. coli* O157:H7 has ever become established and multiplied in a slaughtering/chilling/fabrication operation and contaminated subsequent lots of beef.

Technology is currently available that can:

- minimize carcass contamination during the slaughtering process,
- reduce the likelihood of microbial attachment to exposed tissues, and
- decontaminate carcasses, such as with steam, hot water, or organic acid sprays prior to chilling.

Collectively, the above control measures can significantly reduce, but not eliminate, the likely presence of enteric pathogens on raw beef. Systems for decontamination will likely involve multiple control measures during slaughtering, evisceration, and chilling (Dorsa et al., 1996; Dorsa, 1997; Castillo et al., 1998; Nutsch et al., 1998; Sofos & Smith, 1998; Sofos et al., 1999; Bacon et al., 2000).

It is reasonable to assume that when present, *E. coli* O157:H7 is on the surface of carcasses, not in internal tissues. Rapid chilling of adequately spaced carcasses should retard *E. coli* O157:H7 growth on carcass surfaces. Surface dehydration during chilling is an additional factor that can restrict growth.

After chilling, it is not likely that *E. coli* O157:H7 will multiply during subsequent fabrication because the lower temperature limit for multiplication of *E. coli* O157:H7 is 7–8 °C (ICMSF, 1996). Furthermore, estimates for growth under optimum laboratory conditions in broth indicate that:

- at 10 °C, the time to increase 10-fold is estimated to be 73 h, and
- at 15 °C, the time to increase 10-fold is estimated to be 25 h (Buchanan & Whiting, 1996).

Thus, in controlled chilling/fabrication operations, the concentration of *E. coli* O157:H7 should not increase. Conversely, in operations that do not address chilling/fabrication as important factors requiring control, some degree of multiplication may occur.

Epidemiologic data for outbreaks reported each year by the Centers for Disease Control and Prevention (CDC) indicate the risk of *E. coli* O157:H7 from beef continues to be associated with consumers who have not changed their handling/cooking habits to control the risk of the pathogen. These consumers do not understand the risks and/or do not have the knowledge to deal with a pathogen such as *E. coli* O157:H7. Some may be aware of the risk and have chosen to ignore recommendations for proper handling and cooking of ground beef. A 1996 survey indicated that 19.7% of the population consumed pink (undercooked) hamburger at some time during the previous 12 months (CDC, 1998a).

In its report of FoodNet for 1997, CDC stated: "In contrast with previous investigations, hamburgers eaten at fast food restaurants were not associated with infection, suggesting that recent changes in that industry may have reduced *E. coli* O157:H7 infections from that source" (CDC, 1998b). This is a conservative assessment because a review of the outbreaks reported each year by CDC shows no reported outbreaks from the larger fast food operations between 1993–2000. This verifies that the control measures adopted in those operations have been effective in managing the risk of *E. coli* O157:H7 in ground beef.

The improved controls in fast food operations do not appear to have carried through to other areas. A case-control study conducted at FoodNet sites has determined that undercooked ground beef continues to be a principal food source of *E. coli* O157:H7 infections (CDC, 1998b). Outbreaks have continued to occur in other food service establishments (e.g., restaurants, schools) and among consumers who prepare ground beef at home, on camping trips, or in other settings. Ground beef is the primary vehicle, not roasts, steaks, etc.

This information suggests that a continuing effort should be directed toward educating food handlers and consumers in proper handling and cooking of ground beef before serving. A goal should be established to reduce the percentage of consumers who consume rare ground beef from approximately 20% to some lower value. School systems and other food service establishments should consider additional training/education and other options (e.g., precooked beef products).

17.2.3.1 Survey Results from USDA-FSIS

The US Department of Agriculture (USDA)-Food Safety and Inspection Service (FSIS) established a sampling and testing program for ground beef in 1994 to:

- stimulate industry actions to reduce the presence of *E. coli* O157:H7 in raw ground beef (i.e., to encourage industry to institute and maintain effective control measures),
- encourage industry to sample and test raw ground beef routinely,
- find and remove from commerce product containing *E. coli* O157:H7, and
- expand the agency's information base and understanding in the control of *E. coli* O157:H7 (FSIS, 1994).

Results of the sampling and testing program are summarized in Table 17–2. The methods of sampling and analysis have been modified since 1995 to increase sensitivity of detection. During 1995–1997, a single 25 g sample was analyzed. Since the beginning of fiscal year 1998, five 65 g samples have been analyzed.

During the 1990s, industry implemented new control measures to reduce the occurrence of salmonellae and *E. coli* O157:H7 in ground beef. Due to the increased sensitivity of the methods used by FSIS, it is not possible to assess the effectiveness of changes implemented by industry or to determine whether the prevalence of *E. coli* O157:H7 in ground beef has, in fact, increased since the inception of the sampling program.

Table 17–2 Results from the FSIS Sampling Program Beginning with Fiscal Year 1995

Fiscal Year	No. Samples	No. Positive	% Positive
1995*	5291	3	0.057
1996*	5326	4	0.075
1997*	5919	2	0.034
1998†	7529	14	0.19
1999†	8710	29	0.33
2000‡	>4,000	38	0.95

*25 g analytical units; †325 g analytical units, ‡325 g and improved isolation method (i.e., immunomagnetic beads to concentrate cells). The results for FY2000 are through August 11, 2000.

17.2.3.2 Data from Ground Beef Implicated in Illness and Lots Found Positive by USDA-FSIS

As ground beef has been implicated in outbreaks, samples from some implicated lots have been collected and analyzed. The ability to detect additional positive samples from across implicated lot(s) indicates a relatively high prevalence and concentration. In a few instances, a quantitative analysis was performed to determine the concentration of *E. coli* O157:H7. For example, in the 1993 outbreak that occurred in the northwestern US, frozen ground beef patties produced at a particular facility on November 19 and 20, 1992 were found positive. Positive samples were found in lots 4 and 9–17 in the production of November 19 and in lot 7 in the production of November 20. Quantitative analysis of some of the positive lots conducted by FSIS and CDC resulted in most probable number values in the range of 1–4 cells g^{-1}, with a single high value of 15 g^{-1} (FSIS, 1993).

Other evidence suggesting a relatively high prevalence and concentration of *E. coli* O157:H7 in certain lots of ground beef includes:

- large numbers of cases among consumers,
- multiple outbreaks involving consumers in different locations, and
- ability to detect *E. coli* O157:H7 with just a single sample from an implicated lot.

Follow-up sampling of positive lots of ground beef detected at retail stores has led to a positive sample from coarse ground beef used.

Data from the USDA-FSIS and lots implicated in outbreaks suggest that the vast majority of ground beef has a very low prevalence and concentration of *E. coli* O157:H7. In rare instances, a very small proportion of the lots contains a relatively high prevalence and concentration. It is because of this kind of skewed distribution in the number of bacteria that a log-normal distribution function is often used in microbiology. A hypothetical log-normal distribution of counts with a standard deviation would give the curve shown in Figure 17–1. Baseline studies on the numbers of bacteria in raw meat and poultry in the US and Canada provide further support to the general applicability of a log-normal distribution for the microbial population in these foods (FSIS, 1996; CFIA, 2000). When a log-normal distribution is assumed for these data, it is estimated that the standard deviation would be between 1.3 and 0.55. Figure 17–2 shows an estimated log-normal distribution (mean = 1.2, sigma = 0.8) from counts obtained in the baseline study of the number of Biotype I *E. coli* in raw ground beef in the US compared to the

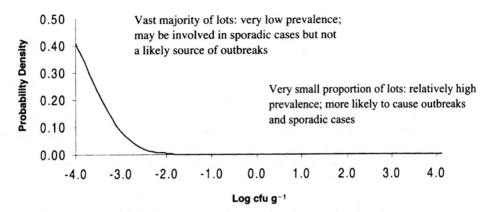

Figure 17–1 Illustration of the likely distribution in prevalence and concentration of *E. coli* O157:H7 in ground beef in North America.

hypothetical distribution for the counts of *E. coli* O157:H7 in raw ground beef. To consider the uses and limitations of sampling for *E. coli* O157:H7 in raw ground beef, we have assumed that the counts for this organism also have a log-normal distribution with a standard deviation value of 0.8, but a mean concentration < 1.2 g^{-1}. Although hypothetical, such a distribution of counts would be consistent with a prevalence of 1% positive in 325 g samples taken.

17.2.3.3 Lots with Low Prevalence and Concentration

The vast majority of ground beef has a very low prevalence and, presumably, low concentration of *E. coli* O157:H7. For example, the results of the FSIS sampling program suggest a background prevalence of about 1% or less when analyzing 325 g. This prevalence is likely influenced by the prevalence of *E. coli* O157:H7 that occurs in cattle at the time of slaughter and also reflects the level that may be normally achievable from slaugh-

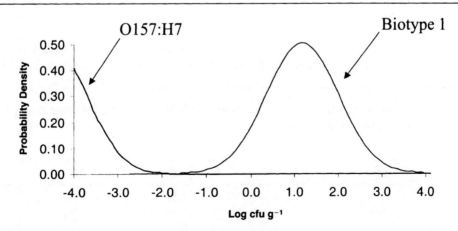

Figure 17–2 Comparison of hypothetical distribution in prevalence and concentration of *E. coli* O157:H to fitted distribution for *E. coli* Biotype I (mean = 1.2, sigma = 0.8) taken from the U.S. Nationwide Raw Beef Microbiological Survey Aug 1993–Mar 1994.

ter to the manufacture of ground beef with existing technology. These lots have such a low prevalence of contamination that defective lots cannot be detected with any degree of confidence through routine microbiological testing. Possibly, these lots may be involved in sporadic cases, but seldom with outbreaks.

Considering the low prevalence, it is questionable that the FSIS sampling program alone, with 5,000–7,000 samples/year, would have a measurable impact on the number of cases/100,000/year attributable to *E. coli* O157:H7 in ground beef. While perhaps ineffective in ensuring consumer protection, the FSIS program does send a strong message to industry and consumers as to its expectations. This has been one factor that has resulted in the implementation of improved control measures in slaughtering operations.

It would be impractical to implement a routine sampling plan to detect and reject contaminated lots with a ≤ 1% prevalence of contamination. For example, if the prevalence rate was 0.7%, the number (n) of samples required to detect the pathogen with a 95% probability would be 428 units. Even sampling to provide a 90% confidence level would require 329 sample units. Additional information describing the difficulty of detecting positive lots with low prevalence of contamination is evident in Table 17–3.

If the defect level is 0.5% and 30 sample units are tested, there is an 86% probability that all 30 samples will be found negative and the lot will be accepted. Even with 100 sample units, there is a 61% probability that all 100 samples will be found negative and the lot will be accepted. Clearly, microbiological sampling and testing are very ineffective for detecting lots with low prevalence of contamination.

17.2.3.4 Lots with Relatively High Prevalence

Experience indicates that a very small proportion of lots of ground beef have a relatively high prevalence (e.g., 5% or higher) and, presumably, higher concentration of cells (e.g., 1–10 g^{-1}). In these lots, it has been possible to find additional positive samples when resampled. Published information does not explain how these lots acquire a higher prevalence or concentration of *E. coli* O157:H7. It is strongly suspected that contamination results when trimmings from one or a few carcasses are introduced into the grinding process with other trimmings and cause a "comet-like" effect as described in Chapter 10 and Figure 10–4c. It is probable that these lots would more likely cause outbreaks as well as sporadic cases, if the product is undercooked or other foods are contaminated.

Due to the higher prevalence of contamination in these lots, it may be possible to apply a statistically valid sampling plan and, to the extent possible, exclude them from the mar-

Table 17–3 Probability of Accepting a Defective Lot with Indicated Proportion of Defective Sample Units

	No. of sample units			
% defective	15	30	60	100
0.1	0.99	0.97	0.94	0.90
0.5	0.93	0.86	0.74	0.61
1	0.86	0.74	0.55	0.37
2	0.74	0.55	0.30	0.13
5	0.46	0.21	0.05	0.01

ket. Since it is not possible to anticipate which lots may be in this category, the intent of the sampling plan would be to detect them as they are produced.

17.2.4 Risk Characterization

The annual per capita consumption of ground beef in the US was estimated to be approximately 12 kg from 1982–1998 (AMI, 2000). If it is assumed that all the ground beef is consumed as hamburger patties weighing 125 g each, this would result in 96 servings/person/year. Since the total population in the US in 1998 was approximately 270 million, the total number of servings on a national basis would be about 26×10^9 servings per year. If the USDA survey data for FY 2000 are used, it can be assumed that the prevalence rate for *E. coli* O157:H7 in ground beef is about 1%. This prevalence rate was determined using five analytical units, each weighing 65 g. A positive result for any of the five analytic units was reported as a positive 325 g.

Data are not available for 125 g samples; however, for this example it will be assumed that the prevalence rate in 125 g samples is 0.7%. Given that assumption, it can be estimated that the number of raw hamburger patties containing *E. coli* O157:H7 each year is 26×10^7 (i.e., 1% of 26×10^9 servings). While a survey indicated that 19.7% of the population had consumed pink (undercooked) ground beef at some time during the previous 12 months, it is not likely that this occurred with each serving. For example, individuals purchasing hamburgers in fast food operations would very likely receive a hamburger that is well done and with no evidence of undercooking. This would significantly reduce the extent of exposure to undercooked hamburgers containing *E. coli* O157:H7. If it is assumed that undercooked hamburgers were consumed 5% of the time among the respondents of the survey, then, it can be estimated that the number of undercooked servings that contained *E. coli* O157:H7 before cooking would be about 26×10^5 [i.e., $5\% \times 19.7\% \times (26 \times 10^7)$]. If it is further assumed that the concentration of *E. coli* O157:H7 in the patties before cooking is 1 g^{-1} (H_0), then the number of cells would be 125/patty and their distribution within the patties would influence whether they are killed during cooking. It has been reported that an expected D value for *E. coli* O157:H7 in ground beef with 30.5% fat is 0.47 min at 62.8 °C (Line *et al.*, 1991). Cooking beef to this temperature typically yields a medium-rare (i.e., not raw) appearance. It is not likely that *E. coli* O157:H7 with an initial level of 1 g^{-1} would survive in a patty that is cooked to 62.8 °C unless, perhaps, all the cells are at the site of slowest heating within the patty. Thus, it may be possible for patties to appear undercooked (medium-rare), but have no viable cells remaining. As with many of the assumptions in this example, data are not available for the frequency of undercooking. Thus, it will be assumed for this example that 10% of the undercooked patties are insufficiently cooked to kill all the cells that are present. This leads to an estimate of approximately 26×10^4 patties/year potentially being consumed in the US that contain viable *E. coli* O157:H7.

The extent to which these patties would lead to illness would be influenced by many factors, in particular the susceptibility of the consumer. It is not likely that the most highly sensitive population, namely children less than five years of age and the elderly, would consume hamburger at a rate of 96 servings/year. In addition, these estimates assume that all the ground beef is consumed as patties. These estimates also ignore the fact that a significant portion of the patties are cooked in USDA inspected facilities and in fast food operations that apply control measures to ensure a minimum internal temperature of 70 °C.

17.3 RISK MANAGEMENT

17.3.1 Level of Consumer Protection

The exposure assessment revealed that about 25% of the *E. coli* O157:H7 outbreaks during 1982–1997 and 33% of illnesses were attributable to ground beef (FSIS, 1998). An estimate for sporadic cases is not available. Using those data, and assuming that there are 62,458 foodborne cases of *E. coli* O157:H7 per year, then the number of cases attributable to ground beef would be approximately 20,611 cases/year. An example of a public health goal for this food-pathogen combination would be to reduce or eliminate the number of cases attributable to ground beef. A 25% reduction in the number of cases attributed to ground beef, for example, would mean a reduction of 5,152 cases/year.

17.3.2 Establishing a Food Safety Objective

In this example, it is difficult to establish a clear relationship between the level of *E. coli* O157:H7 consumed and the probability of illness. However, outbreak data (AGA, 1994; Tarr, 1994) and risk estimates (Cassin *et al.*, 1998) suggest that fewer than 100, to as low as a dozen or so cells, can cause illness among consumers, particularly those at greater risk. Therefore, to establish a food safety objective (FSO) for this example, it is important for the value to be necessarily conservative to reflect the degree of uncertainty, the relatively low infective dose, and the severity of the illness. An example of an FSO would be:

The concentration of *E. coli* O157:H7 in raw ground beef shall not exceed 1 cfu 250 g^{-1}.

Since a concentration of 1 cfu 250 g^{-1} also can be expressed as -2.4 $\log_{10}$ cfu g^{-1}, the FSO also could be stated as the concentration of *E. coli* O157:H7 in raw ground beef shall not exceed -2.4 $\log_{10}$ cfu g^{-1}. The FSO is equivalent to no greater than 1 cell per two patties with each weighing 125 g, a common weight for commercially manufactured ground beef patties.

The proposed FSO should provide a margin of safety considering the thermal destruction that occurs even in undercooked patties. This FSO is only applicable to ground beef for sale in retail stores or for use in food service establishments. Ground beef for manufacturing products cooked in commercial establishments with Hazard Analysis Critical Control Point (HACCP) systems need not meet the proposed FSO because the required cooking processes have been validated to ensure destruction of at least 10^5 g^{-1} salmonellae or *E. coli* O157:H7.

17.4 CONTROL MEASURES

17.4.1 Good Hygienic Practices (GHP)/HACCP

A generic model HACCP plan has been developed for teaching the principles of HACCP to those involved in the manufacture of frozen ground beef patties (Bernard *et* al., 1997). The document discusses the strengths and limitations of GHP and HACCP for control of enteric pathogens in ground beef and provides guidance that may facilitate meeting the FSO by manufacturers of ground beef.

17.4.2 Controlling Initial Number (H_0)

The only means to control the concentration of enteric pathogens in raw ground beef is through raw material selection. The options would be limited to (1) supplier assurance programs and (2) pretesting trimmings before use. In addition, maintaining an effective time-temperature control program is needed to prevent growth of *E. coli* O157:H7 on carcasses during chilling/cutting and through grinding/freezing the ground beef patties.

17.4.2.1 Control through Supplier Assurance Programs

Control through supplier assurance programs entails selecting suppliers who have implemented interventions to minimize carcass contamination (e.g., enclosing bung in plastic bag and tying, an improved method of hoof and hide removal, use of steam vacuum on areas likely to be contaminated) during the slaughtering process and/or to decontaminate carcasses (e.g., hot water or steam pasteurization, organic acid spray) at strategic steps in the slaughtering process. As described earlier, it is likely that these control measures can significantly reduce, but will not likely eliminate, the presence of enteric pathogens (Dorsa et al., 1996; Dorsa, 1997; Castillo et al., 1998; Nutsch et al., 1998; Sofos & Smith, 1998; Sofos et al., 1999; Bacon et al., 2000).

Suppliers should be audited following the guidelines in Chapter 4 and the results used as a basis for approval. Familiarity with the current use of interventions and their effectiveness should be incorporated into the audit. In addition, certain interventions (e.g., steam pasteurization, hot water sprays) may have been validated to achieve a minimum level of reduction under specified conditions of operation.

17.4.2.2 Control through an Incoming Raw Materials Screening Program

Control through an incoming raw materials screening program consists of using microbiological tests of incoming raw materials to identify and select the better, more reliable suppliers and to eliminate suppliers who exercise less control of microbial quality (Bernard et al., 1997; Eisel et al., 1997; Scanga et al., 2000). These tests could involve indicator microorganisms as a general measure of GHP and good manufacturing practices.

Trimmings to be used for ground beef could be pretested for *E. coli* O157:H7 by various sampling plans (AMSA, 1999; Scanga et al., 2000). One option would be to have the supplier sample the trimmings and have them analyzed while the raw materials are in transit. Slaughtering operations that use their own in-house trimmings to manufacture frozen ground beef patties could sample the trimmings before grinding.

Producers of ground beef should consider how much reliance to place on the interventions employed by suppliers and whether it is necessary, or effective, to test for *E. coli* O157:H7. If testing is used as a control measure, then consideration must be given to testing raw materials (i.e., trimmings) or finished product (i.e., frozen patties).

To meet the required FSO through raw material selection, the concentration of *E. coli* O157:H7 in the trimmings would have to be controlled so the ground beef will meet the FSO (i.e., FSO = $-2.4 \log_{10}$ or not exceed 1 per 250 g). This can be expressed in the following simple equation (described fully in Chapter 3).

$$H_0 - \Sigma R + \Sigma I \leq FSO$$
$$H_0 - 0 + 0 \leq -2.4$$
$$H_0 \leq -2.4$$

Thus, the initial concentration of *E. coli* O157:H7 in the frozen ground beef patties must be $-2.4 \log_{10}$ cfu g^{-1} and, if present, there must be no growth ($\Sigma I = 0$).

There has been only one report of beef with a concentration of *E. coli* O157:H7 as high as 1,000/g. That occurred in a very small operation (i.e., a locker plant) that processed individual animals for local customers (Todd et al., 1988). Such operations would be less able to exercise control over the slaughtering/chilling/fabrication process. In larger commercial operations, however, lower numbers of *E. coli* O157:H7 in trimmings and, ultimately, in ground beef would be expected. This would reflect the fact that few carcasses would carry the pathogen and some dilution of the pathogen would occur as trimmings are mixed with the large quantities of negative product. For these reasons, it may be assumed that the concentration of *E. coli* O157:H7 in trimmings from these operations would not likely exceed 100 g^{-1}. Even in the outbreak that occurred in the northwestern US, the highest concentration detected was 15 g^{-1}. Therefore, it is reasonable to assume that the initial level is unlikely to exceed 100 g^{-1} (i.e., $H_0 = 2$).

17.4.3 Reducing Levels (ΣR)

If an initial concentration of 100 g^{-1} is assumed as a worst case, then a performance criterion can be established for the reduction necessary to achieve the FSO of $-2.4 \log_{10}$ cfu g^{-1} (1 cell per 250 g) at the time the patties are consumed. This can be expressed as follows:

$$H_0 - \Sigma R + \Sigma I \leq FSO$$
$$2 - \Sigma R + 0 \leq -2.4$$
$$\Sigma R = 4.4$$

Thus, the performance standard for cooking the frozen patties could be stated as:

a 4.4D reduction of *E. coli* O157:H7 in the coldest area of the patty.

If the initial level (H_0) can be controlled to a lower level (e.g., $H_0 = 1$), then the performance criterion could be met with a milder cooking process (e.g., $\Sigma R = 3.4$).

The only means available to the ground beef producer to ensure that the patties will be cooked sufficiently to meet the FSO in restaurants, institutions, hotels, and other commercial operations would be to provide cooking instructions. This also would be necessary for consumer package labeling, but it is less certain whether consumers follow these instructions. In some regions of the world, regulations specify minimum cooking temperatures for ground beef. In the US, the Food Code (FDA, 1999) recommends that ground beef be cooked to an internal temperature of 66 °C for 1 min, 68 °C for 15 s or 70 °C for < 1 s. These time-temperature relationships provide a 6.5D or greater reduction for salmonellae, a pathogen of comparable heat resistance to *E. coli* O157:H7. Many states have adopted these recommendations. Food service operations that comply with this requirement would be applying a validated cooking process that would exceed the above performance standard. One reason for the greater thermal destruction required in the Food Code is that a value of 1000 g^{-1} was assumed for the estimated initial concentration of *Salmonella* and *E. coli* O157:H7 in ground beef ($H_0 = 3$), and a minimum 6D reduction was desired ($\Sigma R = 6$).

17.4.4 Options Involving Research and Education

Information from the above risk evaluation could lead to several additional risk management options. Examples could include:

- support research that may lead to on-farm control measures (e.g., competitive exclusion, vaccine, control of water supply for cattle). This option should be viewed as a long-term endeavor.
- support research that may lead to additional in-plant control measures to minimize contamination during slaughter and/or decontamination of carcasses.
- provide education to food handlers, particularly those in higher risk situations (e.g., non-fast food service establishments, homes), for proper handling and preparation of ground beef for serving.
- encourage the use of a thermometer and discourage reliance on appearance for estimating doneness when cooking ground beef.
- evaluate the barriers to acceptance of irradiation of ground beef and assess how they may be overcome.
- support research on new technologies to reduce the level of *E. coli* O157:H7 in ground beef while retaining acceptable quality.

17.5 ACCEPTANCE CRITERIA

17.5.1 Organoleptic Criteria

The presence of *E. coli* O157:H7 in raw ground beef cannot be assessed by organoleptic evaluation and, thus, meaningful organoleptic criteria cannot be established.

17.5.2 Chemical and Physical Criteria

The presence of *E. coli* O157:H7 in raw ground beef cannot be assessed by chemical or physical determinations and, thus, meaningful chemical or physical criteria cannot be established.

17.5.3 Microbiological Criteria

The following considers the merits of testing frozen ground beef patties to verify that the FSO has been met by the manufacturer.

While it may not be possible to detect with confidence lots with a low prevalence of contamination, it may be possible to detect some lots of higher prevalence that are more likely to be associated with outbreaks. Thus, one risk management option may be to implement a routine sampling program specifically intended to detect those lots of relatively higher prevalence and, perhaps, concentration. Rare as those lots may be, they are more likely to be associated with illness. It must be emphasized, however, that such a sampling and testing program cannot guarantee the safety of lots that test negative!

The decision to test the frozen ground beef patties as a risk management option can be influenced by many factors, such as:

- the potential public health impact, particularly to young children, in the event *E. coli* O157:H7 is present.
- the intended customer (e.g., food service or home) and intended use (i.e., likelihood of undercooking).
- the effectiveness of product labeling and supporting education to reduce the likelihood of mishandling and undercooking.
- costs associated with illness and liability should product be implicated.

17.5.3.1 Choice of Sampling Plan

The sampling plans recommended in Chapter 8 consider risk to the consumer, including the effect certain microbial agents may have on certain sensitive populations. *E. coli* O157:H7 has been placed into a classification that requires the most stringent sampling plans (i.e., cases 13–15) and 15, 30, or 60 sample units from a lot.

The decision to select 15, 30, or 60 samples depends on whether the hazard would be expected to decrease, not change, or increase between when a food is sampled and when it is consumed (NAS-NRC, 1969; Foster, 1971; NRC, 1985; ICMSF, 1986).

E. coli O157:H7 cannot multiply in frozen ground beef patties; extended holding times above 7–8 °C are needed for even a 10-fold increase. Thus, case 15 does not apply.

Since cooking will kill *E. coli* O157:H7, this suggests that a sampling plan of $n = 15$, $c = 0$ (case 13) would be appropriate. However, approximately 20% of consumers continue to consume rare ground beef. While it would be expected that partial cooking may reduce the number of *E. coli* O157:H7, cooking to a rare state cannot be depended on to eliminate the hazard. This leads to a sampling plan of $n = 30$, $c = 0$ (case 14) as being appropriate for raw ground beef. Recovery of *E. coli* O157:H7 from frozen ground beef implicated in outbreaks is evidence that freezing cannot be relied upon to eliminate the pathogen, further supporting selection of case 14.

It is important to reemphasize the limitations of testing food for the purpose of detecting and eliminating positive lots. For example, even with the proposed sampling plan of $n = 30$, $c = 0$, there is a 21% probability of accepting a lot when the proportion of positive sample units throughout the lot is 5%. There is a 74% probability of accepting a lot when the proportion of positive sample units is 1% (see Table 17–3).

17.5.3.2 Lot

The National Academy of Sciences-National Research Council Salmonella Committee (NAS-NRC, 1969) defined a lot as:

- an identifiable and identified unit of production, distinguishable from other units, that has been packaged
- within a determined period of time
- without major interruptions of flow, shutdowns, or other changes (such as the use of different sources of raw materials) that could be expected to cause one portion of the lot to differ significantly in integrity from another and that is
- complete and
- accessible for inspection and testing.

The ICMSF has defined a lot as follows:

> In the commercial sense, a lot is a quantity of food supposedly produced under identical conditions, all packages of which would normally bear a lot number that identifies the production during a particular time interval, and usually from a particular line, retort, or other critical processing unit. Statistically, a lot is considered as a collection of units of a product from which a sample is drawn to determine acceptability of the lot (ICMSF, 1986).

In this example, the maximum size of a lot consists of all the ground beef produced on common equipment from cleanup to cleanup. Smaller lots could be established (e.g., hourly).

17.5.3.3 Sample and Analytical Units

A sample unit is a representative sample collected at random from a lot that is a small-scale replica of the lot itself and from which a smaller analytical unit may be drawn for analysis.

- The sample unit is one frozen patty.
- The analytical unit is 25 g.

Ideally, the sample units would be submitted to a certified laboratory and analyzed by an official method, or its equivalent, for *E. coli* O157:H7. The method can be adjusted for analysis of 15 analytical units into a single 375 g composite sample. Research has validated that up to fifteen 25 g analytical units of ground beef or trimmings can be composited without significant loss in *E. coli* O157:H7 detection sensitivity (Silliker & Nickelson, 1995). This permits testing at a more reasonable cost and makes it more practical to screen lots to detect those of higher prevalence. The statistical implications of the proposed sampling plan are discussed in the following section.

17.5.3.4 Disposition of Positive Lots

The definition of a lot, the amount of product in a lot, and means of disposition are important considerations in the disposition of a positive lot. In this example, a lot consists of all product produced on common equipment from cleanup to cleanup, including rework. Rework consists of deformed, broken, or underweight patties. Rework can be returned to the grinder or blender on a frequent basis during operation, with the portion remaining at the end of production being discarded or handled separately. None would be added into the next day's production. Possible risk management options for positive lots include diverting the ground beef for use in products to be cooked in a facility that has an effective control system (i.e., GHP, HACCP) to ensure proper handling and destruction of the pathogen. Cooking can provide a substantial margin of safety that *E. coli* O157:H7 will be killed. It is recommended, however, that such lots not be used for the manufacture of fermented meats or in high temperature-short time processes, such as for links, patties, or meatballs.

17.6 STATISTICAL IMPLICATIONS OF THE PROPOSED SAMPLING PLAN

One premise of this example is that sampling can be used to screen out some, but not all, lots that exceed the FSO. A major weakness in this premise is that contamination is more likely to be heterogeneous as illustrated in Figures 10–4b and c. The following discussion assumes, however, that contamination is randomly distributed (i.e., log-normal distribution and standard deviation of 0.8).

The proposed sampling plan involves analysis of 25 g from each of 30 sample units from a lot. A negative result provides 95% confidence that the concentration of *E. coli* O157:H7 in the lot is no more than 1 cell per 250 g (NAS-NRC, 1969; Foster, 1971). In addition, there is a 95% probability that if sampling and testing were extended indefinitely, the positive units found would not exceed 10%, equivalent to a lot average of fewer than one cell in 250 g (NAS-NRC, 1969). The underlying principles for relating the performance of a given sampling plan are explained in Legan *et al.* (2001) and in Chapter 7.

The sampling plans originally proposed by the NAS (NAS-NRC, 1969) and the Interagency-Industry Committee (Foster, 1971) were intended to be applied to suspect lots of food. They were to be used to supplement, not replace, existing routine tests being performed by companies to assess the acceptability of their products. Some companies, however, have adopted the sampling plans for end-product testing where the conditions of operation warranted concern. Similarly, the ICMSF has recommended the use of the appropriate sampling plans as a means to assess incoming lots of food at port-of-entry when no history of the conditions of production and storage is known.

Experience with the foregoing sampling plans has been favorable. Rarely has a food been found to cause foodborne illness, such as salmonellosis, when a lot was sampled and found acceptable. Conversely, foods implicated in foodborne illness normally have been found positive with the sampling plans. For about 30 years, sampling plans involving 15, 30, and 60 sample units from a lot have been accepted and applied by both industry and certain control authorities.

While it may be possible to rationalize the use of 30 sample units for determining the acceptability of a lot, the definition of a lot and the procedure for how and when samples should be collected from a lot are more difficult to justify on a statistical basis. Admittedly, the proposal provides for sampling frequencies that consider the need for procedures to be reasonable in a commercial operation. Confounding the issue is the uncertain distribution of *E. coli* O157:H7 throughout an operating day from cleanup to cleanup. The following from The Committee on *Salmonella* provides some guidance (NAS-NRC, 1969) on this concern.

The matter of *Salmonella* control is complicated by the question of distribution of salmonellae when they do occur in a food product. If this defect is random, it is not controlled by time; i.e., there is an equal opportunity for contamination to occur at any stage of the operation. If, on the other hand, contamination was limited to a certain segment of time during processing, then the defect would not be randomly distributed throughout the lot. If distribution is random, regular sampling procedures will detect the organism; if nonrandom, there is virtually no assurance of control unless the regulatory nature of time is known.

As previously indicated, in sampling a lot, random sampling is recommended. A random sample is one in which any individual tested is as likely to detect contamination as any other. This does not mean that one sample at one given time gives a measure of control of finished product, but rather that selecting samples at random during processing gives information on the lot. It is important that the lot be so defined that the foregoing requirements are met (NAS-NRC, 1969).

In conclusion, the sampling plan proposed above may be useful for screening out lots with a high prevalence of *E. coli* O157:H7 that exceed the FSO ($-2.4 \log_{10}$ cfu g^{-1}). How-

Table 17-4 The Relationship between the Number of 25 g Samples That Would Need To Test Negative To Control a Given Concentration (See Also Figure 17-3)

No. of 25 g samples testing negative	Prevalence (%) needed to detect positives at 95% probability	Log concentration per 25 g controlled at 95% probability	Concentration controlled at 95% probability
60	4.8	−1.33	1/526 g
50	5.8	−1.26	1/455 g
40	7.2	−1.17	1/370 g
30	9.5	−1.05	1/278 g
25	11	−0.97	1/233 g
20	14	−0.87	1/185 g
15	18	−0.73	1/135 g
10	26	−0.52	1/83 g
5	45	−0.10	1/32 g
3	63	−0.27	1/13 g

ever, it is also true that the sampling plan will have a relatively poor ability to detect positive lots where the level of the hazard is not random or is much lower. Another way of explaining this is to consider the likely defect rates. When the mean count is around $-2.4 \log_{10}$ cfu g^{-1}, the defect rate would be around 10% (sigma 0.8). The current defect rate measured in the FSIS sampling program is closer to 1%, which would equate to a mean count of around $-4.5 \log_{10}$ cfu g^{-1}, assuming a similar standard deviation (see Table 17-4 and Figure 17-3). Hence, it is important to again reemphasize the limitations of testing

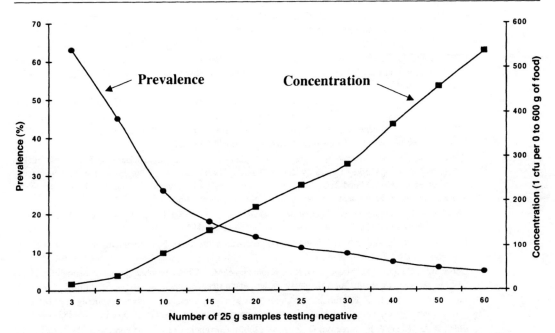

Figure 17-3 Relationship between the number of 25 g samples testing negative and the concentration and prevalence needed before the lot is rejected with 95% probability (assuming sigma = 0.8).

food for the purpose of detecting and eliminating positive lots with a low defect rate. Even with the proposed sampling plan of $n = 30$, $c = 0$, there is a 74% probability of acceptance when the proportion of positive sample units is 1%.

17.7 REFERENCES

Acheson, D. W. K. (2000). How does *Escherichia coli* O157:H7 testing in meat compare with what we are seeing clinically? *J Food Prot* 63, 819–821.

AGA (American Gastroenterological Association) (1994). *Consensus Conference Statement. E. coli O157:H7 Infections: An Emerging National Health Crisis*, July 11–13, 1994.

AMI (American Meat Institute) (2000). *Meat & Poultry Facts* 1999. Washington, DC: American Meat Institute.

AMSA (American Meat Science Association) (1999). *The Role of Microbiological Testing in Beef Food Safety Programs. The Scientific Perspective. Consensus of the 1999 Symposium*. Kansas City, MO: The American Meat Science Association.

Arnold, K. W. & Kaspar, C. W. (1995). Starvation and stationary phase induced acid tolerance in *Escherichia coli* O157:H7. *Appl Environ Microbiol* 61, 2037–2039.

Bacon, R. T., Belk, K. E., Sofos, J. N. *et al.* (2000). Microbial populations on animal hides and beef carcasses at different stages of slaughter in plants employing multiple-sequential interventions for decontamination. *J Food Prot* 63, 1080–1086.

Benjamin, M. M. & Datta, A. R. (1995). Acid tolerance of enterohemorrhagic *Escherichia coli*. *Appl Environ Microbiol* 61, 1669–1672.

Bernard, D. T., Cole, W. R., Gombas, D. E. *et al.* (1997). Frozen, raw beef patties for food service. *Dairy, Food & Environ San* 17, 427–431.

Buchanan, R. L. & Whiting, R. C. (1996). *USDA Pathogen Modeling Program*, version 5.1. Philadelphia, PA: Agricultural Research Service–US Department of Agriculture.

Cassin, M. H., Lammerding, A. M., Todd, E. C. D. *et al.* (1998). Quantitative risk assessment for *Escherichia coli* O157:H7 in ground beef hamburgers. *Int J Food Microbiol* 41, 21–44.

Castillo, A., Lucia, L. M., Goodson, K. J. *et al.* (1998). Comparison of water wash, trimming, and combined hot water and lactic acid treatments for reducing bacteria of fecal origin on beef carcasses. *J Food Prot* 61, 823–828.

CDC (Centers for Disease Control & Prevention) (1998a). Multistate surveillance for food-handling, preparation, and consumption behaviors associated with foodborne diseases: 1995 and 1996 BRFSS food-safety questions. *MMWR Morb Mortal Wkly Rep* 47/No. SS-4.

CDC (Centers for Disease Control & Prevention) (1998b). *FoodNet 1997 Surveillance Results*. http://www.cdc.gov/ncidod/dbmd/foodnet/foodnet.htm

CDC (Centers for Disease Control & Prevention) (2000). Preliminary FoodNet data on the incidence of foodborne illnesses—selected sites, United States, 1999. *MMWR* 49, 201–205.

CFIA (Canadian Food Inspection Agency) (2000). Canadian Microbiological Baseline Survey of Chicken Broiler and Young Turkey Carcasses, June 1997–May 1998. Canadian Food Inspection Agency, Poultry Inspection Programs, Food of Animal Origin, Ottawa, Ontario. (available through website: http://www.inspection.gc.ca/english/anima/meavia/mmopmmhv/chap19/baseline-e.pdf).

Cheville, A. M., Arnold, K. W., Buchreiser, C. *et al.* (1996). rpoS regulation of acid, heat, salt tolerance in *Escherichia coli* O157:H7. *Appl Environ Microbiol* 62, 1822–1824.

Dorsa, W. J. (1997). New and established carcass decontamination procedures commonly used in the beef-processing industry. *J Food Prot* 60, 1146–1151.

Dorsa, W. J., Cutter, C. N., Siragusa, G. R. & Koohmaraie, M. (1996). Microbial decontamination of beef and sheep carcasses by steam, hot water spray washes, and a steam-vacuum sanitizer. *J Food Prot* 59, 127–135.

Eisel, E. G., Linton, R. H. & Muriana, P. M. (1997). A survey of microbial levels for incoming raw beef, environmental sources, and ground beef in a red meat processing plant. *Food Microbiol* 14, 273–282.

Elder, R. O., Keen, J. E., Siragusa, G. R. *et al.* (2000). Correlation of enterohemorrhagic *Escherichia coli* O157 prevalence in feces, hides and carcasses of beef cattle during processing. *Proc Natl Acad Sci* 97, 2999–3003.

FDA (Food and Drug Administration) (1999). *Food Code*. Washington, DC: US Department of Health and Human Services, Public Health Service, Food and Drug Administration.

Foster, E. M. (1971). The control of salmonellae in processed foods: A classification system and sampling plan. *J AOAC* 54, 259–266.

FSIS (Food Safety and Inspection Service) (1993). *Report on the Escherichia coli O157:H7 Outbreak in the Western States*. Washington, DC: US Department of Agriculture.

FSIS (Food Safety and Inspection Service) (1994). Microbiological testing program for *Escherichia coli* O157:H7 in raw ground beef. FSIS Notice 50–94 (12-23-94), US Department of Agriculture, Washington, DC.

FSIS (Food Safety and Inspection Service) (1996). Nationwide Federal Plant Raw Ground Beef Microbiological Survey. August 1993–March 1994. US Department of Agriculture, Science and Technology Division, Microbiology Division, Washington, DC.

FSIS (Food Safety and Inspection Service) (1998). *Preliminary Pathways and Data for a Risk Assessment of E. coli O157:H7 in Beef*. Prepared by the *E. coli* O157:H7 risk assessment team. October 28, Washington, DC.

Hancock, D. D. (1998). *A Summary of* E. coli *O157:H7 Research at Washington State University*. Handout from Governor's Conference on Ensuring Meat Safety, Developing a Research and Outreach Agenda on *E. coli* O157:H7. February 9–10, Lincoln, NB. Pullman, WA: Washington State University.

ICMSF (International Commission on Microbiological Specifications for Foods) (1986). *Microorganisms in Foods 2. Sampling for Microbiological Analysis: Principles and Specific Applications*, 2nd edn. Toronto: University of Toronto Press.

ICMSF (International Commission on Microbiological Specifications for Foods) (1996). Intestinally pathogenic *Escherichia coli*. In *Microorganisms in Foods 5. Characteristics of Microbial Pathogens*, pp. 126–140. Gaithersburg, MD: Aspen Publishers, Inc.

Legan, J. D., Vandeven, M. H., Dahms, S. & Cole, M. B. (2001). Determining the concentration of microorganisms controlled by attributes sampling plans. *Food Control* 12, 137–147.

Leyer, G. J., Wang, L-L. & Johnson, E. A. (1995). Acid adaptation of *Escherichia coli* O157:H7 increases survival in acidic foods. *Appl Environ Microbiol* 61, 3752–3755.

Line, J. E., Fain, A. R., Moran, A. B. *et al*. (1991). Lethality of heat to *Escherichia coli* O157:H7: D-value and z-value determinations in ground beef. *J Food Prot* 54, 762–766.

Mahon, B. E., Griffin, P. M., Mead, P. S. & Tauxe, R. V. (1997). Hemolytic uremic syndrome surveillance to monitor trends in infection with *Escherichia coli* O157:H7 and other Shiga toxin-producing *E. coli*. *Emerging Infect Disease* 3, 409–412.

Mead, P. S., Slutsker, L., Dietz, V. *et al*. (1999). Food-related illness and death in the United States. *Emerging Infect Disease* 5, 607–625.

NAS-NRC (National Academy of Sciences-National Research Council) (1969). *An Evaluation of the Salmonella Problem*. Publication No. 1683, Washington, DC: Committee on *Salmonella*, NAS-NRC.

NRC (National Research Council) (1985). *An Evaluation of the Role of Microbiological Criteria for Foods and Ingredients*. Subcommittee on Microbiological Criteria. Washington, DC: National Academy Press.

Nutsch, A. L., Phebus, R. K., Riemann, M. J. *et al*. (1998). Steam pasteurization of commercially slaughtered beef carcasses: evaluation of bacterial populations at five anatomical locations. *J Food Prot* 61, 571–577.

Scanga, J. A., Grona, A. D., Belk, K. E. *et al*. (2000). Microbiological contamination of raw beef trimmings and ground beef. *Meat Sci* 56, 145–152.

Silliker, J. H. & Nickelson, R. (1995). *Methods for Sampling and Compositing Fresh Red Meat Products for Analysis of Pathogenic and Indicator Bacteria*. Final Report to the National Cattleman's Beef Association (addendum of September 6, 1995). Engelwood, CO: National Cattleman's Beef Association.

Sofos, J. N., Belk, K. E. & Smith, G. C. (1999). Processes to reduce contamination with pathogenic microorganisms in meat. In *Congress Proceedings*, 7-L2 pp. 597–606. 45th International Congress of Meat Science and Technology, August 1–6, Yokohama, Japan.

Sofos, J. N. & Smith, G. C. (1998). Nonacid meat decontamination technologies: Model studies and commercial applications. *Int J Food Microbiol* 44, 171–188.

Tarr, P. I. (1994). *Testimony to Washington State Senate, Department of Agriculture*. January 20, 1994.

Todd, E. C. D., Szabo, R. A., Peterkin, P. *et al*. (1988). Rapid hydrophobic grid membrane filter-enzyme-labeled antibody procedure for identification and enumeration of *Escherichia coli* O157 in foods. *Appl Environ Microbiol* 54, 2536–2540.

Tompkin, R. B. & Bodnaruk, P. W. (1999). A proposed food safety management plan for *E. coli* O157:H7 in ground beef. In *The Role of Microbiological Testing in Beef Food Safety Programs*. Edited by C. Calkins and M. Koohmaraie. Kansas City, MO: American Meat Science Association.

USDA (US Department of Agriculture–Animal and Plant Health Inspection Service–Centers for Epidemiology and Animal Health) (1994). *Escherichia coli O157:H7 Issues and Ramifications*. Fort Collins, CO: US Department of Agriculture, Animal and Plant Health Inspection Service, Centers for Epidemiology and Animal Health.

WHO (World Health Organization) (1997). *Prevention and Control of Enterohaemorrhagic Escherichia coli (EHEC) Infections* (WHO/FSF/FOS/97.6). Geneva: Food Safety Unit, Programme of Food Safety and Food Aid, World Health Organization.

Appendix A

Glossary

Acceptance criteria (for a food operation)
    Statements of conditions that differentiate acceptable from unacceptable food operations.

Acceptance criteria (for lot acceptance)
    Statements of conditions that differentiate acceptable from unacceptable lots (batches) of food.

Control (verb)
    To take all necessary actions to ensure and maintain compliance with established criteria.

Control (noun)
    The state wherein correct procedures are being followed and criteria are being met.

Control measure
    Any action and activity that can be used to prevent or eliminate a food safety hazard or reduce it to an acceptable level.

Corrective action
    Any action to be taken when the results of monitoring at the critical control point indicate a loss of control.

Critical control point (CCP)
    A step at which control can be applied and is essential to prevent or eliminate a food safety hazard or reduce it to an acceptable level.

Critical limit
    A criterion which separates acceptability from unacceptability.

Default criterion
    A conservative criterion established to ensure the safety of a food under worst-case conditions.

Deviation
    Failure to meet a critical limit.

Dose-response assessment
> The determination of the relationship between the magnitude of exposure (dose) to a chemical, biological, or physical agent and the severity and/or frequency of associated adverse health effects (response).

Expert panel
> A group of individuals who collectively have knowledge or experience with a hazard or food and the conditions that can lead to foodborne illness, and who have the ability to provide advice based on available scientific information.

Exposure assessment
> The qualitative and/or quantitative evaluation of the likely intake of biological, chemical, and physical agents via food, as well as exposures from other sources if relevant.

Flow diagram
> A systematic representation of the sequence of steps or operations used in the production or manufacture of a particular food item.

Food operation
> A location along the food chain where food is handled or prepared for commercial reasons.

Food safety objective (FSO)
> A statement of the maximum frequency and/or concentration of a microbiological hazard in a food at the time of consumption that provides the appropriate level of protection.

Hazard Analysis Critical Control Point (HACCP)
> A system that identifies, evaluates, and controls hazards which are significant for food safety.

HACCP plan
> A document prepared in accordance with the principles of HACCP to ensure control of hazards that are significant for food safety in the segment of the food chain under consideration.

Hazard
> A biological, chemical, or physical agent in, or condition of, food with the potential to cause an adverse health effect.

Hazard analysis (in HACCP)
> The process of collecting and evaluating information on hazards and conditions leading to their presence to decide which are significant for food safety and therefore should be addressed in the HACCP plan.

Hazard characterization
> The qualitative and/or quantitative evaluation of the nature of the adverse health effects associated with the hazard. For the purpose of microbiological risk assessment the concerns relate to microorganisms and their toxins.

Hazard identification
    The identification of biological, chemical, and physical agents capable of causing adverse health effects and which may be present in a particular food or group of foods.

Monitor
    The act of conducting a planned sequence of observations or measurements of control parameters to assess whether a CCP is under control.

Performance criterion
    The required outcome of a step, or combination of steps, that contribute to assuring a food safety objective is met.

Process criteria
    The control parameters of a step, or combination of steps, that can be applied to achieve a performance criterion.

Product criteria
    A parameter of a food that can be used to assess the acceptability of a lot or consignment.

Qualitative risk assessment
    A risk assessment based on data which, while forming an inadequate basis for numerical risk estimations, nonetheless, when conditioned by prior expert knowledge and identification of attendant uncertainties permits risk ranking or separation into descriptive categories of risk.

Quantitative risk assessment
    A risk assessment that provides numerical expressions of risk and indication of the attendant uncertainties.

Risk
    A function of the probability of an adverse health effect, and the severity of that effect, consequential to a hazard(s) in food.

Risk analysis
    A process consisting of three components: risk assessment, risk management, and risk communication.

Risk assessment
    A scientifically based process consisting of the following steps: (i) hazard identification, (ii) hazard characterization, (iii) exposure assessment, and (iv) risk characterization.

Risk characterization
    The process of determining the qualitative and/or quantitative estimation, including attendant uncertainties, of the probability of occurrence and severity of known or potential adverse health effects in a given population based on hazard identification, hazard characterization, and exposure assessment.

Risk communication
    The interactive exchange of information and opinions throughout the risk analysis process concerning hazards and risks, risk-related factors and risk perceptions, among

risk assessors, risk managers, consumers, industry, the academic community, and other interested parties, including the explanation of risk assessment findings and the basis of risk management decisions.

Risk estimate
Output of risk assessment.

Risk management
The process, distinct from risk assessment, of weighing policy alternatives, in consultation with all interested parties, considering risk assessment and other factors relevant for the health protection of consumers and for the promotion of fair trade practices, and, if needed, selecting appropriate prevention and control options.

Safe food (the Codex document gives this as the definition for food safety – not safe food)
Food that does not cause harm to the consumer when it is prepared and/or eaten according to its intended use.

Sensitivity analysis
A method used to examine the behavior of a model by measuring the variation in its outputs resulting from changes in its inputs.

Step
A point, procedure, operation, or stage in the food chain including raw materials, from primary production to final consumption.

Tolerable level of risk (TLR)
The level of risk proposed following consideration of public health impact, technological feasibility, economic implications, and that which society regards as reasonable in the context of and in comparison with other risks in everyday life.

Transparent
Characteristics of a process where the rationale, the logic of development, constraints, assumptions, value judgements, decisions, limitations, and uncertainties of the expressed determination are fully and systematically stated, documented, and accessible for review.

Uncertainty analysis
A method used to estimate the uncertainty associated with model inputs, assumptions and structure/form.

Validation
Obtaining evidence that the elements of the HACCP plan are effective.

Verification
The application of methods, procedures, tests, and other evaluations, in addition to monitoring to determine compliance with the HACCP plan.

Appendix B

Objectives and Accomplishments of the International Commission on Microbiological Specifications for Foods

HISTORY AND PURPOSE

The International Commission on Microbiological Specifications for Foods (ICMSF, the Commission) was formed in 1962 through the action of the International Committee on Food Microbiology and Hygiene, a committee of the International Union of Microbiological Societies (IUMS). Through the IUMS, the ICMSF is linked to the International Union of Biological Societies (IUBS) and to the World Health Organization (WHO) of the United Nations.

In the 1960s, there was growing recognition of foodborne disease and greatly increased microbiological testing of foods. This, in turn, created unforeseen problems in international trade in foods. Different analytical methods and sampling plans of doubtful statistical validity were being used. Furthermore, analytical results were interpreted using different concepts of biological significance and acceptance criteria, creating confusion and frustration for both the food industry and regulatory agencies.

In this environment, ICMSF was founded to: (1) assemble, correlate, and evaluate evidence about the microbiological safety and quality of foods; (2) consider whether microbiological criteria would improve and assure the microbiological safety of particular foods; (3) propose, where appropriate, such criteria; and (4) recommend methods of sampling and examination.

Thirty years later, the primary role of the Commission remains to give guidance on: (1) appraising and controlling the microbiological safety of foods and (2) microbiological quality, since this influences consumer acceptance and losses due to spoilage. Meeting those objectives assists international trade, national control agencies, the food industry, international agencies concerned with humanitarian food distribution, and consumer interests.

FUNCTIONS AND MEMBERSHIP

The ICMSF provides basic scientific information through extensive study and makes recommendations without prejudice based on that information. Results of the studies are published as books, discussion documents, or refereed papers. Major publications of the Commission are listed in Appendix D.

The ICMSF functions as a working party, not as a forum for the reading of papers. Meetings consist largely of discussions within subcommittees, debating to achieve consensus, editing draft materials, and planning. Most work is done between meetings by the Editorial Committee and members, sometimes with the help of nonmember consultants.

Since 1962, 33 meetings have been held in 19 countries (Australia, Brazil, Canada, Denmark, Dominican Republic, Egypt, England, France, Germany, Italy, Mexico, South Africa, Spain, Switzerland, The Netherlands, the United States, the former USSR, Venezuela, and the former Yugoslavia). During its meetings, Commission members frequently participate in symposia organized by microbiologists or public health officials of the host country.

Currently, the membership consists of 15 food microbiologists from 9 countries, with combined professional interests in research, public health, official food control, education, product and process development, and quality control, from government laboratories in public health, agriculture, and food technology; from universities; and from the food industry (see Appendix C). The ICMSF is also assisted by consultants, specialists in particular areas of microbiology who are critical to the success of the Commission (see Appendix C for lists of the consultants, contributors, and reviewers of this volume). New members and consultants are selected for their expertise, not as national delegates. All work is voluntary without fees or honoraria.

Three Subcommissions (Latin American, South-East Asian, and Balkan and Danubian) promote ICMSF's activities among food microbiologists in their regions and facilitate communication worldwide (see Appendix C).

The ICMSF raises its own funds to support its meetings. Support has been obtained from government agencies, WHO, IUMS, IUBS, and the food industry (over 80 food companies and agencies in 13 countries). Grants for specific projects and seminars/conferences have been provided by a variety of sources. Some funds are received from the sale of its books.

RECENT PROJECTS

Microorganisms in Foods 5: Characteristics of Microbial Pathogens (1996) is a thorough, but concise, review of the literature on growth, survival, and death responses of foodborne pathogens. It is intended as a quick reference manual to assist in making decisions in support of Hazard Analysis Critical Control Plans (HACCP) and to improve food safety. (Detailed publication information on the major documents mentioned here is available in Appendix D.)

Microorganisms in Foods 6: Microbial Ecology of Food Commodities (1998) updates and extends an earlier volume in the ICMSF series (1980). For 16 commodity areas, Volume 6 describes the initial microbial flora and the prevalence of pathogens, the microbiological consequences of processing, typical spoilage patterns, episodes implicating those commodities with foodborne illness, and measures to control pathogens.

Discussion Documents Prepared for the Joint Food and Agriculture Organization (FAO) and World Health Organization (WHO) Food Standards Programme, Codex Alimentarius Commission

1. Establishment of Sampling Plans for Microbiological Safety Criteria for Foods in International Trade

2. Discussion of Sampling Plans for *L. monocytogenes, Salmonella, Campylobacter* and Verocytotoxin-Producing *E. coli* in Foods in International Trade
3. Recommendations for the Future Management of Microbiological Hazards for Foods in International Trade
4. Principles for the Establishment of Food Safety Objectives and Related Control Measures

ICMSF's recommendations for sampling foods and acceptance criteria for *Listeria monocytogenes* were subsequently published as "Sampling plans for *Listeria monocytogenes*" (1994, *Int J Food Microbiol* 22, 89–96), as was "Establishment of microbiological safety criteria for foods in international trade" (1997, *World Health Stats Quarterly*, 50, 119–123).

At the request of the Secretariat of Codex, the ICMSF developed recommendations for revision of the document Principles for the Establishment and Application of Microbiological Criteria for Foods, originally published in the Procedural Manual of Codex.

Addressing the need for a scientific basis in risk assessment, a working group of the ICMSF published "Potential application of risk assessment techniques to microbiological issues related to international trade in food and food products" (1998, *J Food Prot* 61, 1075–1086).

PAST AND FUTURE

For almost 25 years, the ICMSF's major efforts were devoted to methodology. This resulted in improved comparisons of microbiological methods and better standardization (17 refereed publications). Among many significant findings, it was established that when analyzing for salmonellae, analytical samples could be bulked (composited) into a single test with no loss of sensitivity. This made practical the collection and analysis of the large number of samples recommended in some sampling plans.

With the rapid development of alternative methods and rapid test kits, and the ever expanding list of biological agents involved in foodborne illness, the Commission reluctantly discontinued its program of comparison and evaluation of methods in 1986, recognizing that issues of methodology were being addressed effectively by other organizations.

A long-term objective of the Commission has been to enhance the microbiological safety of foods in international commerce. This was initially addressed through books that recommended uniform analytical methods (ICMSF, 1978) and sound sampling plans and criteria (ICMSF, 1974, 1978, 1986). The Commission then developed a two volume set on the microbial ecology of foods (ICMSF, 1980a, b) intended to familiarize analysts with processes used in the food industry and microbiological aspects of foods submitted to the laboratory. Knowledge of the microbiology of the major food commodities and the factors affecting the microbial content of these foods helps analysts to interpret analytical results.

At an early stage, the Commission recognized that no sampling plan can ensure the absence of a pathogen in food. Testing foods at ports-of-entry or elsewhere in the food chain cannot guarantee food safety. This led the Commission to explore the potential value of HACCP for enhancing food safety. A meeting in 1980 with the WHO led to a report on the use of HACCP for controlling microbiological hazards in food, particularly in developing countries (ICMSF, 1982). The Commission then developed a book on the principles of HACCP and procedures for developing HACCP plans (ICMSF, 1988), covering the importance of controlling the conditions of producing/harvesting, preparing, and handling foods. Recommendations were given in that volume for the application of HACCP from production/harvest to consumption, together with examples of how HACCP can be applied at each step in the food chain.

The Commission next recognized that a major weakness in the development of HACCP plans is the process of hazard analysis. It has become more difficult to be knowledgable about the many biological agents recognized as responsible for foodborne illness. ICMSF (1996) summarizes important information about the properties of biological agents commonly involved in foodborne illness and serves as a quick reference manual when making judgments on the growth, survival, or death of pathogens.

Subsequently, the Commission updated its volume on the microbial ecology of food commodities (ICMSF, 1998).

The current book, *Microorganisms in Foods 7: Microbiological Testing in Food Safety Management* (2001), illustrates how systems such as HACCP and Good Hygienic Practices (GHP) provide greater assurance of safety than microbiological testing, but also identifies circumstances in which microbiological testing still plays a useful role.

We believe that the original objectives of the Commission are still relevant today. The European Union, the many other political changes occurring throughout the world, the growth of developing countries seeking export markets, and the increased trade in foods worldwide, as evidenced by the passage of the General Agreement on Tariffs and Trade (GATT) and the North American Free Trade Agreement (NAFTA) all point to the continuing need for the independent recommendations on food safety, such as those of the Commission. It is essential that import/export policies be established as uniformly as possible and on a sound scientific basis. The Commission's overall goal will continue to be to enhance the safety of foods moving in international commerce. The Commission will continue to strive to meet this goal through a combination of educational materials, promoting the use of food safety management systems using microbiological food safety objectives, HACCP, and GHP, and recommending sampling plans and microbiological criteria where they have been developed according to Codex principles and offer increased assurance of microbiological safety. The future success of ICMSF will continue to depend upon the efforts of members, support from consultants who generously volunteer their time, and those who provide the financial support so essential to the Commission's activities.

Appendix C

ICMSF Participants*

OFFICERS

Chairman

Dr. Martin Cole (from 2000), Group Manager, Food Safety & Quality, Food Science Australia, P.O. Box 52, North Ryde, NSW 1670, Australia

Dr. Terry A. Roberts (1991–2000), Food Safety Consultant, 59 Edenham Crescent, Reading RG1 6HU, UK

Secretary

Prof. Mike van Schothorst, Vice President. Food Safety Affairs, Nestle, Avenue Nestle 55, CH-1800 Vevey, Switzerland; and Food Safety Microbiology, Wageningen University, P.O. Box 8129, Wageningen 6700 EV, The Netherlands

Treasurer

Dr. Jeffrey M. Farber (from 2000), Health Canada, Food Directorate, Microbiology Research Division, Banting Research Centre, Postal Locator 2204A2, Tunney's Pasture, Ottawa, Ontario K1A OL2, Canada

Prof. Frank F. Busta (1998–2000), Department of Food Science and Nutrition, University of Minnesota, 2168 Ferris Lane, St. Paul, MN 55113–3876, USA

Dr. A. N. Sharpe (1989–1998), Head of Automation Section, Bureau of Microbial Hazards, Health Protection Branch, Health Canada, Tunney's Pasture, Ottawa, Ontario K1A 0L2, Canada

Members

Dr. A. C. Baird-Parker, Consultant, Food Microbiology, 2 Pagnell Court, Hardingstone, Northampton NN4 6EF, UK (retired 1999)

Dr. Robert L. Buchanan, U.S. Food and Drug Administration, Centre for Food Safety and Applied Nutrition, HFS-500, Room 3832, 200 C-Street, SW, Washington, DC 20204, USA

Dr. Jean-Louis Cordier, Head, Industrial Hygiene, Senior Microbiologist, Food Safety & Quality Assurance, Nestlé Research Centre, Vers-chez-les-Blanc—P.O. Box 44, CH-1000 Lausanne 26, Switzerland

Dr. Susanne Dahms, Fachbereich Veterinärmedizin, Institut für Biometrie und Informationsverarbeitung, Freie Universität Berlin, Oertzenweg 19b, D-14163 Berlin, Germany

---

*Titles and locations during preparation of Book 7.

Prof. M. P. Doyle, Center for Food Safety & Quality Enhancement, University of Georgia, Georgia Station, Griffin, GA 30223, USA (resigned 1999)

Dr. M. Eyles, CSIRO, Division of Food Science & Technology, P.O. Box 52, North Ryde, NSW 2113, Australia (resigned 1999)

Prof. J. Farkas, Vice Rector, Department of Refrigeration and Livestock Products Technology, Faculty of Food Industry, University of Horticulture and Food Industry, H-1118 Budapest, Menesi ut 45, Hungary (retired 1998)

Dr. R. S. Flowers, Silliker Laboratories, 900 Maple Road, Homewood, IL 80430, USA

Prof. Bernadette D. G. M. Franco, Departamento de Alimentos e Nutricao Experimental, Faculdade de Ciencias Farmaceuticas, Universidade de São Paulo, Av. Prof. Lineu Prestes 580, 05508–900, São Paulo, SP, Brazil.

Prof. Lone Gram, Danish Institute for Fisheries Research, Department of Seafood Research, Soltafts Plads, Danish Technical University, Bldg 221, DK-2800 Kgs. Lyngby, Denmark

Dr. F. H. Grau, CSIRO, Division of Food Science & Technology, Brisbane Laboratory, P.O. Box 3312, Tingalpa DC, QLD 4173, Australia (retired 1999)

Prof. Jean-Louis Jouve, European Commission, DG Health and Consumer Protection, Rue Belliard, 232-Room 6/14, B-1040 Brussels, Belgium

Dr. Anna M. Lammerding, Food Safety Risk Assessment Unit, Health of Animals Laboratory, 110 Stone Road West, Guelph, Ontario N1G 3W4, Canada

Dra. Silvia Mendoza, Division of Biological Sciences, Department of Biological and Biochemical Process Technology, Simon Bolivar University, P.O. Box 89.000, Caracas 1080 A, Venezuela (retired 1998)

Ms. Zahara Merican, Technical Services Centre, Malaysian Agricultural Research & Development Institute (MARDI), P.O. Box 12301 GPO, 50774 Kuala Lumpur, Malaysia

Dr. John I. Pitt, Chief Research Scientist, Food Science Australia, P.O. Box 52, North Ryde, NSW 1670, Australia

Prof. Dr. F. Quevedo, Food Quality & Safety Assurance International, F. Villareal National University, Las Petunias 140, Dpt. 201, Urb. Camacho, Lima 12, Peru (retired 1998)

Dr. Paul Teufel, Director and Professor, Institute for Hygiene and Food Safety, Federal Dairy Research Centre, Hermann-Weigmann Strasse 1, D-24103 Kiel, Germany

Dr. R. Bruce Tompkin, Vice President Product Safety, ConAgra Refrigerated Prepared Foods, 3131 Woodcreek Drive, Downers Grove, IL 60515–5429, USA

PAST MEMBERS OF THE ICMSF

NAME	COUNTRY	MEMBERSHIP
Dr. A. C. Baird-Parker	UK	1974–1999
Dr. M. T. Bartram	USA	1967–1968
Dr. H. E. Bauman	USA	1964–1977
Dr. F. L. Bryan	USA	1974–1996[1]
Dr. L. Buchbinder*	USA	1962–1965
Prof. F. F. Busta	USA	1985–2000[2]
Dr. R. Buttiaux	France	1962–1967
Dr. J. H. B. Christian	Australia	1971–1991[3]
Dr. D. S. Clark	Canada	1963–1985[4]
Dr. C. Cominazzini	Italy	1962–1983

Dr. C. E. Dolman*	Canada	1962–1973
Dr. M. P. Doyle	USA	1989–1999
Dr. R. P. Elliott*	USA	1962–1977
Dr. Otto Emberger	former Czechoslovakia	1971–1986
Dr. M. Eyles	Australia	1996–1999
Dr. J. Farkas	Hungary	1991–1998
Mrs. Mildred Galton*	USA	1962–1968
Dr. E. J. Gangarosa	USA	1969–1970
Dr. F. Grau	Australia	1985–1999
Dr. J. M. Goepfert	Canada	1985–1989[5]
Dr. H. E. Goresline*	USA/Austria	1962–1970
Dr. Betty C. Hobbs*	UK	1962–1996
Dr. A. Hurst	UK/Canada	1963–1969
Dr. H. Iida	Japan	1966–1977
Dr. M. Ingram*	UK	1962–1974[6]
Dr. M. Kalember-Radosavljevic	former Yugoslavia	1983–1992
Dr. K. Lewis*	USA	1962–1982
Dr. John Liston	USA	1978–1991
Dr. Holger Lundbeck*	Sweden	1962–1983[7]
Dr. S. Mendoza	Venezuela	1992–1998
Dr. G. Mocquot	France	1964–1980
Dr. G. K. Morris	USA	1971–1974
Dr. D. A. A. Mossel*	The Netherlands	1962–1975
Dr. N. P. Nefedjeva	USSR	1964–1979
Dr. C. F. Niven, Jr.	USA	1974–1981
Dr. P. M. Nottingham	New Zealand	1974–1986
Dr. J. C. Olson, Jr.	USA	1968–1982
Dr. H. Pivnick	Canada	1974–1983
Dr. T. A. Roberts	UK	1978–2000[8]
Dr. F. Quevedo	Peru	1965–1998
Dr. A. N. Sharpe	Canada	1985–1998[9]
Dr. J. Silliker	USA	1974–1987[10]
Mr. Bent Simonsen	Denmark	1963–1987
Dr. H. J. Sinell	Germany	1971–1992
Dr. G. G. Slocum*	USA	1962–1968
Dr. F. S. Thatcher*	Canada	1962–1973[11]

*Founding member
[1]Secretary, 1981–1991
[2]Treasurer, 1998–2000
[3]Chairman, 1980–1991
[4]Secretary-Treasurer, 1963–1981
[5]Treasurer, 1987–1989
[6]Ex-officio member, 1962–1968
[7]Chairman, 1973–1980
[8]Chairman, 1991–2000
[9]Treasurer, 1989–1998
[10]Treasurer, 1981–1987
[11]Chairman, 1962–1973

MEMBERS OF THE LATIN AMERICAN SUBCOMMISSION

Chairperson

Dra. Maria Alina Ratto, General Manager, Microbiol S.A., Joaquin Capello 222, Lima 18, Peru, e-mail:microbl@terra.com.pe

Secretary/Treasurer

Lic. Ricardo A. Sobol, Director Tecnico, Food Control S.A., Santiago del Estero 1154, 1075 Buenos Aires, Argentina, e-mail: 50601@foodcontrcl.com

Honorary Members

Prof. Fernando Quevedo, Food Quality and Safety Assurance International, Buenos Aires 188, Miraflores, Lima 18, Peru, e-mail:fquevedo@amauta.rcp.net.pe

Prof. Sebastião Timo Iaria, Av. Angelica 2206, apto 141, 01228–200, Sao Paulo, SP, Brazil, e-mail: stiaria@aol.com.br

Prof. Silvia Mendoza, Conjunto Residencial E1, Av. Washington Torre 1A, piso 12 apto 123, Caracas, Venezuela, e-mail: silmendoza@cantr.net

Prof. Nenufar Sosa de Caruso, Alimentarius, Tomas de Tezanos 1323, Montevideo, Uruguay, e-mail: alimenta@adinet.com.uy

Members

Prof. Bernadette D. G. M. Franco, Departamento de Alimentos e Nutricao Experimental, Faculdade de Ciencias Farmaceuticas, Universidade de São Paulo, Av. Prof. Lineu Prestes 580, 05508–900, São Paulo, SP, Brazil, e-mail: bfranco@usp.br

Dra. Eliana Marambio, Coventry 1046, Depto 405, Ñuñoa, Santiago, Chile, e-mail: emarambio@entelchile.net

Profa. Janeth Luna Cortéz, Universidad de Bogota, Carrera 4 No. 22–61 Of 436, Santafé de Bogotá, DC, Colombia, e-mail: ingeneria.alimentos@utadeo.edu.co

Dra. Dora Martha González, Sarmiento 2323, Montevideo, Uruguay, e-mail: dmgonzal@adinet.com.uy

Profa. Pilar Hernandez S., Universidad Central de Venezuela, Apartado 40109, Caracas 1040-A, Venezuela, e-mail: hernands@camelot.rect.ucv.ve

Former Members of the Latin-American Subcommission

Dra. Ethel G.V. Amato de Lagarde, Argentina
Dr. Rafael Camperchioli, Paraguay
Dr. Cesar Davila Saa, Ecuador
Dr. Mauro Faber de Freitas Leitao, Brazil
Dra. Josefina Gomez-Ruiz, Venezuela (former chairperson)
Dra. Yolanda Ortega de Gutierrez, Mexico
Dr. Hernan Puerta Cardona, Colombia
Dra. Elvira Regus de Pons, Dominican Republic

MEMBERS OF THE SOUTH-EAST ASIAN SUBCOMMISSION

Chairperson

Dr. Zahara Merican, Food Technology Research Centre, Malaysian Agricultural Research and Development Institute, P.O. Box 12301, GPO 50774 Kuala Lumpur, Malaysia

Secretary/Treasurer

Dr. Pho Lay Koon, Section Head, Plant Biotechnology & Agrotechnology, Chemical Process & Biotechnology Dept., Singapore Polytechnic, Dover Road, Singapore 0513.

Members

Dr. Srikandi Fardiaz, Head of Food Microbiology Laboratory, Inter University Centre for Food & Nutrition, Bogor Agricultural University, P.O. Box 220, Bogor, Indonesia (deceased 2000)

Ms. Quee Lan Yeoh, Food Technology Research Centre, Malaysian Agricultural Research and Development Institute, PO Box 12301, GPO 50774 Kuala Lumpur, Malaysia

Dr. Reynaldo C. Mabesa, Assoc. Professor, Institute of Food Science and Technology, University of the Philippines at Los Banos, Los Banos, Laguna 4031, Philippines

Ms. Chakamas Wongkhalaung, Deputy Director, Institute of Food Research and Product Development (IFRPD), Kasetsart University, P.O. Box 1043 Kasetsart, Bangkok 10903, Thailand

Dr. Lor Kim Loon, Senior Manager, Food Research and Development, SATS Catering Pte Ltd., SATS Inflight Catering Centre, P.O. Box 3 Singapore Changi Airport, Singapore 918141

MEMBERS OF THE BALKAN-DANUBIAN SUBCOMMISSION

Chairperson

Dr. Hajnalka Domjan Kovacs, Food Bacteriologist, National Food Investigation Institute, Pf. 1740, H-1465 Budapest 94, Hungary

Secretary/Treasurer

Dr. Vladimir Spelina, CSc. Center of Hygienes of Nutrition, Institute of Hygiene and Epidemiology, Srobarova 48, CZ-100 42 Praha 10 - Vinohrady, Czech Republic

Members

Dr. Milica Kalember-Radosavljevic, Military Medical Academy, Institute of Hygiene, Crnotravska 17, 11000 Beograd, Ljermonrova 22, Yugoslavia

Prof. Livio Leali, Professore Associato di Igene del Latte, Universita di Milano, Via Celoria 10, I-20133 Milano, Italy

Doz. Dr. vet. Ivan Kaloyanov, Central Veterinary Research Institute, 15 Pencho Slaveikov Blvd., BG-Sofia 1606, Bulgaria

Former Members of the Balkan-Danubian Subcommission

Dr. Vladimir Bartl, former Czechoslovakia (former chairperson)

Dr. Zora Bulajic, former Yugoslavia
Dr. Deac Cornel, Romania
Dr. Corneliu Ienistea, Romania
Dr. John Papavassilliou, Greece
Prof. Dr. Oscar Prandl, Austria
Prof. Dr. Mirko Sipkaformer, Yugoslavia (first chairperson)
Dr. H.J. Takacs, Hungary (former chairperson)
Dr. Zenai Muammer Tancman, Turkey
Dr. S. Tzannetis, Greece
Doc. Dr. Muammer Ugar, Turkey
Dr. Fuad Yanc, Turkey
Prof. Dr. Z. Zachariev, Bulgaria

MICROORGANISMS IN FOOD 7

Consultants

Dr. J. Braeunig, Germany, 2000
Dr. S. Dahms, Germany, 1997, 1998 (member since 1998)
Dr. P. Desmarchelier, Australia, 1999
Dr. J. M. Farber, Canada, 1998 (member since 1998)
Dr. B. D. G. M. Franco, Brazil, 1998, 1999 (member since 2000)
Dr. W. Garthwright, USA, 1999
Dr, L, G, M, Gorris, Netherlands, 2000
Dr. L. Gram, Denmark, 1997, 1998 (member since 1998)
Dr. H. Kruse, Norway, 1999, 2000
Dr. A. M. Lammerding, Canada, 1997, 1998 (member since 1998)
Dr. B. Shay, Australia, 1997, 1999
Dr. K. M. J. Swanson, USA, 2000
Dr. A. von Holy, South Africa, 1997

Contributors to Microorganisms in Food 7

Dr. D. Kilsby (UK), Dr. R. B. Smittle (USA), Dr. J. H. Silliker (USA)

Reviewers

Dr. V. N. Scott (USA), Dr. D. Kilsby (UK), Dr. P. Bodnaruk (USA), Dr. J. N. Sofos (USA), Dr. W. P. Pruett (USA)

Appendix D

Publications of the ICMSF

BOOKS

    Food and Agriculture Organization & International Atomic Energy Agency/ICMSF (1970). *Microbiological Specifications and Testing Methods for Irradiated Foods.* Technical Report Series No. 104, Vienna: International Atomic Energy Agency.

    ICMSF (1978). (Reprinted 1982, 1988 with revisions). *Microorganisms in Foods 1. Their Significance and Methods of Enumeration,* 2nd edn. Toronto: University of Toronto Press. (ISBN 0–8020–2293–6).

    ICMSF (1980). *Microbial Ecology of Foods. Volume 1. Factors Affecting Life and Death of Microorganisms,* New York: Academic Press. (ISBN 0–12–363501–2).

    ICMSF (1980). *Microbial Ecology of Foods. Volume 2. Food Commodities,* New York: Academic Press. (ISBN 0–12–363502–0).

    ICMSF (1986). *Microorganisms in Foods 2. Sampling for Microbiological Analysis: Principles and Specific Applications,* 2nd edn. Toronto: University of Toronto Press. (ISBN 0–8020–5693–8) (available outside North America from Blackwell Scientific Publications, Ltd., Osney Mead, Oxford OX2 0EL, UK) (First edition: 1974; revised with corrections, 1978).

    ICMSF (1988). *Microorganisms in Foods 4. Application of the Hazard Analysis Critical Control Point (HACCP) System to Ensure Microbiological Safety and Quality.* Oxford: Blackwell Scientific Publications (ISBN 0–632–02181–0). Also published in paperback under the title *HACCP in Microbiological Safety and Quality* (1988) (ISBN 0–632–02181–0).

    ICMSF (1996). *Microorganisms in Foods 5. Characteristics of Microbial Pathogens.* Gaithersburg, MD: Aspen Publishers Inc. (ISBN 0–412–47350-X).

    ICMSF (1998). *Microorganisms in Foods 6. Microbial Ecology of Food Commodities.* Gaithersburg, MD: Aspen Publishers Inc. (ISBN 0–8342–1825–9). Gaithersburg, MD: Aspen Publishers Inc.

    (Aspen Publishers Inc., 200 Orchard Ridge Dr, Suite 200, Gaithersburg MD 20878, USA, 800–638–8347, http://www.aspenpublishers.com. Outside the USA, contact Plymbridge Ltd., Plymouth, UK, tel. (44) 1752 202 301.)

WHO PUBLICATIONS

    ICMSF (authors: J. H. Silliker, A. C. Baird-Parker, F. L. Bryan, J. C. Olson, Jr., B. Simonsen, & M. van Schothorst,)/WHO (1982). *Report of the WHO/ICMSF meeting on Hazard Analysis: Critical Control Point System in Food Hygiene.* WHO/VPH/82.37. Geneva: World Health Organization. (also available in French).

    Christian, J. H. B. (1983). *Microbiological Criteria for Foods.* (Summary of recommendations of FAO/WHO expert consultations and working groups 1975–1981). WHO/VPH/83.54. Geneva: World Health Organization.

    ICMSF (authors: B. Simonsen, F. L. Bryan, J. H. B. Christian, T. A. Roberts, J. H. Silliker, & R. B. Tompkin) (1986). *Prevention and Control of Foodborne Salmonellosis through*

Application of the Hazard Analysis Critical Control Point System. Report, International Commission on Microbiological Specifications for Foods (ICMSF). WHO/CDS/VPH/86.65. Geneva: World Health Organization.

OTHER ICMSF TECHNICAL PAPERS

Thatcher, F. S. (1963). The microbiology of specific frozen foods in relation to public health: Report of an international committee. *J Appl Bacteriol* 26, 266–285.

Simonsen, B., Bryan, F. L., Christian, J. H. B., Roberts, T. A., Tompkin, R. B., & Silliker, J. H. (1987). Report from the International Commission on Microbiological Specifications for Foods (ICMSF). Prevention and control of foodborne salmonellosis through application of hazard analysis critical control point (HACCP). *Int J Food Microbiol* 4, 227–247.

ICMSF (1994). Choice of sampling plan and criteria for *Listeria monocytogenes*. *Int J Food Microbiol* 22, 89–96.

ICMSF (1997). Establishment of microbiological safety criteria for foods in international trade. *World Health Stats Quarterly* 50, 119–123.

ICMSF (1998). Potential application of risk assessment techniques to microbiological issues related to international trade in food and food products. *J Food Prot* 61, 1075–1086.

ICMSF [M. van Schothorst, Secretary] (1998). Principles for the establishment of microbiological food safety objectives and related control measures. *Food Control* 9, 379–384.

TRANSLATIONS

Thatcher, F. S. & Clark, D. S. (1973). *Microorganisms in Foods 1. Their Significance and Methods of Enumeration.* (in Spanish, translated by B. Garcia). Zaragoza, Spain: Editorial Acribia.

ICMSF (1981). *Microorganismos de los Alimentos 2. Métodos de Muestreo para Análisis Microbiológicos: Principios y Aplicaciones Específicas,* (in Spanish, translated by J.A. Ordonez Pereda. & M.A. Diaz Hernandez). Zaragoza, Spain: Editorial Acribia.

ICMSF (1983). *Ecología Microbiana de los Alimentos 1. Factores que Afectan a la Supervivencia de los Microorganismos en los Alimentos,* (in Spanish, translated by J. Burgos Gonzalez et al.). Zaragoza, Spain: Editorial Acribia.

ICMSF (1984). *Ecología Microbiana de los Alimentos 2. Productos Alimenticios,* (in Spanish, translated by B. Sanz Perez et al.). Zaragoza, Spain: Editorial Acribia.

ICMSF (1988). *El Sistema de Análisis de Riesgos y Puntos Críticos. Su Aplicación a las Industrias de Alimentos,* (in Spanish, translated by P.D. Malmenda & B.M. Garcia). Zaragoza, Spain: Editorial Acribia.

ICMSF (1996). *Microorganismos de los Alimentos: Caraterísticas de los Patógenos Microbianos,* (in Spanish, translated by M. R. Vergés). Zaragoza, Spain: Editorial Acribia.

ABOUT THE ICMSF

Bartram, M. T. (1967). International microbiological standards for foods. *J Milk & Food Technol* 30, 349–351.

Saa, C. C. (1968). The Latin American Subcommittee on Microbiological Standards and Specifications for Foods. *Revista de la Facultad de Quimica y Farmacia* 7, 8.

Cominazzini, C. (1969). The International Committee on Microbiological Specifications for Foods and its contribution to the maintenance of food hygiene. (In Italian). *Croniche Chimico* 25, 16.

Saa, C. C. (1969). El Comité Internacional de Especificaciones Microbiológicas de los Alimentos de la IAMS. *Revista de la Facultad de Química y Farmacia* 8, 6.

Mendoza, S. & Quevedo, F. (1971). Comisión Internacional de Especificaciones Microbiológicas de los Alimentos. *Boletín del Instituto Bacteriológico de Chile* 13, 45.

Thatcher, F. S. (1971). The International Committee on Microbiological Specifications for Foods. Its purposes and accomplishments. *J AOAC* 54, 836–814.

Clark, D. S. (1977). The International Commission on Microbiological Specifications for Foods. *Food Technol* 32, 51–54, 67.

Clark, D. S. (1982). International perspectives for microbiological sampling and testing of foods. *J Food Prot* 45, 667–671.

Anonymous (1984). International Commission on Microbiological Specifications for Foods. *Food Lab Newsl* 1, 23–25. (Box 622, S-751 26 Uppsala, Sweden)

Quevedo, F. (1985). Normalización de alimentos y salud para América Latina y el Caribe. 3. Importancia de los criterios microbiológicos. *Boletín de la Oficina Sanitaria Panamericana* 99, 632–640.

Bryan, F. L. & Tompkin, R. B. (1991). The International Commission on Microbiological Specifications for Foods (ICMSF). *Dairy, Food & Environ San* 11, 66–68.

Anonymous (1996). The International Commission on Microbiological Specifications for Foods (ICMSF): update. *Food Control* 7, 99–101.

Appendix E

Table of Sources

CHAPTER 2

Figure 2–1. *Source:* Data from Center for Disease Control and Prevention, Atlanta, Georgia.

Figure 2–2. *Source:* Data from Federal Institute for Health Protection of Consumers and Veterinary Medicine (FAO/WHO Collaborating Centre for Research and Training in Food Hygiene and Zoonoses), Berlin, Germany and International Commission on Microbiological Specifications for Food, University of Toronto Press, Toronto, Canada.

Figure 2–3. *Source:* Reprinted with permission from *Journal of Food Protection*. Copyright held by the International Association for Food Protection, Des Moines, Iowa, U.S.A. Robert L. Buchanan, William G. Damert, Richard C. Whiting, and Michael van Schothorst, authors, U.S. Department of Agriculture, Agricultural Research Center, Eastern Regional Research Center, 60 E. Mermaid Lane, Wyndmoor, PA 19038, USA and Nestle, Avenue Nestle 55, CH-1800, Vevey, Switzerland.

Table 2–1. Courtesy of World Health Organization, 2000, Geneva, Switzerland and the Food and Agriculture Organization of the United Nations from "Exposure Assessment of Listeria Monocytogenes in Ready-to-Eat Foods, *Preliminary Report for Joint FAO/WHO Expert Consultation on Risk Assessment of Microbiological Hazards in Foods*, 2000.

Table 2–2. *Source:* Data from Center for Disease Control and Prevention, 2000a, Annual Reports from the Center for Disease Control and Prevention, Preliminary FoodNet Data on the Incidence of Foodborne Illnesses—Selected Sites, United States, 1999, *Morbidity and Mortality Weekly Reports,* Vol. 49, pp. 201–205, Table 2–2.

CHAPTER 3

Figure 3–1. *Source:* Data from Center for Disease Control and Prevention, Atlanta, Georgia.

Figure 3–2. *Source:* Data from Center for Disease Control and Prevention, Atlanta, Georgia.

Figure 3–3. *Source:* Reprinted from Center for Disease Control and Prevention, Atlanta, Georgia.

Figure 3–4. *Source:* Reprinted from Center for Disease Control and Prevention, 2000a, Preliminary FoodNet Data on the Incidence of Foodborne Illnesses—Selected Sites, United States, 1999, *Morbidity and Mortality Weekly Reports,* Vol. 49, pp. 201–205, and Center for Disease Control and Prevention, 2000b, Outbreaks of Salmonella Serotype Enteritidis Infection Association with Eating Raw or Undercooked Shell Eggs—United States, 1996–1998, *Morbidity and Mortality Weekly Reports*, Vol. 49, pp. 73–79.

Table 3–1. *Source:* Reprinted with permission from R. B. Tompkin and T. V. Kueper, Microbiological Considerations in Developing New Foods. How Factors Other Than Temperature Can be Used to Prevent Microbiological Problems, In *Microbiological Safety of Foods in Feeding Systems,* pp. 100–122, ABMPS Report No. 125, © 1982, Advisory Board on Military Personnel Supplies, National Research Council, National Academy Press.

Table 3–2. *Source:* Reprinted from United States Department of Agriculture, 1996, Pathogen Reduction; Hazard Analysis and Critical Control Point (HACCP) Systems; Final Rule, *Federal Register,* Vol. 61, pp. 38806–38989.

CHAPTER 4

Figure 4–1. *Source:* Adapted with permission from CAC (Codex Alimentarius Commission) 2000. Proposed Draft Framework for Determining the Equivalency of Sanitary Measures Associated with Food Inspection and Certification Systems. *CCFICS/CX/FICS 00/6, Attachment 1.* Joint FAO/WHO Food Standards Programme, Codex Committee on Food Import and Export Inspection and Certification Systems, Food and Agriculture Organization of the United Nations.

CHAPTER 5

Table 5–1. Courtesy of International Commission on Microbiological Specifications for Foods, *Microorganisms in Foods 2, Sampling for Microbiological Analysis: Principles and Specific Applications, 2nd Ed.,* Table 25, p. 178 © 1986, University of Toronto Press, Toronto, Canada.

CHAPTER 6

Figure 6–1. Courtesy of International Commission on Microbiological Specifications for Foods, *Microorganisms in Foods 2, Sampling for Microbiological Analysis: Principles and Specific Applications, 2nd Ed.,* p. 20, © 1986, University of Toronto Press, Toronto, Canada.

CHAPTER 7

Figure 7–1. *Source:* Reprinted from *Food Control,* Vol. 12, No. 3, J. D. Legan, M. H. Vandeven, S. Dahms, M. B. Cole, Determining the Concentration of Microorganisms Controlled by Attributes Sampling Plans, pp. 137–147, © 2001, with permission from Elsevier Science.

Figure 7–2. Courtesy of International Commission on Microbiological Specifications for Foods, *Microorganisms in Foods 2, Sampling for Microbiological Analysis: Principles and Specific Applications, 2nd Ed.,* Figure 2, p. 33, © 1986, University of Toronto Press, Toronto, Canada.

Figure 7–3. Courtesy of International Commission on Microbiological Specifications for Foods, *Microorganisms in Foods 2, Sampling for Microbiological Analysis: Principles and Specific Applications, 2nd Ed.,* Figure 4, p. 69 © 1986, University of Toronto Press, Toronto, Canada.

Figure 7–5. *Source:* Reprinted from *Food Control,* Vol. 12, No. 3, J. D. Legan, M. H. Vandeven, S. Dahms, M. B. Cole, Determining the Concentration of Microorganisms Controlled by Attributes Sampling Plans, pp. 137–147, © 2001, with permission from Elsevier Science.

Figure 7–6. *Source:* Reprinted with permission from *Journal of Food Protection.* Copyright held by the International Association for Food Protection, Des Moines, Iowa, U.S.A. G. Hildebrandt, L. Bohmer and S. Dahms, authors, Institut fur Lebensmittelhygiene, Freie Universitat Berlin, Konigsweg 69, Berlin 14163 Germany.

Table 7–1. Courtesy of International Commission on Microbiological Specifications for Foods, *Microorganisms in Foods 2, Sampling for Microbiological Analysis: Principles and Specific Applications, 2nd Ed.,* Table 2, p. 34, © 1986, University of Toronto Press, Toronto, Canada.

Table 7–2. Courtesy of International Commission on Microbiological Specifications for Foods, *Microorganisms in Foods 2, Sampling for Microbiological Analysis: Principles and Specific Applications, 2nd Ed.,* Table 3, p. 35, © 1986, University of Toronto Press, Toronto, Canada.

Table 7–4. Courtesy of International Commission on Microbiological Specifications for Foods, *Microorganisms in Foods 2, Sampling for Microbiological Analysis: Principles and Specific Applications, 2nd Ed.,* Table 4, pp. 38–39, © 1986, University of Toronto Press, Toronto, Canada.

Table 7–5. Courtesy of International Commission on Microbiological Specifications for Foods, *Microorganisms in Foods 2, Sampling for Microbiological Analysis: Principles and Specific Applications, 2nd Ed.,* Table 16, p. 110, © 1986, University of Toronto Press, Toronto, Canada.

Table 7–6. Courtesy of International Commission on Microbiological Specifications for Foods, *Microorganisms in Foods 2, Sampling for Microbiological Analysis: Principles and Specific Applications, 2nd Ed.,* Table 17, p. 111, © 1986, University of Toronto Press, Toronto, Canada.

Table 7–7. Courtesy of International Commission on Microbiological Specifications for Foods, *Microorganisms in Foods 2, Sampling for Microbiological Analysis: Principles and Specific Applications, 2nd Ed.*, Table 18, p. 112, © 1986, University of Toronto Press, Toronto, Canada.

Table 7–8. Courtesy of International Commission on Microbiological Specifications for Foods, *Microorganisms in Foods 2, Sampling for Microbiological Analysis: Principles and Specific Applications, 2nd Ed.*, Table 5, p. 40, © 1986, University of Toronto Press, Toronto, Canada.

CHAPTER 8

Table 8–1. Courtesy of International Commission on Microbiological Specifications for Foods, *Microorganisms in Foods 2, Sampling for Microbiological Analysis: Principles and Specific Applications, 2nd Ed.*, © 1986, University of Toronto Press, Toronto, Canada.

Table 8–2. Courtesy of International Commission on Microbiological Specifications for Foods, *Microorganisms in Foods 2, Sampling for Microbiological Analysis: Principles and Specific Applications, 2nd Ed.*, © 1986, University of Toronto Press, Toronto, Canada.

Table 8–3. Courtesy of International Commission on Microbiological Specifications for Foods, *Microorganisms in Foods 2, Sampling for Microbiological Analysis: Principles and Specific Applications, 2nd Ed.*, © 1986, University of Toronto Press, Toronto, Canada.

Figure 8–2. Courtesy of International Commission on Microbiological Specifications for Foods, *Microorganisms in Foods 2, Sampling for Microbiological Analysis: Principles and Specific Applications, 2nd Ed.*, Figure 3, p. 66, © 1986, University of Toronto Press, Toronto, Canada.

Appendix 8–A. Courtesy of International Commission on Microbiological Specifications for Foods, *Microorganisms in Foods 2, Sampling for Microbiological Analysis: Principles and Specific Applications, 2nd Ed.*, © 1986, University of Toronto Press, Toronto, Canada.

CHAPTER 9

Exhibit 9–1. Courtesy of International Commission on Microbiological Specifications for Foods, *Microorganisms in Foods 2, Sampling for Microbiological Analysis: Principles and Specific Applications, 2nd Ed.*, Table 11, p. 77, © 1986, University of Toronto Press, Toronto, Canada.

Table 9–1. Courtesy of International Commission on Microbiological Specifications for Foods, *Microorganisms in Foods 2, Sampling for Microbiological Analysis: Principles and Specific Applications, 2nd Ed.*, Table 12A, p. 78, © 1986, University of Toronto Press, Toronto, Canada.

Table 9–2. Courtesy of International Commission on Microbiological Specifications for Foods, *Microorganisms in Foods 2, Sampling for Microbiological Analysis: Principles and Specific Applications, 2nd Ed.*, Table 12B, p. 78, © 1986, University of Toronto Press, Toronto, Canada.

Table 9–3. Courtesy of International Commission on Microbiological Specifications for Foods, *Microorganisms in Foods 2, Sampling for Microbiological Analysis: Principles and Specific Applications, 2nd Ed.*, Table 13, p. 79, © 1986, University of Toronto Press, Toronto, Canada.

Table 9–4. *Source:* Data from R. M. Cannon and R. T. Roe, *Livestock Disease Surveys: A Field Manual for Veterinarians,* Australian Bureau of Animal Health, Department of Primary Industry, Canberra, Australia, 1982, Australian Government Publishing Service.

CHAPTER 10

Figure 10–1. Courtesy of International Commission on Microbiological Specifications for Foods, *Microorganisms in Foods 2, Sampling for Microbiological Analysis: Principles and Specific Applications, 2nd Ed.*, Figure 5, p. 82, © 1986, University of Toronto Press, Toronto, Canada.

Figure 10–2. Courtesy of International Commission on Microbiological Specifications for Foods, *Microorganisms in Foods 2, Sampling for Microbiological Analysis: Principles and Specific Applications, 2nd Ed.*, Figure 6, p. 88, © 1986, University of Toronto Press, Toronto, Canada.

Figure 10–3. Courtesy of Silliker Laboratories, 1993, Homewood, Illinois.

Figure 10–4a. Courtesy of Silliker Laboratories, 1993, Homewood, Illinois.
Figure 10–4b. Courtesy of Silliker Laboratories, 1993, Homewood, Illinois.
Figure 10–4c. Courtesy of Silliker Laboratories, 1993, Homewood, Illinois.
Figure 10–5. *Source:* Reprinted with permission from J. H. Silliker and D. A. Gabis, ICMSF Methods Studies, I. Comparison of Analytical Schemes for Detection of Salmonella in Dried Foods, *Canadian Journal of Microbiology,* Vol. 19, pp. 475–479, © 1973, NRC Research Press.
Table 10–1. Courtesy of International Commission on Microbiological Specifications for Foods, *Microorganisms in Foods 2, Sampling for Microbiological Analysis: Principles and Specific Applications, 2nd Ed.,* Table 14, p. 85, © 1986, University of Toronto Press, Toronto, Canada.
Table 10–2. Courtesy of International Commission on Microbiological Specifications for Foods, *Microorganisms in Foods 2, Sampling for Microbiological Analysis: Principles and Specific Applications, 2nd Ed.,* Table 15, p. 89, © 1986, University of Toronto Press, Toronto, Canada.
Table 10–3. Courtesy of Silliker Laboratories, 1993, Homewood, Illinois.

CHAPTER 13

Figure 13–1. *Source:* Reprinted with permission from *Journal of Food Protection.* Copyright held by the International Association for Food Protection, Des Moines, Iowa, U.S.A. Robert L. Buchanan, author, U.S. Food and Drug Administration, Center for Food Safety and Applied Nutrition, 200 C St., SW, Washington, DC 20204.
Figure 13–2. *Source:* Reprinted with permission from *Journal of Food Protection.* Copyright held by the International Association for Food Protection, Des Moines, Iowa, U.S.A. Robert L. Buchanan, author, U.S. Food and Drug Administration, Center for Food Safety and Applied Nutrition, 200 C St., SW, Washington, DC 20204.
Figure 13–3. *Source:* Reprinted with permission from *Journal of Food Protection.* Copyright held by the International Association for Food Protection, Des Moines, Iowa, U.S.A. Robert L. Buchanan, author, U.S. Food and Drug Administration, Center for Food Safety and Applied Nutrition, 200 C St., SW, Washington, DC 20204.
Figure 13–4. *Source:* Reprinted with permission from *Journal of Food Protection.* Copyright held by the International Association for Food Protection, Des Moines, Iowa, U.S.A. Robert L. Buchanan, author, U.S. Food and Drug Administration, Center for Food Safety and Applied Nutrition, 200 C St., SW, Washington, DC 20204.
Figure 13–5. *Source:* Reprinted with permission from *Journal of Food Protection.* Copyright held by the International Association for Food Protection, Des Moines, Iowa, U.S.A. Robert L. Buchanan, author, U.S. Food and Drug Administration, Center for Food Safety and Applied Nutrition, 200 C St., SW, Washington, DC 20204.
Table 13–3. *Source:* Reprinted from Food and Safety Inspection Service, 1996, Pathogen Reduction; Hazard Analysis and Critical Control Point (HACCP) Systems; Final Rule, *Federal Register,* Vol. 61, pp. 38806–38989.

CHAPTER 15

Table 15–1. *Source:* Reprinted from *Food Control,* Vol. 12, No. 3, J. D. Legan, M. H. Vandeven, S. Dahms, M. B. Cole, Determining the Concentration of Microorganisms Controlled by Attributes Sampling Plans, pp. 137–147, © 2001, with permission from Elsevier Science.

CHAPTER 16

Figure 16–2. *Source:* Reprinted with permission from *Journal of Food Protection.* Copyright held by the International Association for Food Protection, Des Moines, Iowa, U.S.A. Robert L. Buchanan, William G. Damert, Richard C. Whiting, and Michael van Schothorst, authors, U.S.

Department of Agriculture, Agricultural Research Center, Eastern Regional Research Center, 60 E. Mermaid Lane, Wyndmoor, PA 19038, USA and Nestle, Avenue Nestle 55, CH-1800, Vevey, Switzerland.

Table 16–1. *Source:* Reprinted with permission from J. M. Farber, E. Daley, M. T. Mackie et al., A Small Outbreak of Listeriosis Potentially Linked to the Consumption of Imitation Crab Meat, *Letters in Applied Microbiology,* Vol. 31, pp. 100–104, Blackwell Science Ltd.

Table 16–2. *Source:* Reprinted from United States Department of Agriculture (USDA), 1995, Nationwide Raw Ground Chicken Microbiological Survey, March 1995–May 1995, Food Safety and Inspection Service, Microbiology Division, Washington, DC and USDA Nationwide Raw Ground Turkey Microbiological Survey, January 1995–March 1995, Food Safety and Inspection Service, Microbiology Division, Washington, DC and USDA Nationwide Federal Ground Beef Microbiological Survey, August 1993–March 1994, Food Safety and Inspection Service, Microbiology Division, Washington, DC.

Table 16–3. *Source:* Reprinted from R. L. Buchanan and R. C. Whiting, USDA Pathogen Modeling Program, Version 5.1, United States Department of Agriculture, Agricultural Research Service, Philadelphia, Pennsylvania.

Table 16–5. *Source:* Data from J. M. Farber and P. I. Peterkin, *Listeria monocytogenes,* in *The Microbiological Safety and Quality of Food,* p. 1205. Edited by B. M. Lund, A. C. Baird-Parker, & G. W. Gould, © 2000, Gaithersburg, MD: Aspen Publishers, Inc.

CHAPTER 17

Table 17–1. *Source:* Data from World Health Organization (WHO), 1997, Prevention and Control of Enterohaemorrhagic *Escherichia coli* (EHEC) Infections (WHO/FSF/FOS/97.6). Food Safety Unit, Programme of Food Safety and Food Aid, World Health Organization, Geneva.

Table 17–2. *Source:* Reprinted from Food Safety and Inspection Service, 1993, Report on the *Escherichia coli* O157:H7 Outbreak in the Western States, United States Department of Agriculture, May 21, Washington, DC.

APPENDIX B

Courtesy of International Commission on Microbiological Specifications for Foods, Toronto, Canada.

APPENDIX C

Courtesy of International Commission on Microbiological Specifications for Foods, Toronto, Canada.

APPENDIX D

Courtesy of International Commission on Microbiological Specifications for Foods, Toronto, Canada.

Index

A

Acceptable level of risk, 25
Acceptance criteria, 79–97
    application of, 84–85
    definition, 17, 333
    equivalence, 80–81
    establishment, 81–84
    FSO and, 81, 85
    guideline, 83
    lot acceptance, 87–90
    selection and use, 79–96
    specification, 83, 84
    standard, 83
    types of, 82–84
Acceptance of a lot
    auditing 90–96
    examples, 87–89
    parameters to assess, 87
    probabilities, 160–161
    procedures, 86
    role of FSO, 85
    supplier approval and, 85
Aflatoxins, 168
    in peanuts, 263–270
        acceptance criteria, 269–270
        control measures, 267–269
        food safety objective, 266
        GHP, 267
        HACCP, 267
        microbiological testing, 270
        process criteria, 268
        risk assessment, 263–265
        risk management, 265–266
        tolerable level of risk, 266
Appropriate level of protection, definition, 25
Arcobacter spp., 171

Attribute charts, in process control, 257
Attributes sampling plans, 123–131
    comparison, 134–139, 141–142
    selection of, 145–165
    three-class, 125–131
        construction of, 139–141
    two-class, 123–127
    variables, 131–134
Auditing lot acceptance, 90–96

B

Bacillus spp., 172
Baseline
    ingredient data, 241
    microbiological data, 240
    in process control, 240
    variability, causes, 242
    variability, types, 242
Beef patties, raw ground frozen and *E. coli* O157:H7, 313–330
Biofilms, 209–211
Biogenic amines, 172
Botulinum toxin, 167
Botulism, 10, 11
Bovine spongiform encephalopathy, 168
Brucella spp., 167
Brucellosis, in humans, 49
Burkholderia cocovenenans, 167

C

c
    definition, 123
    selecting, 161–162
Campylobacter spp., as cause of illness, 9, 10, 30, 150, 168

Cases
    choice of plan, 153, 157
    choosing two-class or three-class plan, 157–158
    definition, 152–153
    examples, 156–157
    selection of, 145–165
    susceptibility of consumers influencing, 155
Challenge testing, 62–63
Clostridium botulinum, 150, 169
Clostridium perfringens, as cause of illness, 9, 11, 150, 168, 172
Consumer protection, 26–28
Contamination
    environmental examples, 203–204
    post-process, 200–202
Control
    definition, 333
    of environment, 199–220
    of food processes, 237–261
        baseline establishment, 241
    of foodborne diseases, 71–78
    of pathogens, 207–210
Control charts
    attribute charts, 257
    cumulative sum charts, (CUSUM) 248, 258–259
    moving range charts, 256
    moving sum charts (MOSUM), 248, 258–259
        moving window, 260
    in process control, 246, 248–259
    selecting limits, 254
    software, 259
    variable, 249
Control measures, 45–68, 71–78
    definition, 17
    effectiveness, examples, 49–54
    establishment, 17
    importance, 49
    monitoring, 65
    verification, 65
Control options, 66–68
    equivalency, 68
Criteria, *see* Acceptance criteria, Microbiological criteria
Critical control point (CCP)
    definition, 333
    in process control, 244, 246, 247
Cryptosporidium parvum, 169, 171
Cyclospora cayetanensis, 171

D

Default criterion, 61
    definition, 17, 333

Discrimination, sampling plans, 117
Distribution, of microorganisms in a lot, 191–194
    heterogeneous, 191–194
    homogeneous, 191–194
    nonrandom, 192–194
    random, 192–194

E

Egg products, microbiological criteria, examples, 111
Enterobacter sakazakii, 169
Enterohaemorrhagic *E. coli*, 10, 167
Enteropathogenic *E. coli*, 168, 172
Enterotoxigenic *E. coli*, 168, 172
Environment, control, 199–220
    environmental contamination, 201
        examples, 203–204
    in food processing, 202–220
        biofilms, 209
        niches, 210
        plant design, 208
        resident microorganisms, 205
        sampling, 210
        transient microorganisms, 202
    post process, 200–202
Epidemiological data, 28–31
Equivalence, of processes, 80–81
    flow diagram, 82
Escherichia coli
    as cause of illness, 10, 150
    control options, 66–68
Escherichia coli O157:H7
    acceptance criteria, 325–327
    control measures, 322–325
    exposure assessment, 315–317
    food safety objective, 322
    GHP, 322
    HACCP, 322–325
    incidence in meat and cattle, 314–318
    level of consumer protection, 322
    risk characterization, 321–322
    in raw ground beef patties, 313–330
    risk evaluation, 314–321
    risk management, 322
    sampling plans, 326–330
Evaluation of risk, 31–33
Expert panels
    definition, 334
    for evaluating risk, 32
Exposure assessment, 36
    definition, 37–38, 334

F

Flow diagram
 definition, 334
 frankfurter manufacture, 286
Food processes, 51
 control, 237–261
Food processing, environment control, 202–220
 control measures, 207–210
  biofilms and niches, 209–210
  GHP, 207–208
  plant design, 208–209
  resident microorganisms, 205–207
  transient microorganisms, 202, 205
 sampling, 210–220
  investigational sampling, 216–220
  protocols, 212–215
  zone concept, 211–212
Food safety management system, 4–6
Food safety objective
 achievability, 48
 for aflatoxins in peanuts, 266
 characteristics, 100
 concept, 33–34
 control measures to meet, 45–68
 definition, 25, 334
 effectiveness, 14
 for *Escherichia coli* in raw ground beef patties, 322
 establishment of, 35–36
 for *Listeria monocytogenes* in frankfurters, 294
 microbiological criteria, as basis for, 99–101
 quantitative risk assessment basis, 40–41
 risk evaluation and, 33–34
 for *Salmonella* in dried milk, 275
 stringency, 41–42
 uses of, 15–16, 34–35
 value in food safety management, 14–16
Foodborne hazards
 control, 7–8, 13, 71–78
 surveillance, 12–13
Foodborne illness, 9–13
 etiolgic agents or contaminants
  bacteria, 9–11
  protozoa, 11
  seafood toxins, 11–12
  toxigenic fungi, 12
  viruses, 11
Frankfurters, *Listeria monocytogenes* and, 285–309
Fumonisin mycotoxins, 171
Fungi, toxigenic, as cause of illness, 12, 167, 171

G

Good hygienic practice (GHP), 14, 47
 effectiveness, 14
 in food processing, 207–208
 principles of, 200
 and ready-to-eat foods, 199

H

HACCP, 14, 48
 definition, 334
 process control and, 237–238, 244, 247
Hazard characterization, 36
 curves, 37, 39, 40
 definition, 38–39, 334
Hazard groups, foodborne pathogens, toxins, 166–172
Hazard identification, 36
 definition, 37, 335
Hazards
 control, initial levels, 46
 increasing levels, 46
 microbial, according to risk, 151–152
  moderate, 151–152
  serious, 152
  severe, 152
 reducing levels, 46
Hepatitis A virus, 169–171
Hepatitis virus, 264, 265
Heterogeneous distribution, 191–194
Homogeneous distribution, 191–194

I

ICMSF, xi–xiii, 1–4, 337–349
 publications of, 347–349
Indicator tests, microbiological criteria and, 146
Indicators, microbial, 108
 factors in choosing, 109
Injured cells, recovery of, 232–233

L

Laboratories, accreditation, 233
 proficiency testing, 234
 validation of methods, 235
Level of protection, appropriate, definition, 25
Listeria monocytogenes
 cause of illness, 9, 11, 150, 169, 170
 foodborne outbreaks, 290–291
 in frankfurters,, 285–309
  acceptable level of protection, 293–294

acceptance criteria, 308–309
control measures, 295–296
flow diagram for manufacture, 286
food safety objective, 294
GHP, 308
HACCP, 308
performance criteria, 296–304
process and product criteria, 304–308
risk characterization, 292–293
risk evaluation, 287–293
risk management, 293–304
hazard characterization, 37
thermal destruction values, 306, 307
Listeriosis, incidence, 24
Lot
definition, 118
distribution of microorganisms in, 191–194
Lot acceptance
auditing, 90
examples, 87–89
microbiological criteria, 99–112
parameters to assess, 87
procedures for, 86
role of FSO, 85
supplier approval and, 85
testing to ensure, 89, 91
Lot rejection, disposal of, 110–111

M

M
definition, 127
m
determining values for, 158–159
definition, 123, 125
determining values for 158–159
Microbial hazards, categorizing risk, 151–152, 166–172
Microbial indicators, 108
Microbial populations
charting in process control, 253
selection of limits in process control, 254
Microbiological criteria, 18, 60, 102
application to foods, 101–104
characteristics, 100
components, 106–111
default, 61
definition, 101
examples, 111–112
food safety objectives, comparison, 99–101
indicator tests, 146
for lot acceptance, 99–112

pathogen tests, 147
principles for establishment, 104–106
sampling plans for, 109
types, 102
utility tests, 146
Microbiological guidelines, 102
Microbiological limits, 106
Microbiological sampling, 59–62, 123–142
Microbiological specifications, 102
Microbiological standards, 102
Microbiological testing, 18, 19, 145–165
Milk, dried, *Salmonella* in, 273–282
Mycobacterium bovis, 168
Mycotoxins, aflatoxins, 168
deoxynivalenol, 171
fumonisins, 171
nivalenol, 171
ochratoxin A, 171
T-2 toxin, 171
trichothecenes, 171
zearalenone, 171

N

n
definition, 123
selecting, 161–162
Niches, 210
Nonrandom distribution, 192–194
Norwalk virus, 172

O

Ochratoxin A mycotoxin, 171
Operating characteristics, curve, 116, 117, 118
function, 115

P

Pathogen tests
decision tree, 148
epidemiological considerations, 148
factors affecting risks, 147
microbiological criteria and, 147
Pathogenic microorganisms
clinical features, 150
ecology, 149
ranking, 167–172
Peanuts, aflatoxins in, 263–270
Performance criterion, 25, 54–59
control options and, 66

definition, 16, 335
examples of, 55–59, 60
Population, sampling, 114
Probability, concepts, 113–121
Process control, 237–261
    baseline establishment, 240
        end product testing, 241
        equipment and environmental tests, 241
        ingredient data, 241
        in-line sampling, 241
        shelf-life sampling, 241
    capability study, 243
        criteria, 244
        variability, sources, 244–246
        variability, types, 242
    control charts, 246, 248–259
    critical control points, 244, 246, 247
    ecology, 149
    in food production, 246–249
    in food regulation, 259
    HACCP, 244, 247
    options, 66
    during production, 246–249
    regulatory tool, 259
    requirements, 238–240
    statistical process control, 244
    stringency, 244
    systems, 65
Process criterion
    definition, 16, 335
    examples, 59
Process validation, 61
    challenge testing and, 62–63
Process variability, 63, 242
Product criterion, definition, 16, 335
Protozoa, as cause of illness, 11

Q

Qualitative risk assessment, definition, 335
Quantitative risk assessment, 36–41
    definition, 335

R

Random distribution, 192
Ready-to-eat foods, 199, 201
Representative sample, of a lot, 119–120
Risk, consumer, 117
    definition, 335
    producer, 117

Risk, tolerable level of, definition, 25
Risk analysis, definition, 335
Risk assessment, quantitative, 36
    definition, 335
    in food processing environment, 210
Risk characterization, 36
    definition, 39–40, 335
Risk communication, definition, 335
Risk estimate, 39
    definition, 336
Risk evaluation, 31–33
    by quantitative risk assessment, 36
Risk management, definition, 336

S

Safe food, definition, 336
Salmonella
    in dried milk, 273–282
        acceptable level of protection, 275
        acceptance criteria, 281–282
        control measures, 276
        exposure assessment, 274
        food safety objective, 275
        GHP, 280
        HACCP, 280
        hazard characterization, 274–275
        hazard identification, 273–274
        performance criterion, 277
        process criteria, 279–280
        risk characterization, 275
        risk evaluation, 273–275
        risk management, 275–279
    incidence in meat, 54
    performance standards, 260
    two-class attributes plans, examples, 186–188
Salmonella enteritidis, 53, 170
    as cause of illness, 9, 29, 150, 167
Salmonella typhi, 52
Salmonella typhimurium, 150, 170
Salmonellosis
    in Germany, 29
    in USA, 53
Sample
    analysis, 231–232
    collection of sample units, 226
    event samples, definition, 228
    injured cells in, 232–233
    representative, of a lot, 119
    storage and transportation, 229, 231
Sampling, 173–182, 225–235
    food, special considerations, 230

in food processing environment, 210–220
    examples, 214
    investigational sampling, 216–220
    protocols, 212–215
    zone concept, 211–212
  principles, 113–121
  procedures, 228–229
Sampling plans, 115, 123–142, 162–165
  case, definition, 152, 153
  choosing two-class or three-class, 157–158
  comparison, 134–139, 141–142, 163
  interpretation of results, 120
  investigational, 173–177
  performance, 124
  performance and case, 162
  reduced, 174, 182
  stringency, 117, 152, 180
  three-class, 125–131
    construction of, 139–141
  tightened, 173–180
  two-class, 123–127, 186–188
  variables, 131–134
Sausage, *Listeria monocytogenes* in, 285–309
Seafood toxins, as cause of illness, 11
Set point, in food processing, 64
Shigella, as cause of illness, 10, 29, 167, 170
Shigellosis, in US, 53
Special foods, susceptible consumers and, 155
Staphylococcus aureus, as cause of illness, 9, 11, 172
Stringency, of sampling plans, 117
  problems in implementation, 188
  relation to commercial practice, 189–191

T

Three-class attributes plans, 125–131
Tolerable level of risk
  for aflatoxins in peanuts, 266

definition, 25, 26–28, 336
economic impact, 27
epidemiology, 28
Transient microorganisms, 202, 205
Trichinosis in humans, 50
Trichothecene mycotoxins, 171
Two-class attributes plans, 123–127, 183–197
  application of 183–197
  discrepancies between tests, 191–196
  pathogens (*Salmonella*), 186
  relation to commercial practice, 189–191
  stringent sampling plans, 188
  zero tolerance, 184–185
Typhoid fever, 52

U

Utility tests, microbiological criteria and, 146

V

Variables, plans, 131–134
Verotoxins, 10
Vibrio cholerae, as cause of illness, 168, 172
Vibrio parahaemolyticus, as cause of illness, 9, 10, 172
Vibrio vulnificus, 169
Viruses, as cause of illness, 11

Y

Yersinia, as cause of illness, 11, 170

Z

Zearalenone mycotoxin, 171
Zero tolerance, 184–185
Zone concept, 211–212

Lightning Source UK Ltd.
Milton Keynes UK
UKOW041421040212

186627UK00008B/2/A